Give Space My Love

Give Space My Love

● ● ●

**An Intellectual Odyssey with
Dr. Stephen Hawking**

TERRY BRISTOL

Institute for Science, Engineering and Public Policy
Affiliated with Portland State University

For permissions to reproduce selections from this book, write to
Permissions, Institute for Science, Engineering and Public Policy
3941 SE Hawthorne Blvd
Portland OR 97214

Author website: terrybristol.org

Library of Congress Control Number: 2015914066
LCCN Imprint Name: Institute for Science, Engineering and Public Policy
Portland Oregon

ISBN: 0940530015 (hardcover)
ISBN 13: 9780940530010 (hardcover)

Published by the Institute for Science Engineering and Public Policy
Portland Oregon Website: www.isepp.org
Affiliated with Portland State University

Dedication

To Suzanne, Caitlin, Gavin and Kris whose loves repeatedly rescued me from my excesses, and kept me balanced on the path of the middle way.

Love is never-ending.

Plato's *Symposium*

Acknowledgements

During the 'emergence' of this book through serial revisions over more than a decade, I have naturally gained from numerous dialogues and critiques. As host of the Linus Pauling Memorial Lecture Series for more than twenty years, I have had the unique opportunity to discuss ideas with an amazing range of leading-edge thinkers. I have benefited enormously from their critiques and encouragements beginning with BBC interdisciplinary journalist James Burke, followed by physicist Nick Herbert, Ilya Prigogine, Fred Alan Wolf, Philip Morrison, Frank Tipler and John Barrow (The Cosmological Anthropic Principle), F. David Peat, Steven Weinberg, Freeman Dyson, Kip Thorne, Brian Greene, Murray Gell-Mann, Lawrence Krauss, Paul Davies, Roald Hoffmann, Fritjof Capra, Robert Laughlin, Gibor Basri, Roger Penrose, Mario Livio, Lisa Randall, Sean Carroll and Ray Jayawardhana.

My understanding of evolution was repeatedly challenged and enhanced through discussions with Stephen Jay Gould, Richard Dawkins, Lynn Margulis, Stuart Kauffman, Robert Ulanowicz, Francisco Varela, Mary Catherine Bateson, Richard Leakey, Jane Goodall, Tim White, Dorion Sagan and Eric Schneider (Into the Cool), Evelyn Fox Keller, Craig Venter, Michael Ruse, Michael Russell, Robert Hazen, David R. Cox, Robert G.B. Reid, and James A. Shapiro. In philosophy of mathematics I have gained from discussions with Ian Stewart, Ralph Abraham, Benoit Mandelbrot, Bart Kosko, Tom Siegfried, Keith Devlin.

In the history and philosophy sphere I have gained from discussions with Jerry Ravetz, Daniel Dennett, John Casti, David Albert, Nancey Murphy, David Boersema, Terence Love, George Johnson, George Lakoff, Steven Goldman, Susan Haawk, Philip Clayton, Rebecca Goldstein, Nancy Cartwright, and John Dupre.

In the new engineering and economic paradigms I have gained considerably from interactions with Henry Petroski, Scott Borg, George Bugliarello, Louis Bucciarelli, Moira Gunn, William Wulf, Paul Alivasatos, Ramez Naam, David Pearce Snyder, Paul Romer, David Warsh, and Joseph Stiglitz.

Fellow faculty and colleagues in a dozen professional societies have offered valuable reviews and encouragements. Appreciation is also due to my students at University of California at Santa Barbara, Marylhurst College, Portland Community College, Linfield College and Portland State University who after tolerating my sometimes rambling lectures still managed to offer novel penetrating questions.

My intellectual and academic mentors Paul Feyerabend and Imre Lakatos above all others served to help me define the path of my inquiries. And it was not just the integrity of their ideas that I found compelling but, perhaps, most of all, their humanity.

Story editor Dave Malone coached me in making numerous improvements in the narrative continuity. Copy editor Pam Raschio Brown saved me from the embarrassment of grammatical grafts and contributed to major enhancements in overall expression.

And finally this book would not have been possible without my more than 25 year friendship with Stephen William Hawking, who nonetheless initially tried to dissuade me from writing it. Having been assured that the book 'wasn't just about him,' he added that it should be made clear that it is not an authorized text. Be it so clarified.

Stephen's graduate assistants have all been delightful comrades and valued friends. And yet in the end it was two of Stephen's nurses — exceptionally wonderful people and great friends — Joan Godwin and Pam Benson, who repeatedly encouraged me to go ahead, insisting that 'these stories of Stephen's life and in particular his meetings with the students with disabilities need to be told.'

Contents

**Part Two - The Two Paths to
Complementarity in the 20ᵗʰ Century**

GIVE SPACE MY LOVE

Prologue

Hawking's Meeting with the Shuttle Astronauts

Who is the real Stephen Hawking? How are we to make sense of this great scientist, for many the quintessential representative of Modern Science, who turns out to be such a lovable and loving character? What is his real place, his real role, in the universe? The question of course isn't just about Stephen Hawking. The question is for all ostensible scientists. Are scientists detached Spectators seeking to understand a fixed, fully determined, objective reality 'out there'? Or are we embodied Participants, inquirers actively experimenting and exploring – developing reality as we learn? How are we to make sense of scientific inquiry into the nature of reality as part of the nature of reality? Who are we – really? These self-reflexive questions pertain to everyone who asks questions, to everyone who wonders about humanity's place in the universe, to everyone who wonders about how he/she fits into the grand scheme.

Stephen's nurses take the opportunity to serve him tea – about half of it cascades down into his bib-pocket. As Stephen's host and organizer of the Science, Technology and Society lecture series I had arranged a small private jet to take Hawking and the crew down from Seattle to Portland. The private jet gave us greater scheduling flexibility and was actually cheaper than flying commercial for the six of us. Admittedly, the real deciding reason was that with our own private jet we had a chance on the flight down to buzz the crater of the recently volcanic Mt. St. Helens.

On our arrival at Galvin's Flying Service at Boeing Field where we were to meet our plane, I had noticed two small, military-like fighter jets parked outside with prominent blue NASA markings on their tails. As I was peeking outside toward the runways for a closer look, two guys approached me. They were wearing powder blue jumpsuits with NASA insignia patches above their hearts.

"Is that Dr. Hawking?" one of them asked me. "Yes," I answered.

"We are shuttle astronauts and we were wondering if we might introduce ourselves – very briefly – and thank Dr. Hawking for his contribution to our program of space exploration?" one of them queried. After clearing this with Stephen, I motioned to the astronauts. They approached and introduced themselves. The lead was Captain Frank Culbertson an impressive individual – clear-eyed, physically fit, perfect posture, polite, friendly, self-effacing – a perfect human specimen with manners.

After Culbertson and his partner made their formal introductions, and Stephen returned a "Hello, how are you?" Culbertson offered, "Thank you for your work in cosmology" and emphasized how important Hawking's support had been in the development of NASA's Hubble Space Telescope project.

Stephen listened. The rest of Stephen's entourage had gathered around so I introduced the crew to Culbertson and his partner. This morphed into one of those uncomfortable moments where no one was really quite sure what was next. Stephen's graduate assistant at the time, Tom Kendal, offered that Stephen had just come from Cal Tech and the Jet Propulsion Lab. Culbertson acknowledged JPL's prominent role in several NASA projects.

Then suddenly Stephen boomed out: "How's weightlessness?" Smiles crept across everyone's face. Culbertson, smiling, made a dramatic pause – looked down and sort of pawed the ground with his foot. Then looking up straight at Stephen, he said, "Well, it's one of the two greats." Tom laughed and said, "Ah, weightlessness. Yeah, Stephen would like that. He would do very well in weightlessness." But Culbertson had us. All attention focused on him. He started again. "Yes, there are two greats of space flight: weightlessness and the view of the Earth – its beauty."

As the communication-delay anxiety lifted, Joan Godwin, one of Stephen's nurses, asked what Culbertson and his partner were doing in Seattle. Culbertson related that he was the ground commander of NASA's portion of the historic joint US-Russian space flight. Fellow astronaut Shannon Lucid had just come aboard the Russian Space Station, Mir. It was in the early days of the flight and Culbertson was coordinating communication systems with facilities shared with Boeing.

Gradually the conversation drifted once again into the 'I-don't-really-know-what-is-happening-in-this-conversation' realm. People were shuffling. Culbertson began to move like he was preparing to disengage. We all started the 'Well – Thank Yous.' "Thank you for introducing yourselves." "Well, thank you for obliging us." "And thank you Dr. Hawking, it was a great honor to meet you."

Culbertson and his partner had just begun to turn, to exit the group, when Hawking boomed out: "Give space my love." There were smiles all around. Culbertson and his partner acknowledged and promised that they would.

That parting comment was a turning point for me – the moment when I thought I needed to write a book to tell the stories of the four-city lecture tour. The first thought was something like 'Adventures with Stephen Hawking'. Over the years and on reflection I came to recognize that the 'stories of the road' were far more fascinating, and their larger significance could be presented, when framed in the context of the intellectual odyssey of the 20th century. A prominent feature of our time together had been the initially casual but increasingly intense dialogue between Hawking the scientist and me, the philosopher of science.

The new 20th century physics had raised enigmatic questions about the classical Spectator conception of 'objective' reality 'out there'. Niels Bohr's embrace of complementarity meant that 'making sense of reality' required an irreducible reference to the observer as a Participant in reality. Heisenberg's uncertainty principle suggested that the inquirer's choices – how to observe, how to proceed – were formally under-determined – constrained and yet, to some irreducible extent, 'free' – under-determined.

In parallel, in 20th century philosophy of science, the Logical Positivist representation of the history of science as logico-mathematical, as conceptually continuous, had been discarded. Thomas Kuhn's careful historical studies showed that 'real' learning was logically discontinuous – conceptually revolutionary! Karl Popper's insistence that all meaningful theories must be falsifiable led to a parallel embrace of complementarity in philosophy of science. Philosopher of science, Willard Quine established that the inquirer's choice of what to believe was, again, in parallel, under-determined by all possible evidence – perhaps constrained and yet, to some irreducible extent, 'free'.

Complementarity seemed to suggest that the perennial disputes between ideological opponents might be productively understood in a new way. In the broader intellectual milieu, the new limitations on 'objectivist' science entailed a limitation on all 'one right answer' ideological belief systems. Such was the stuff of our dialogues. Everyone participated: Hawking and I, his graduate assistants and his nurses. The four-city lecture tour became an intellectual odyssey through 20th–21st century culture – seeking to find our home in the universe.

Although we all inhabit diverse roles in our lives, Dr. Stephen Hawking operates in two dramatically incompatible worlds. As traditional scientist he searches for the timeless laws that govern the inevitable course of events in the universe, almost as if he were a detached Spectator. Yet Hawking is actively promoting a liberal politics and a new empowering vision of the disabled. Understood as Participant in a developing universe, he is working to alter the course of events, to steer the universe toward a more desirable future. His meetings with students with disabilities, officially collateral events on the lecture tour, were remarkable in and of themselves and yet more thought-provoking in the context of the question of the nature of the universe and our role in it.

Part One

The Journey of a Thousand Miles is Under Your Feet

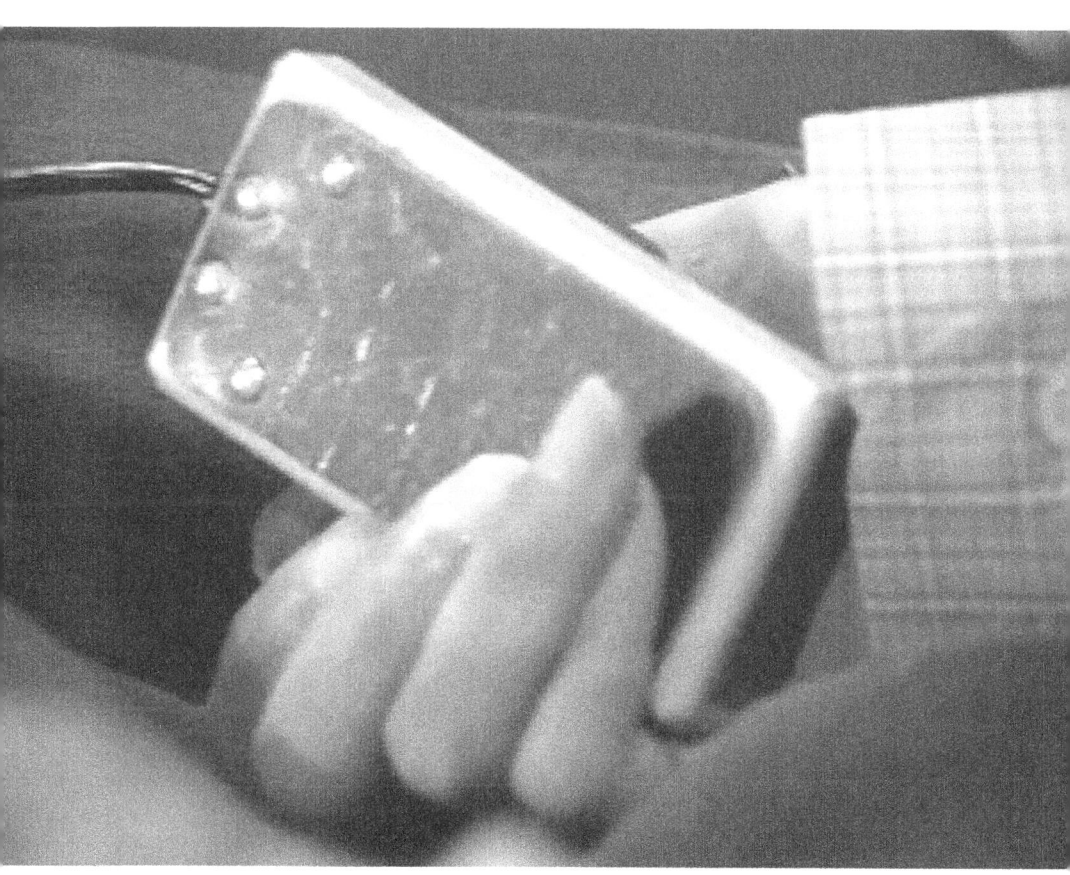

1

The Scientific Litany and
the Rebels at Berkeley

Reflecting back on my own story, I can now understand, but only with hindsight, the path that led me to my encounters with Stephen Hawking… not inevitable but with an emerging narrative coherence. When I entered the University of California at Berkeley in the fall of 1964, my clear and unshakable intention was to major in Astronomy. In high school I had led a sort of split life. On the one hand I played lots of sports, received more attention than I deserved from the ladies and was elected Senior Class President. On the other hand, I had been something of a standout in math and the hard sciences. I had a separate group of geeky, intellectual and artistic friends. Since I was ten years old, my longtime hobby had been amateur astronomy, pursued with a telescope purchased with funds from my morning newspaper delivery route. I'm not entirely clear about my initial attraction to astronomy, but I recall having been intrigued by both the beauty and mystery of the cosmos. In choosing Berkeley, I consciously thought of myself as escaping – avoiding the social temptations of my many 'popular' friends attending Oregon universities. It was an escape to my intellectual side. When I arrived in Berkeley I knew no one there. I reveled in the anonymity – gradually re-inventing my social self.

As I advanced through my undergraduate classes in chemistry, biology, physics and math, I began to appreciate that *my* idea of astronomy, and, more specifically, *my* idea of science, was somehow different from the 'official' litany of the 'Scientific Research Program' (viz. recitation of the official tenets). The more I learned about the formal agenda of scientific inquiry, the more I explored the sciences, the more I felt a need to seek a broader framework of inquiry. My curiosity somehow was calling for a more general, a more comprehensive inquiry into the nature of the universe.

The questions I had started asking about the cosmos when I was ten years old, the questions that had fueled my initial curiosity, were certainly 'compatible' with the official representation of the scientific worldview, and yet they seemed to reach further.

What I took to be everyone's natural-born sense of inquiry included questions about the meaning of life and our place in the cosmos and, at least plausibly, included normal day-to-day questions about justice and beauty. It seemed to me common sense, really, that these questions, along with the scientific questions about how the world worked, should be part of one overall complete and consistent understanding. I imagined that all inquiry would naturally be included, constituting one self-referentially coherent research program.

I was rather surprised – and more than a little dismayed – to find that when I raised the broader questions in scientific circles at Berkeley, I was told that they were not meaningful questions. Even the question about the place of scientific inquiry itself in the cosmos was considered to be a 'philosophical' question. Real science, it was emphasized, was concerned with understanding objective reality – 'out there' – independent of the observer, independent of the inquirer, independent of inquiry itself. I was told that questions about values, about justice, about what was desirable, about what was 'really' beautiful 'in itself out there', were not meaningful questions. Social and political philosophies were not meaningful – by their very nature – because their unique validity could not be empirically established. Social, political and aesthetic questions were not decidable. Values were subjective – in the mind – not part of objective material reality – 'out there'.

These 'subjective' value questions, when addressed at all, were characterized as analogous to preferences for one or another flavor of ice cream.

Some like chocolate, some like strawberry. And these personal preferences were most likely to be explained, eventually, by analysis of one's genes.

In stark contrast, questions about facts, about objective reality, about one's preferences for one or another scientific theory, were treated quite differently. All beliefs about objective reality, it was presupposed, could be rationally decided – at least eventually – on the basis of the results of carefully designed scientific experiments. To arrive at a complete and consistent understanding of objective reality was the agenda of the Scientific Research Program

This 'official scientific litany' was somehow hard-wired. All meaningful questions were scientific. Everything else was some sort of primitive pre-scientific, philosophical-religious fantasizing. In the Scientific Research Program there could be only 'one right answer' to every meaningful question, to every empirically testable question; that is, to every scientific question. Answers, meaningful statements, were statements of fact about objective reality. I had never imagined any such ultimate division between factual and value questions and claims – particularly with the former being meaningful and the latter being completely meaningless.

What actually disturbed me the most was that to question this scientific litany, to question these presuppositions of the Scientific Research Program, was itself inherently heretical. To question the tenets of science, from 'outside', was inconsistent with the tenets of science. The question as to whether *only* scientific questions are meaningful, it turns out, is not itself a scientific question. To ask such a question required 'thinking outside the box', venturing into non-scientific foreign territory, into meaningless philosophical territory.

When thinking about this over the years, I often recalled an encounter when I was quite young. A Catholic friend was telling me that it was a mortal sin to question or to think outside the Catholic litany (viz. tenets, doctrine). Even to raise questions was, by its very nature, by the very nature of the questioning, heretical and meant you had strayed from the true path. A true Catholic, I was told, was also not supposed to listen to or discuss alternative viewpoints. You would be thinking outside the box. My young friend, I suspect, was as dumb-founded by this policy, by what he had been told, as I was. I do not think the analogy with the scientific litany is spurious. In both communities there are efforts to actively enforce an 'official' litany. To question the litany is to question what defines what each community takes to be the nature of rational thought and meaningful inquiry. To raise questions outside the core defining tenets was not just denigrated as heretical but in the final analysis was characterized, essentially by definition, as irrational.

Only much later did I come to see these self-justifying 'inside the box' 'rationalizations' as characteristic of all ideologies – all belief systems that imagined that they have 'the one right answer'. Other belief systems were not just wrong – they could not be 'made sense of' in terms of the official core defining tenets of the ideology. Alternatives were, in the final analysis, nonsense, irrational – and not to be respected, not to be tolerated.

Ironically, since the question about values was not a 'scientifically' meaningful question, it couldn't actually come up *within* the Scientific Research Program; it couldn't arise *within* respectable rational thought and inquiry. As a consequence the actual question was never 'officially' raised and never actually given a rational, a scientific answer. Such questions simply didn't make sense. That value questions were meaningless was somehow a hard-wired, logical consequence. The regular response to those who raised such questions was dismissive: 'That is just another of those meaningless philosophical questions.'

The bad guy, I gradually realized, the culprit, at least in the case of science, was what I came to call the Scientific Hypothesis, the core hypothesis that defines the Scientific Research Program. The Scientific Hypothesis is that all phenomena in the universe are governed by One universal, time-space invariant order. That's a mouthful! Consider a simpler formulation. Virtually everyone buying into the Scientific Hypothesis understands it as equivalently expressed in terms of 'repeatability'. The mark of scientific knowledge is that it is demonstrably repeatable – with changes in time and location. This means that there must be One universal (viz. time-space invariant) order over changes in time and location. Galileo's experiment dropping the balls from the Leaning Tower of Pisa circa 1600 can be repeated here in Oregon in 2015. This is a repetition of experimental knowledge, over changes in time and location. The repeatability criterion is often alternatively expressed by the scientific maxim 'same cause, same effect'.

When I suggested that presupposing the Scientific Hypothesis was itself a 'philosophical' position, I was told: 'we don't do philosophy here.' But they did. The Scientific Hypothesis that all phenomena are uniquely linked by One universal causal order is also, equivalently, referred to in the more general context as the Mechanical Philosophy. Modern Science was born with the Mechanical Philosophy starting most explicitly with Galileo, traveling up through to perhaps its most simple, canonical modern expression in Newtonian Mechanics.

The problem with the Scientific Hypothesis and the Mechanical Philosophy is that there doesn't seem to be any way to make sense of the normal day-to-day, common sense value dialogues. In a fully deterministic Mechanical Philosophy the traditional humanities are characterized, at best, as unnecessary and, at worst, as simply meaningless, based in childish, pre-scientific mythological fantasies. For those like me, who continued to take value questions to be meaningful, it was made clear that overtly pursuing such questions was not only inappropriate in the scientific community, but that it would not to be tolerated. Respectable scientific dialogue, the standards of scientific rationality, needed to be protected and enforced – from the classroom to the research laboratory, including strict policies on what could be presented and discussed in scientific journals.

I was finding it increasingly difficult to see myself as comfortable with a career and a life within the enforced tenets of the Scientific Hypothesis and the Mechanical Philosophy. There was something missing – something serious, something important. Outside the formal cloistered confines of the modern scientific research community, in the real world, most people seemed to think that questions about values and about our place and role in the universe were meaningful.

My broader questioning led one of my Berkeley professors to suggest that I talk to Professor Paul Feyerabend in the Philosophy Department. Feyerabend was a major player in a field I had never heard of – philosophy of science. We connected. Here in the philosophy of science the questions I was asking were being taken seriously – not only the questions about the nature of the universe but also the self-reflexive questions about our place in the universe and about what counted as a meaningful question.

About the same time I took a philosophy course from the famous Berkeley professor of philosophy John Searle. That was an important influence. Searle made it clear that these 'philosophical' questions about questions, about inquiry, about what we know and how we know, and about our place in the universe were not new. Historically these questions had not been so rigorously banned from the active research community. Before William Whewell's introduction of the term 'scientist' in the 19th century – associating it with the Mechanical Philosophy (*The Philosophy of the Inductive Sciences*, 1840), those engaged in inquiry into the nature of reality had referred to themselves as 'natural philosophers' and their inquiry more broadly as Natural Philosophy. Indeed, many of the developers of the Modern Scientific Tradition including Newton, Descartes and Leibniz were equally honored and renowned for their contributions in the broader philosophical tradition. Unfortunately, in the 20th century, the philosophical contributions of these giants are completely absent from the narrowed scientific curriculum of the modern university.

Similarly the history of science itself has been deemed unnecessary for scientists to learn since, if all the laws governing phenomena are timeless, everything worth studying, everything we are trying to understand, can be found in the present. 'The present is the key to understanding both the past and the future.' With the dominance of the Scientific Hypothesis and the Mechanical Philosophy, the history of science, along with history in general, have become foreign territory, relegated to the humanities.

My initial impression of Feyerabend's philosophy of science community was of a rebel stronghold made up of scientifically and mathematically sophisticated folks questioning the Mechanical Philosophy. You could raise the more general questions under the umbrella of a philosophy department – questions not allowed or tolerated in science departments.

Feyerabend himself was a unique figure. Born and raised in Vienna Austria, he developed an interest in theater and operatic singing. Graduating high school in 1942 during Hitler's regime he enrolled in officer school hoping that the war would be over by the time he finished. Instead he was eventually sent to fight on the Russian front. As a lieutenant, while directing traffic during the German retreat he was hit by three bullets. One bullet lodged permanently, inoperably, in his lower spine, so that he would always walk with difficulty, awkwardly, requiring a stick for support.

Feyerabend once told me, what he took to be a sort of humorous story, about visiting a new hospital for a checkup. He mentioned that he had persistent pain in his lower back, but didn't mention the bullet. The doctor came back with the x-ray – very excited about his discovery of the bullet. For Feyerabend it was a little macabre entertainment.

Paul Feyerabend

After the war, Feyerabend trained briefly to sing in the opera, studied with Bertold Brecht and wrote plays for theatre. At university he studied history and sociology, but was dissatisfied and soon transferred to physics. After graduating, he received a scholarship to study in England with one of the most famous 20th century intellectual rebels, Ludwig Wittgenstein, but Wittgenstein died before he arrived. Feyerabend had met Karl Popper in 1948 at a conference. Popper was now at the University of London's London School of Economics, and Feyerabend chose to switch to Popper as his Ph.D. supervisor. Popper's influence on Feyerabend was both 'positive' and 'negative'. Over the years Feyerabend was both an important champion and a severe critic of Popper's voluminous work. In Berkeley, Feyerabend, identified as rebel, was a renowned teacher whose courses were packed.

Feyerabend was culturally erudite, reflected for instance in his regular attendance at the world famous San Francisco Opera, always with front row seats. He told me that he particularly liked to follow the development of the voices of the singers. On the other hand, Feyerabend also regularly attended the seemingly preposterous 'Texas Wrestling' – where characters such as 'Gorgeous George' would fight 'The Destroyer'. When asked about the contrast and why someone of his sophistication would attend 'Texas Wrestling', Feyerabend said, "It's the only place where you can really tell the good guys from the bad guys."

I began to align with the rebel forces. However, it soon became clear that the philosophy of science community was not made up exclusively of rebels. There were conservative forces who saw the Mechanical Philosophy as providing the correct representation of scientific inquiry and knowledge. These were the Logical Positivists who expanded on Galileo's posit that the language of nature was mathematics. The Positivists, noting that the reasoning in mathematics was 'formally logical' argued that all successful inquiry in a mechanical universe should proceed logico-mathematically. Scientific method, they proposed, must be completely and consistently systematic – logico-mathematical. Teaching students how to learn was to teach them scientific method, to teach them how to think logico-mathematically.

The curious, ironic twist is that many of the original Logical Positivists, after only a few years, abandoned this defining hypothesis of a logico-mathematical scientific method. However, the hypothesis had leaked out into the broader scientific community and is still, and perhaps increasingly, popular there. The reason for this popularity, I think, is that the Scientific Research Program is defined by the presuppositions of the Scientific Hypothesis. If you begin with the assumption that the universe is completely and consistently mechanical, governed by universal mathematical laws, 'it stands to reason' that investigating such a universe should be done by means of a universal logico-mathematical scientific method. One plausible implication of this line of reasoning is that scientific research could eventually be turned over to logico-mathematically programmed mechanical computers. Stephen Hawking even suggested this in his 1976 blockbuster, *A Brief History of Time*. This concept is still the driving theme in much of modern artificial intelligence research.

When I entered the philosophy of science community the rebel forces had really just begun to coalesce. Popper had written his *Logic of Scientific Discovery* in 1934, but hardly anyone knew about it until Feyerabend began to reference it in the 1960s. It was the 1962 publication of Thomas Kuhn's *The Structure of Scientific Revolutions* that served as the intellectual cornerstone of the rebel movement. Kuhn wasn't actually saying anything entirely new or unique. Others had made similar points. But Kuhn brought it all together with his rigorous scholarship and carefully focused reasoning – and it had penetrated. Kuhn was raising serious questions about how 'actual scientific inquiry' was being represented. Kuhn's most powerful arguments were from physics, supported by the history of science. He showed that the evidence didn't support the Positivists' expectations of uniform logico-mathematical advances. Kuhn argued that advances were 'revolutionary' – logico-mathematically discontinuous (non-uniform). Kuhn challenged the Positivist's representation of inquiry focusing in large part on their attempt to interpret the new physics (quantum theory and relativity) as making sense within the classical understanding of the Scientific Research Program.

Over the decades since *Structure* there have been many dialogues. And our understanding of the nature of 'real' inquiry has matured. I asked Hawking some twenty years after the first publication of *A Brief History of Time* about his statement that science could eventually be turned over to computers. He said, "It's embarrassing. I wish I had never said it." Hawking had by then read Kuhn and Popper and had begun to understand my concerns – at least that there was something to be concerned about.

It took me some time to realize another curious overarching irony – that the rebels didn't actually have anything *positive* to say. They were almost exclusively critics. Over the years, it seemed to me, they were 'backing into their positions'. They were 'learning through hindsight', by pointing out the ways in which the Positivist theory didn't work. They didn't have any clear idea of how inquiry did work – but they could see that the Positivist logico-mathematical representation wasn't correct. The rebel's criticisms pointed out that the Positivist representation of scientific theories, scientific method and the history of science actually *misrepresented* scientific theories, scientific method and the historical process. It was only through these careful, relentless critiques that the rebel cause advanced. Kuhn helped us to distinguish between how 'real' advances actually occurred and the Positivist's after-the-fact 'rationalized' logico-mathematical accounts. 'Real' inquiry wasn't a systematic, uniform, logico-mathematical discovery process. It was revolutionary – logico-mathematically discontinuous.

Thomas Kuhn

Kuhn's historical critiques reached back to Copernicus and Galileo and the beginnings of what we think of as Modern Science. He also pointed out that Ancient Science and Modern Science were based on the same presuppositions about a universal objective order. In important ways Galileo had simply resurrected the Ancient Scientific Hypothesis and its Ancient Scientific Research Program.

Kuhn and the other rebels in the philosophy of science community were not alone in their criticisms of the standard attempts to make sense of both quantum theory and relativity within the Scientific Research Program. Nobel Laureate physicist Niels Bohr, in particular, one of the founders of quantum theory, led the forces within the formal physics community, insisting that the complementarity of wave and particle phenomena in the new physics required a revolutionary new type of theory, a more general, post-mechanical, post-scientific theory of how the universe works.

So here is the final irony that continues to define the current moment. Even with decades of serious accumulating critiques of the Positivist's representation of inquiry and the paradoxical wave-particle complementarity as viewed from within the classical Scientific Hypothesis and Mechanical

Philosophy – the expected, passionately anticipated and sought after revolutionary new, more general, post-mechanical, post-scientific theory just isn't popping out at us. We are stuck with what must be characterized as a paradoxical view of reality and our place in it.

One of Kuhn's historical examples clarifies the situation. The retrograde motion of the planets in the sky was well known and well documented for centuries before Copernicus. The planets move forward with great regularity in their orbits, and then, occasionally, they just, for a brief period, reverse the direction of their motion. The Ptolemaic Earth-Centered astronomy was even able to predict these occasional periodic 'backward' motions with considerable accuracy. We now understand that these 'apparent' retrograde motions are actually due to the motion of the Earth, our observing platform, around the Sun. The Ptolemaic system had tacitly presupposed 'the obvious', that we – the observers on the Earth – were not moving, located in the unchanging, stable center of the universe. Copernicus's move to a Sun-Centered astronomy offered *a new way* to understand the well-established, well-confirmed existing observations of retrograde motion – as they had been understood in the Ptolemaic system.

The point is, by analogy, that even though we have been aware for some time of serious inadequacies of the Scientific Hypothesis, of the Mechanical Philosophy – such as the irreversibility of many phenomena, the fundamental complementarity of the order(s) governing Newtonian particle physics and Maxwell's electromagnetic wave physics, and the obvious conflict with our normal day-to-day understanding of human action and inquiry, nonetheless, the new way of understanding, the new more advanced theory hasn't come forth. Where is our Copernicus? Einstein's theories have captured crucial insights, but, as both Kuhn and Bohr emphasized, they have also generated serious problems in our understanding of the universe and our place in it.

Throughout the 20th century, for many of us, it has felt like the new post-scientific, post-mechanical paradigm was struggling to be born, fighting to emerge. And yet – nothing. We have been 'stuck' for an uncomfortably long period of time. Leading quantum cosmologist Lee Smolin, addressing a group of incoming graduate students in theoretical physics in 2010, expressed the feeling well. 'In the 1960s when my generation was coming into theoretical physicists, we were enormously optimistic. The previous generation, the actual founders of quantum theory and relativity, had all failed to resolve the question of the nature of reality in light of the new physics. – My generation was excited about bringing forth the new paradigmatic understanding of reality. ... It is now 2010, and it has become rather Kafkaesque [viz. seriously weird and unsettling] that we haven't made any progress whatsoever.'

For those still pursuing the Mechanical Philosophy there have been heroic efforts to 'cram' quantum theory and relativity into a single unified Mechanical Research Program, resulting in what Jim Baggott, in his recent book, *Farewell to Reality*, has characterized as an increasingly 'fairy-tale physics.' In philosophy of science there has been a painful, ongoing parallel effort to 'cram' Kuhn's revolutionary history of 'science' and Feyerabend's rejection of a universal 'scientific method' backwards into a conceptually continuous, universally logico-mathematical framework.

The difficulty, that only gradually dawned on me, is that in order to find the new post-scientific paradigm you need to think outside the box; you need to think outside the scientific paradigm; you need to think outside the presuppositions of the Scientific Hypothesis. To make novel discoveries you need to be an explorer in foreign territory. At the same time, it is always crucial to remain cognizant of, to remember, where you came from. If you forget, then you are no longer expanding your understanding, but merely wandering in the mire. This last constraint means that in shifting to a new paradigm you must not reject the previous, highly successful paradigm, you must not reject the Scientific Paradigm as completely wrong. As the common metaphor expresses it you don't want to throw the scientific baby out with the bath water. *Completely* rejecting the hypothesis of a universal Mechanics because of its inadequacies is untenable. The successes of the Newtonian and Maxwellian Research Programs have been enormous and dramatic and are simply undeniable. Any new post-mechanical, post-scientific paradigm must be able to understand and explain their successes, albeit, perhaps, in a new way.

The trick is to find a new More General Paradigm that both subsumes the successes of the Scientific Research Program and supersedes them by explaining the successes in a new way – as having limited validity.

Perhaps it was coincidental that I entered Berkeley in 1964, the beginning of the baby boomer student rebellions. The dominant California political establishment had decided that 'free' political speech and organizing could not happen on a state government funded campus. They came up with the argument that this new policy followed reasonably from the ban on political activity by government employees in government buildings. As the conservative (right-wing-dominated) establishment, their real aim was to suppress oppositional political dialogue coming from the (left-wing-dominated) campuses. This attempt to enforce a right-wing litany on a predominantly left-wing community gave birth to the infamous Berkeley student rebellion known as the 'Free Speech Movement'. The mid-to-late-1960s was also the period of the flower children and the baby boomer's 'generational rebellion'. Boomers weren't happy about their parents' life-style. Politically lots of things didn't seem right, didn't seem to embody the values of the American Constitution and the American Dream – civil rights, the Vietnam War. As a generation we began a serious questioning of authority.

One common attitude among my contemporaries was a *complete* rejection of the established order. In my personal heated dialogue with the older generation, particularly with my father, I kept getting the seemingly diversionary question: What is your alternative? That hadn't seemed to be my point. The point was that you were wrong and that your system was corrupted. Many of my friends simply took the question about an alternative as a defensive obfuscation. But the more I thought about it – the more it seemed to be a reasonable question.

What was our new alternative paradigm? It was a good question. The Beatles captured our dilemma in their song *Revolution*: "You say you want a revolution. Well, you know we all want to change the world. You tell me that it's evolution. Well, you know we all want to change the world. But when you talk about destruction. Don't you know that you can count me out. ... You say you got a real solution. Well, you know we'd all love to see the plan."

I sometimes think that the most important lesson I learned at Berkeley was to question authority. But the concomitant lesson was that critical questioning and rebellious resistance isn't enough. To really move forward, to advance, you need a new better plan, a new better theory. You need a better understanding that both subsumes (viz. includes and incorporates) what was right and successful in the previous understanding, and yet, also supersedes it – understanding reality in a new, better way. Our parents' generation, for all its flaws, had brought forth the opportunity space for us to be able to question the authority of their current order. They had brought forth the opportunity for us to be able to seek, and hopefully bring about, a more desirable future. A particularly attractive Berkeley-love-child-like entailment of the supersession approach was that it sought to honor, respect and to sympathetically understand both our parents' generation as well as all those in our generation who were actually not so happy with the rebellion strategy. They were both wondering about 'our plan', passionately anticipating our new paradigm. As soon as I began trying to formulate the new social-economic political 'plan', I quickly realized that this was no easy task. Indeed, it wasn't at all clear even how to proceed.

These, perhaps coincidental, Berkeley experiences of critically questioning authority served to hone and mature my research agenda in critiquing the Scientific Hypothesis – in seeking to find the More General Theory that could properly subsume and supersede both the Scientific Hypothesis and the associated Mechanical Philosophy.

Like my rebel mentors in the philosophy of science, lacking a clear vision of how to bring forth a new post-scientific understanding, I learned the most by critically challenging my traditionalist physics and Positivist-oriented philosophy of science professors and colleagues. Raising uncomfortable foreign-sounding questions isn't a collegially endearing path. In most cases it was difficult to get scientists to take us seriously. Feyerabend once responded rather bluntly to my bellyaching that: "If you are looking for acceptance and adoration you are in the wrong business."

As I moved into the philosophy of science community I did so still thinking of myself as a scientist in some sense, insisting with the other rebels that however we were to advance it must be firmly evidence-based and be able to make sense of the 'apparent' successes of mechanics, albeit in a new way.

As my thinking matured, I felt that there must be some empirical evidence that would conclusively demonstrate the limits of the Scientific Hypothesis. There were logical and mathematical arguments about the limits of consistency and completeness of any formal, mechanical system. However, in the end I felt strongly that the limit must be empirically, experimentally demonstrable – beyond question.

Central to my approach was to ask what I came to refer to as Popper's Question: What evidence, if it were to occur, would force you to conclude that the Scientific Research Program was inherently limited, an inevitably incomplete approach to a comprehensive understanding of reality? I imagined that the identification of such evidence would at least clarify the path forward toward the new paradigm for the new physics.

All through the period of my research I was somehow aware that the evidence already existed, just as retrograde motion was well-known for centuries prior to the Copernican Revolution. The problem, perhaps, was to make sense of existing evidence for and against the Mechanical Worldview, in a new way. Coming from physics, like the other rebels, Kuhn, Popper, Feyerabend and Lakatos, I naturally focused on Bohr's complementarity thesis as it presented itself in both quantum theory and relativity.

There was a point, a mark in time, in Berkeley, when I stopped looking back, stopped trying to make sense of the new physics within the Scientific Research Program. My journey forward really began with my first real extended one-to-one meeting with Paul Feyerabend in his office. I poured my heart out to him, my frustrations with 'official' science, my burning sense that I thought maybe I could see another way – another way to understand ourselves and our place in the universe. It was intense, emotional. I just unloaded on him for about fifteen minutes without a pause. He listened – quiet and attentive. As I fumbled to a finish he finally spoke. "I understand," he said. I was a little shocked by his simple straightforward expression of a recognition of what others seemed to find so difficult to grasp. A door opened, here was a kindred soul – and a mentor. Before I left his office that day he gave me three books right off the bookshelf in his office: Karl Popper's *The Logic of Scientific Discovery,* and *Conjectures and Refutations* and Thomas Kuhn's book, *The Structure of Scientific Revolutions.*

These Berkeley experiences were the beginnings of my involvement in the 20th-21st century journey to discover a better, more general post-scientific understanding of reality and a new better understanding of our role in the universe and its evolution. The journey, humanity's modern journey, has, by its very nature, been an exploration, experimental, marked by surprising intellectual unfoldings. My personal path has been just one path amidst many in this broad and ongoing cultural revolution. Throughout this book, covering Hawking's four-city lecture tour, as well as our dialogues about and recountings of the ongoing revolution, I am your humble guide.

This isn't a book about Stephen Hawking – and yet it is. Stephen embodies the modern dilemma in dealing with the Scientific Litany. On the one hand, he seems to be the quintessential scientist, symbolic representative of those who imagine they are seeking the universal laws that uniquely determine the course of events in the universe. On the other hand, his life exemplifies the socio-political activist, champion of the disabled, seeking through his deliberate efforts to bring about a more desirable future. Stephen serves here to personify, and thereby to make more accessible to a wider audience, what are otherwise abstract technical arguments. The effort here is to present the questions about the nature of universe and our place in it by means of the questions about 'Who is the real Stephen Hawking?'. The narrative tension is between Stephen Hawking the Spectator on a fuller deterministic, classically scientific universe and Stephen Hawking the Participant, seeking to bring about a more desirable future in a developing universe. Stephen, of course, is symbolic, representing every scientist. Even more generally he symbolically represents everyone who questions, and anyone seeking to bring about a more desirable future.

2

Lunch at the End of the Universe

"Don't."

It was just one word. The sound from the voice synthesizer delivered to us as if by some sorcerer through his apprentice.

What had been a raucous party atmosphere suddenly just stopped – mid-thought, midsentence – everyone and everything frozen in time.

All eyes shifted to the head of the table. I caught a brief look from Pam, sitting close on Stephen's right. 'Dublin Pam', the most medically skilled of the nurses, had been working through the lunch menu with Stephen. She knew what he meant sooner and surer than the rest of us.

I was the first to move – to unfreeze, perhaps because I was feeling more immediately uncomfortable than the others. We had all been co-conspirators in dishing out a little justice to our ill-mannered waiter. Now, once Stephen had spoken, we epitomized abandonment; responsibility for our enterprise evaporated.

It had been just over two decades since my journey had begun in Berkeley. Following Hawking's first spectacular lecture event in Portland the night before, we were now headed south to Eugene for a lecture this evening in Eugene.

En route, to save travel time, I suggested drive-through at McDonald's. This brought me an immediate encounter with one of the Nurses' Rules: "Stephen WILL eat." Skipping or even rushing a meal is unacceptable. They had medical orders. Stephen was in charge, but he had delegated elements of his lifestyle to the nurses. Nurses Rule! – at least on this issue. No compromise. "Dr. Hawking will have a proper lunch." … "And he will not be rushed… Thank you very much."

Being new to Stephen I might have been the last one in those few moments of stillness to understand, but I was the first to speak. With my pencil in one hand, I held up the small comment card in the other… "You mean…?" Eyes darted back and forth, noticeably meeker than a moment before, glancing at Stephen and at each other over the white cloth-covered table, set formally with silverware, flowers and monogrammed napkins.

The stark emptiness of the huge dining room accentuated 'the word' – and equally accentuated the ensuing, penetrating silence.

Stephen didn't say anything more. He just looked at us, no obvious emotion in his face. But that was the emotion. Heads started to hang. There were audible sighs of resignation.

"Well, yeah…" "Right." "Sure." Everyone was falling into line. Within a few seconds our righteous little tribunal had transformed into a repentance party.

The dining room was illuminated by a white radiance of early afternoon sunlight streaming through a gigantic window that revealed a spectacular view south down the Oregon coastline. Spray from the Pacific Ocean had lightly coated the window with a crystalline mist, filtering, perhaps polarizing, the light. Since we were the only ones, the only table, being served in the entire dining room – there was an eerie quiet.

Everyone had been so intent on our moral mission that we now struggled for conversation. "Nice weather we're having." Smiles. Giggles. "Yes! Hasn't rained in hours." Contagious, understated, laughter brought us back into a comfort zone.

The occasion was our lunch at the Inn at Spanish Head in Lincoln City on the Oregon Coast. The Inn at Spanish Head is built on a 600-foot high outcropping where the mountains of Oregon's Coast Range descend to the seashore. Only a small, anchoring segment of the Inn can be properly described as built 'on' the cliff. Most of the hotel and restaurant hangs dramatically seaward over the edge of the cliff, facing due west to the vastness of the Pacific Ocean. From any one of the numerous floor-to-ceiling windows you can look out to the ocean, or down – way down – to the sandy beach, and, depending on the tides, to the waves softly, slowly, with a timeless reliability, caressing the shore.

The maître d' had seated us and politely handed us each a menu. Conforming to Nurses' Rules meant that lunch here was bound to take longer. I swallowed hard, thinking about the sold out crowd of 2500 people who would be sitting and waiting in Eugene's Hult Center if we didn't make it there in time. It was nearly 2:00 pm and the restaurant did not normally serve mid-afternoon. I had encouraged the manager to seat us. He had agreed, despite the fact that the cooks and wait staff were ready to leave for the afternoon.

Soon conversation lapsed as we all buried our faces in contemplation of the list of culinary offering. Pam held the menu up in front of Stephen while he looked it over. Pam Benson, who lives most of the time in Dublin, Ireland, is a neurological surgical nurse. Neurological surgery is the specialty in which they open people's skulls and remove brain tumors or do repairs following traumatic head injuries. She often travels with Stephen. Pam is attractive, slim and fit, probably in her in mid 30s but prematurely graying. Pam is also a runner – marathons. Head nurse on this trip, Pam is reserved most of the time, unless Stephen is having any sort of problem, at which point you see a valued professional jump into action. She is quiet but attentive, fully aware of everything happening around her – ready. Pam Benson is also just one of the nicest people you could ever hope to meet.

Pam Benson

After a few minutes the waiter arrives. He is in his mid-thirties, noticeably chubby and slightly effeminate. Immediately, just in the way he approaches the table I am sensing that he is a little annoyed with a world that requires him to be in this job. From the tone of his greeting and his ill-at-ease demeanor it was apparent that he had a problem. This was not your happy and cheerful culinary service provider. When someone asked for a glass of water he gave a loud sigh of exasperation. There was a notable tone of disgust in his assent to the second request for water. There was no third request. In the beginning we were cowed. There was a natural hope that maybe he just had a bad moment and everything would get better from now on. But it didn't.

Tensions began to escalate when Pam, who was 'on duty' with Stephen, asked about the soup de jour. "Clam Chowder," the waiter snapped. Pam asked whether it had been made with gluten. The waiter appeared to just ignore the question. Whatever she had said must not have been directed at him. "Excuse me!" And Pam repeated the query. This didn't elicit a polite response. "What's gluten? What does it matter?" he snapped. In an attempt to dismiss the question entirely the waiter offered: "Look, it's just regular food." Pam remained perfectly composed – that was Pam – and persistent. "I understand that it's regular food. But some people use wheat flour to thicken a soup and some use cornstarch. It is important for us to know how the food was prepared since some people have a mild allergy to the gluten in wheat flour."

"How should I know?" the waiter snapped.

"Well, perhaps you could ask the chef," said Pam, very calmly, politely, with an impeccable, educated Irish accent, looking him straight in the eye. His annoyance was now more postural than verbal. His weight shifted to one side and his eyes turned toward the ceiling – perhaps seeking strength from above in dealing with this latest imposition. It was written all over him: 'Not only do I have to stay late, I have to put up with this!'

In the mean time the rest of us, heads bowed, intensified our study of the menu, beginning to dread our turn with the waiter. At first, some of this seemed funny – quaint waiter. But it soon became apparent that this guy really was just being rude.

Although his snippy, understated anger had temporarily frightened us into obedience, gradually we were becoming angry. We felt *actively* abused. Offended! When he walked away from the table our silence ended.

"Can you believe how rude this guy is?" said Jonathan, who is the newest in a long line of graduate students serving as Stephen's personal assistant for nonmedical matters. It is not a cushy job. Tall and lean with dark hair, Jonathan Brenchley is studying mathematical physics in Stephen's department in Cambridge University in England. Jonathan's priority tasks are to make sure Stephen's electronically sophisticated wheelchair, computer and voice-synthesizer are working.

Being the official Oregon host I felt that decorum called for me to apologize for the waiter's behavior. "This is NOT typical of Oregon," I pleaded.

Mild reactions slowly but surely morphed into outrage, and into a spirit of rebellion. Our final psychological transition begat an aggressive stance.

"Did you see how he just snatched the menu out of my hand?" complained Diana.

"I was still reading it." Diana Briscoe, the accompanying nurse, is a piece of work. Mid-thirties, she is strikingly beautiful – not cute – more gorgeous and sultry. With her long dark hair, she dresses stylishly with make-up and noticeable jewelry. Unlike Pam, Diana might not be the perfect candidate for who you would want to take home to meet Mom and Dad. It crosses my mind that if you had to be in Stephen's physical situation, the constant care and presence of nurses like Pam and Diana would certainly make it all a bit easier to take.

I had noticed, at the entrance, at the maître d' station, on the way into the dining room, some 'comment cards'. So it was decided – through a sort of anonymous agreement – that we should obtain a batch of these cards in order to 'responsibly inform' the management about the behavior of this waiter.

Even Sue Masey, the final member of our party, had joined in. Sue is the cheery administrator in Stephen's department – DAMTP, Department of Applied Mathematics and Theoretical Physics at Cambridge University. This was an uncommon step to the dark side for Sue, who is in her forties, dark hair, slim, divorced with two teenage daughters – and always smiling. Departmental administrators who last for more than a year or two must be undaunted by bureaucratic complexity, daily frustrations and professorial prima donnas. Sue is most affectionately the perfect accomplish in Stephen's ongoing comedy show. She never misses an opportunity to tease him – never allowing anyone to take him too seriously. Stephen responds to her in kind.

Mind you Stephen has not said anything throughout the build-up of tensions with the waiter. He has been communicating with Pam in a series of quick 'yes' and 'no' expressions working through the menu. "Would you like cordon bleu or the roast beef?" Eyes-brows up is 'yes.' Eyes or lower jaw side-ways left is 'no'. Eye signals are much quicker than composing on the computer. For the uninitiated it isn't easy to read these gestures. His communications are complex. Sometimes he is just looking around, and sometimes responding to queries from others.

There was a certain joy in our self-righteousness. While the waiter was still in the kitchen, I slipped over to grab a small stack of the comment cards, passing them out. "Who has something to write with?" Composition. The poetry of justice. This was right. This was good. And we got to return the insult at the same time. Glorious! Giggles of joy. Stern words hit paper.

It was then that Stephen interjected: "Don't."

Full stop!! Stephen had spoken. But what did he mean? He wasn't responding to Pam about anything on the menu. He was looking straight at us.

Since Stephen spoke through his voice synthesizer there was no unique intonation to the single word. No extra clues there. And since he doesn't move much – if at all – there was no body language to read. Just the word, "Don't," from the robotic voice synthesizer.

In a very few moments it became clear that he was suggesting that we call off the 'let's get back at the waiter' agenda. But why? And *how* was he saying this? The way I first took it, understood it, might be translated as, "Hey, group, let's not do this." A second thought was, "Maybe this isn't such a great idea."

It wasn't like a command, although it could have been taken that way. I was aware that he was in charge. Stephen ran the group – moment to moment. Any decisions to be made? Ask Stephen. His comment in this case was more like a vote – a little delayed – in the formation of the consensus.

We could have done it anyway. At least I certainly could have – not being under his direct employ.

What Stephen had offered was a thought. Stephen had suggested and was indeed pushing us toward what he intuited as a better, more beautiful and more just universe. Stephen had offered us a different vision of time-future – not a prediction, but a prescription offered by a Participant in a universe that could improve, could become better. 'The word' had been a moral act of some sort – an affirmation of value – redirecting the universe, ever so slightly in a new, and for the rest of us, unexpected direction.

On the other hand 'the word' and the events surrounding it could be described scientifically. There had been a rise in energy followed by an exponential dampening – and perhaps a little scattered light. Thinking in terms of the Mechanical Philosophy it was just another rearrangement of molecules energetically dispersing themselves toward a cold, meaningless equilibrium at the end of time.

Stephen hadn't said anything further on the matter. He had just glanced around the table. He looked at us. We looked at him. It was over. You could have cut the atmosphere with a knife. There was some serious backpedaling, people looking at their napkins and rearranging their silverware in front of them. Gradually, what had seemed like such a good idea, on reflection, now seemed like a rather bad idea, small and petty. The topic changed and the lunch moved on.

And yet, on that rocky outcrop several hundred feet above the Pacific Shoreline on that day in January, 1992, waves ceaselessly moving toward us, scattering the myriad particles of sand on the beach beneath us, all this drenched in sunlight, another side of Stephen Hawking had been revealed to me.

It had been just one word, yet the universe had stirred, just slightly, reorienting on the open waters of reality. Stephen's 'word' had opened an unexpected door to an unanticipated side of Stephen William Hawking, presenting a dilemma. Who was the real Stephen Hawking?

Officially, of course, Stephen is, Dr. Stephen William Hawking, Lucasian Professor of Mathematics at the University of Cambridge in England, the same prestigious professorship held earlier by Sir Isaac Newton. Hawking was born on January 8, 1942 in Oxford, 300 years to the day after Galileo's death in 1642, a fact he loved to emphasize whenever his birth date was mentioned. Surely a coincidence. Or was it? Hawking took his undergraduate degree at Oxford University and then moved to Cambridge for his

Ph.D. in Cosmology under Dennis Sciama. His scientific celebrity was, arguably, launched in a collaboration with Roger Penrose. Hawking extended Penrose's work on singularities in black holes, and applied it to the Big Bang – to the singularity at the beginning of the universe. Hawking's scientific and personal biographies are well told in several books and movies so they need not be repeated here. The best personal and intellectual biography is Kitty Ferguson's authorized: *Stephen Hawking: An Unfettered Mind.*

After the death of Albert Einstein in 1955, the question arose in the scientific community as to who would replace Einstein as the symbolic genius of science, the informal Pope of science, the final word – 'he who could bless and condemn.' The natural choice for a variety of reasons had been John Archibald Wheeler, whose students included Richard Feynman and Kip Thorne, the latter being one of Hawking's closest friends. Wheeler had casually coined the phrase 'black hole' as an alternative to the more cumbersome 'gravitationally completely collapsed star.' Wheeler was impressive, but he had refused to perpetuate, what is referred to by detractors, behind close doors, as 'the Einstein cult.'

Both inside and outside of the science community, perhaps primarily through the characterization of Hawking in the popular press as 'the disabled genius,' Stephen has, at least in many people's minds, come to wear the 'mantle of Einstein.' Theoretical physicist and world famous cosmologist; that was one side of Stephen William Hawking, the most public face. What I experienced on this four-city lecture tour was something different, something more.

Stephen Hawking now seemed to me to live in two incompatible worlds. As a scientist he was searching for the universal, timeless laws that determine the inevitable course of all events in the universe. As scientist the image was of an abstracted, detached observer, a Spectator intent on explaining the workings of objective reality – 'out there'. Yet Hawking was also inside the universe – a Participant. And as became increasing clear he was a morally deliberate activist. As a Participant in a historically developing universe, Hawking seemed to be working to alter the course of events; he seemed to be working to steer the universe toward a more just future.

By choosing amongst real, alternative possibilities, Stephen Hawking, the embodied Participant, seemed to be trying to improve the universe, trying to bring about a more morally desirable future.

I had entered Berkeley as a scientist – thinking of myself as a scientist. I never left that behind. And yet it seemed increasingly clear to me that the dominant representation of science, as engaged in a detached Spectator-like inquiry about objective realty – 'out there', was seriously incomplete. The humanities, many recently attempting to transform themselves into sciences, seemed equally incomplete. The political, economic, social, and psychological 'sciences' had ironically embraced the Spectator's detached paradigm, concerning themselves at best, at least 'officially', with mechanistic explanations of human activities – 'out there'. Philosophy was no better, having become dominated by the new 'analytic' tradition of the Logical Positivists.

My initial frustration was that I could not find any obvious way forward both for myself and, hopefully, to lead the scientific and the philosophy of science communities to think 'outside the box', to think outside the Mechanical Paradigm. In seeking a more general perspective, I had to explore in foreign territory. None of the presuppositions of the Scientific Research Program would help guide me here. There was no map. How could I navigate?

The path forward was inherently problematic, uncertain and unclear. The approach had to be exploratory, self-consciously experimental and regularly self-critical.

3

In the Beginning... a Serendipitous Breakthrough

My attempts to engage Hawking to be part of the lecture series had started two years earlier. One glorious day my persistent, previously unproductive phone calls and letters to Hawking's office in Cambridge made a serendipitous breakthrough. The discovery was the key to finally penetrating the fence lines, hedgerows and communication filters surrounding Stephen. My invitation to undertake a four-city public lecture tour in the Pacific Northwest was now to receive serious consideration.

Following the last of my five years of graduate work in England at the University of London I had migrated back to Oregon, taking up teaching positions for seven years at Linfield College and Portland State University. With the encouragement of the Provost at Portland State, I had created the Institute for Science, Engineering and Public Policy, a freestanding non-profit institute to be affiliated with the university. One of the main activities of the Institute was a Public Understanding of Science, Technology and Society Program that developed into a major local cultural enterprise, drawing an average of 2000 attendees per lecture.

The invitation to Hawking had made it through the first gate – credibility – only because I could point out that I had successfully organized public lectures for celebrity scientists such as Richard Leakey and Dr. Carl Sagan. At the time I first contacted Hawking's office, the Institute was presenting its first full six-part series, including ethnologist Jane Goodall, environmentalist David Suzuki, evolutionary biologist Stephen Jay Gould and MIT physicist Philip Morrison.

By virtue of my accumulating, yet still unsuccessful phone calls and emails, by midsummer of 1991 I had begun to develop a speaking acquaintance with Sue Masey, the cheery departmental administrator at DAMTP – Department of Applied Mathematics and Theoretical Physics at Cambridge University. Sue Masey had pointed out the hedgerows… "Stephen isn't easy to work with," she said. "I want to schedule his visit a year in advance," I said. Sue laughed, pleasantly warning me, "You'll be lucky if you have sixty days. Stephen just doesn't like to arrange things very far out." Sue and I came to describe this as Hawking's policy of 'making decisions on a need-to-decide basis.' If it was possible to delay making a scheduling commitment, it was delayed. This, of course, gave him maximum flexibility if something more appealing came up. And as far as appealing options in his life, the proposed lecturing visit to the Pacific Northwest wasn't a high priority.

Another hedgerow was that Hawking traveled with at least two nurses, a graduate assistant and occasionally an administrative person, like Sue Masey. This last inclusion depended on the complexity of the travel plans, possible press involvement, and so forth. The nurses took care of Stephen's medical needs on a 24-hour, 7 days-a-week basis. The graduate assistant was responsible for all technical matters, primarily Hawking's high-tech, battery-powered wheelchair and the computer system attached to the all-important voice synthesizer. Hosting Hawking for a public lecture meant transporting and accommodating four to six people from the United Kingdom to the United States and back.

Another early conversant at DAMTP was Hawking's then graduate assistant, Stuart Jamieson. "Hi, it is me again – Terry Bristol in Oregon. How are you today? Just calling to see if we've made any further progress on the latest proposal." Stuart wasn't optimistic.

The real key to gaining serious consideration arose unexpectedly in a conversation with Stuart following the rejection of my third revised proposal. At the height of my frustrations, I fell into a sort of rambling, free-wheeling inquiry: "What does he want?" I asked, meekly adding, "just wondering?" Then I started probing with more focus. "What are his buttons? What would it take to persuade him?"

In my early experiences running the Institute's lecture series, I had come to understand that adding a little extra incentive was often what convinced the speaker-invitee to accept. For instance, 'How about staying over for the weekend in a resort in Central Oregon?' And I had learned the importance of being personal. Nothing turns off a public lecturer like being treated like a commodity. Treat them like a colleague, partners in the overall effort of enhancing the intellectual culture, the Public Understanding of Science, Technology and Society. And you definitely ask – "Is it OK if I call you Stephen?"

"Well," said Stuart, "if your lecture tour had extended further south into California, that might have made a difference."

The invitation had been for lectures in the Pacific Northwest – Portland, Eugene, Seattle and Vancouver, British Columbia. Gradually it became clear to me that Hawking only rarely did anything that couldn't somehow be related to his academic responsibilities and research interests. One might have expected this sort of focus and commitment from a leading-edge research professor at Cambridge University. There were expectations of, and consequent pressures on, Stephen William Hawking, the highly visible Lucasian Professor. California was appealing to Hawking, not for the sunshine and beaches, but for the presence of his close colleague, Dr. James Hartle at the University of California at Santa Barbara.

As soon as Stuart had outlined the relationship, my proposal began to transform: popular public lectures in the Pacific Northwest might make sense to Hawking as part of a larger schedule that included seeing and working with Jim Hartle in Santa Barbara.

"Is that all?" I blurted back in amazement.

What had come to mind as Stuart talked was a little background knowledge about international flights. On international flights, the cost of extending your flight to a further destination as long as it was more or less in the same geographical line was relatively minor. I envisioned Hawking and his crew flying from London via the north polar route to Vancouver,

British Columbia. This was the route I had first taken, in the reverse direction, to attend graduate school at the University of London. Extending south from Vancouver in a line through Seattle, Portland and Eugene was straightforward. Santa Barbara was, from a north polar route perspective, a simple extension along roughly the same geodesic.

This newly imagined invitation broke through the communication barriers, and Hawking himself actually took it seriously. At the time, I assumed that Hawking's interest was reasoned as follows: 'Bristol is willing to pay for me and my entourage to visit Hartle in Santa Barbara, and all I have to do is a couple of public lectures on the way.' The question of a visit to Santa Barbara had simply never come up in any of the original conversations. Why would it?

"So, the Institute will fly him to Santa Barbara as part of the deal?" Stuart was noticeably surprised and, I think, a little suspicious of the simplicity of my solution. "You mean you'll fly him to Santa Barbara even though you aren't doing a lecture down there?" he asked. "Yes," I said, and explained that the extension of the international flight wasn't very expensive.

And there was light.

We finally began to discuss particulars, hypothetically, to understand what Stephen could and would undertake as a visiting lecturer. Sue Masey rejoined the conversation. "We're really not very easy to work with. Are you sure you want to do this?" she reemphasized. Sometimes in life you just gotta say, 'what the hell.' I figured that if we lined up everything except the specific date – a corporate co-sponsor, publicity materials without printing them, and a detailed understanding of Stephen's special needs, then we could, in fact, make all the final arrangements within a sixty day window. A little crazy, but doable.

Sue let me know, two days before Thanksgiving, 1991, that Stephen was open to two, not four, public lecture dates in January, 1992 – for Portland and Eugene. – Yikes!

Hawking was initially conflicted about taking a lecture fee for a public program where tickets were sold. As Lucasian Professor at Cambridge University he was not just any professor. He was, just by virtue of the title, a principal academic role model in a leadership position. Much of his early concern with doing popular public lectures where tickets were sold arose from a natural conservatism born of a deep sense of cultural responsibility seated in university traditions. But I also know that Stephen, in contrast to many of his academic colleagues, admired Dr. Carl Sagan. Sagan had become the quintessential popularizer of modern science – who, by the way, had no problem charging what the market would bear. On the other hand, Sagan did at least as much for free – pro bono. Stephen admired Sagan for communicating the current state of modern cosmology to a broad public and perhaps even more for his courage in standing up and speaking out politically about the possible Global Winter that would result from an all out nuclear exchange.

Again the activist side of Stephen Hawking can be seen in his admiration for the political activity of scientists like Dr. Linus Pauling. My office is now in Linus Pauling's boyhood home in Portland. When I mentioned this to Hawking he was impressed with my efforts to promote Pauling's legacy and told me that Pauling was his cultural hero. Pauling had received two unshared Nobel Prizes, the first in chemistry, in 1954, for his work on the quantum nature of the chemical bond, the foundation stone of much of 20th century chemistry. Then in 1962, Pauling was awarded the Nobel Peace Prize for his leadership in rallying the global scientific community in support of nuclear disarmament. Hawking was surely impressed by Pauling's scientific work leading to the first Nobel Prize, but it was clearly his moral activism leading to the second Nobel Prize – to the Nobel Peace Prize – that gave Pauling first rank with Hawking.

The altogether unexpected spectacular success of *A Brief History of Time* raised in Hawking an awareness of his potential influence in the broader cultural setting outside the academic box. "Stephen has a 'calling'," I pointed out to Sue and Stuart and eventually to Stephen, regularly, whenever I had the chance. "You have a duty to communicate with the public, to follow up on the book, to fulfill the role of public intellectual – to be a cultural leader." Hawking of course understood all this way before my encouragements.

Earlier Hawking had found himself and his place in the community of academic scientific inquiry, in the search for the hypothesized universal mechanical laws governing the universe. Like all scientists he tacitly understood his 'official' role here as some sort of Spectator on a fully deterministic reality. The scientific inquirer is detached from the subject of study. Scientific inquiry never alters 'objective' reality – 'out there'. Reality is understood, presupposed, to be, by its very nature, independent of the observer, independent of the inquirer. This separation between the inquirer and objective reality had at least tacitly been accepted as one of the defining characteristics of Modern Science.

Yet the real, full Stephen Hawking had now come into, and had accepted, a new, unexpectedly wider role as a Participant in altering the nature of reality. Beyond communicating the latest cosmology to the general public, Hawking had become an inspiring figure and a role model for millions of people with disabilities. No longer merely the detached scientist, Hawking had now willfully begun to explore and engage reality in a new way – at least now more powerfully – to move reality toward a more desirable future.

The Spectator and the Participant offer two different representations of the nature of inquiry and learning. It surprised me, and surprised me that it surprised me, to realize that each of these contrasting representations of inquiry carries with it *parallel entailments* about the nature of the universe being investigated. This link between one's representation of the nature of inquiry and one's theory of the nature of reality became a cornerstone of my research, what I later came to refer to as the Parallel Hypothesis.

The Spectator framework can be characterized in terms of the following thought experiment. Here we are – incarnate in the universe. Imagine that we have enough food and shelter, and now things are getting a little boring. So one of us proposes: "Why don't we try to figure out how the universe works?... Just for fun." The illuminating realization is that in order for this enterprise to be successful, for us to be able to move from our current state of ignorance to an understanding of the nature of reality, it is important – essential – that the nature of reality remain the same over time. If the nature of reality was changing arbitrarily we would be unable to converge to a final complete understanding. There would be no time-invariant target. Similarly, the nature of the reality must be the same everywhere, in different locations. If it weren't it would not be One reality. The somewhat surprising parallel entailment of the Spectator representation of inquiry is that the nature of reality must be time-space invariant. Another way to say this is that the order governing what happens in the universe must be time-space invariant, must be the same for all time from the beginning to whatever future and must be the same (consistent) everywhere. This seems to be equivalent to what I have called the Scientific Hypothesis and defines the Mechanical Worldview: all phenomena in the universe are governed by One time-space invariant order, by universal (viz. time-space invariant) laws.

One further implication of the Spectator representation of inquiry is that our actions as inquirers must not disturb, must not causally alter, the nature of the reality. This has led to the 'detached observer' characterization of the Spectator representation. We are studying the nature of 'objective' reality – 'out there' – as it operates independent of our inquiry, independent of being observed. Einstein, who always opted for this classical representation put it succinctly: "Physics is an attempt conceptually to grasp reality as it is thought, independently of its being observed. In this sense one speaks of 'physical reality.'"

The Spectator is a passive observer. In the Spectator's representation of observation, information and causal influence flow in only one direction – to the Spectator. If the Spectator's activities as an inquirer were causally to alter the 'objective' course of events, then the 'target' – the nature of reality – would be altered in ways the inquirer would be unable to discover. For the Spectator enterprise to be successful, to be able to arrive at a complete and consistent understanding of how the universe works, a complete and consistent understanding of the nature of reality – the order governing reality, must be, and remain, time-space invariant.

And finally it is important to keep in mind that in the Spectator representation inquiry is 'just for fun.' The Spectator's research enterprise does not, by its very nature, imagine or claim that there is or will be any practical benefit resulting from the successes of the research program. Pressing the point a little, it will be difficult, if not impossible to account for any 'benefit' – since the Spectator representation has the parallel entailment that the nature of reality is governed by One space-time invariant order and that the course of events is completely and consistently causally determined by that order everywhere from the beginning of time.

The Participant representation of inquiry is quite different and can be characterized as follows. Participant inquiry is embedded and embodied in reality. The Participant representation immediately accepts that inquiry and learning alter both the course of events as well as the structure and function of reality. The Participant representation of inquiry has the parallel entailment that Participant inquiry must actually alter the nature of reality – in the sense of and as conceived in terms of the nature of the Spectator's uniquely determining mechanical reality. The Participant representation of inquiry entails that the nature of reality itself must develop along with the inquiry. As the Participant learns – discovering and implementing new innovative ways of doing things – the universe, the very nature of the universe – that is, how it works – develops, evolves. Real learning for the Participant is not a theoretical convergence, to what the Spectator presupposes, is a fixed reality. Real learning for the Participant is emergent and involves a self-inclusive progressive development.

In the Spectator's fully deterministic time-space invariant reality, it is unclear how to make sense of inquiry as either an essential, or even possibly coherent, aspect of objective reality 'out there'. Spectator inquiry can't make sense of itself as anything more than a completely passive, meaningless entertainment – 'just for fun'. In the Participant understanding of reality, successful inquiry has an integral, beneficial, role in the development, in the progressive evolution, of reality.

In the Spectator perspective, Hawking, the scientist, is represented as detached from the world. Even if imagined 'somehow' self-reflexively to be part of the fixed time-space invariant order, everything he does, as well as everything we think about him, is supposed to be understood as completely determined from the beginning – everywhere and always. On the other hand, in his very down-to-earth, in-the-world human reality, marked in part by his courageous response to his disability, Hawking is an activist for persons with disabilities, as well as other social and political causes. Understood as a Participant, he is trying to alter the course of events, perhaps altering the structure and function of reality, trying to steer the universe toward a more desirable future.

I laid out these contrasting Spectator and Participant representations of inquiry and their associated entailments about the nature of reality to one of my physicist become engineer colleagues, Bruce Adams. He responded in the 'affirmative' saying, "Yes, there are two realities. The 'real' deterministic reality and then the other reality where we live our lives." I responded, "Bruce there is only one reality and these are two incompatible representations. What do you really believe?" Bruce responded, "There are two realities. The 'real' deterministic reality and then the other reality where we live our lives." I said, Bruce, there is only one reality – take your pick." Bruce responded, "There are two realities. The 'real' deterministic reality and then the other reality where we live our lives." This frustrating back and forth 'dialogue' went on for a while, neither of us giving way to the other and then we went our ways.

As my journey unfolded, frequently encountering these dialogic impasses, I came to reflect on a moment in Berkeley where I picked up a wisdom about engaging people with whom you disagree. I guess my natural tendency had been to be a little combative. I mean, 'obviously', when we disagree I am right and you are wrong. That is what a disagreement is all about – or maybe not. Being combative makes sense if there is only 'one right answer' to all questions. The strategy, as in the debate model, is to defeat the other's position – show him that he is wrong, and to (hopefully, supposedly) convince him of the truth of your position. It is easy to become angry when debating those with whom you disagree, to question their motives and to think bad thoughts about them.

The moment of wisdom was offered to me at what was a major rally in Free Speech Movement in Berkeley's central Sproul Plaza. The immediate plan was for the crowd of perhaps 1500 students to march about a quarter mile across campus to where the Regents of the University of California were meeting to decide what to do about our demands for unrestricted political speech – free speech – on campus. We were just going to sit quietly outside their meeting, so they could see us if they looked out the window or, at least, as they went in and out of the building. Several of those speaking to the crowd had us pretty riled up – stirring a righteous indignation about the proposed oppressive policies of the Regents. Obviously, we were right and they were wrong. As we were organizing, locking arms, into lines of 16-across – forming an impressive spectacle – I remember having this sense of anger and aggression. "Stupid Regents!" "Yeah, let's get 'em." As the procession took some few minutes to form up, folk singer Joan Baez had begun to entertain us with uplifting protest songs – "We Shall Overcome" and such. Then, as she still had the microphone, just as we were departing Sproul Plaza, she paused for a moment and then said, "Do this thing with love in your heart."

What?

A natural challenge of my research agenda to understand the limits of science and to find a More General Paradigm was to 'convince' my scientific and humanities friends and colleagues, to get them, perhaps by force of argument, to think 'outside the box'. How should I deal with those who simply disagreed with me, who opposed me, sometimes vigorously? As my journey matured over time, that moment of wisdom in Sproul Plaza kept coming back to me.

4

What It's Like Being with Stephen Hawking

Most people find it difficult to be with Hawking for the first time. It takes some adjustment to get used to the pace. Not that everything is just slower, some things are faster, and some things are slower. The harmony of the overall ensemble is different. Synchronization is not straightforward.

Besides his computer and voice synthesizer Hawking has evolved a system of quick "yeses" and "nos" with his staff. 'Yes' is signaled by eyebrows up. The actual motion is subtle, and the novice can easily miss it. 'No' is the eyes and/or lower jaw moving to his left, also subtle.

Those who spend more time with Hawking learn to read his face, his expressions. This of course opens up a much wider range of communications – signs and signals. However, as the range of non-verbal expressions increases, so too does the opportunity for miscommunication, misunderstanding, misinterpretation.

In the beginning, the greatest initial difficulty is the relativistic time delay between the greeting – "hello" – and Stephen's response. What do you do for the five to ten seconds while Stephen is composing his response? At receptions when he is expecting a series of polite exchanges – 'Hello, how are you?' 'Pleased to meet you.' – Stephen is ready and fairly quick. He has preselected the screen that has a few such phases so he can quickly trigger them with a touch of his finger on the gravity switch sending the appropriate pattern of electrons from his computer to the voice synthesizer.

If you greet him with a substantive question, the response might take one to two minutes. And here is where one of the more complex discomforts begins. After asking your question, how do you know he is even in the process of responding? Perhaps he didn't hear you. Perhaps he is composing a response to someone else who asked him a question before you entered.

At first I expected, I think quite reasonably, that the staff would help by looking over Stephen's shoulder to see if he is composing a response – and to somehow let you know. His computer system requires that he compose his remarks on the screen, facing Stephen, not you, and only then send the sentence or two to the voice synthesizer. By personally peeking over his shoulder one could, theoretically at least, reduce the uncertainty and associated anxiety by seeing that he is indeed composing – something. One could then return to the waiting position and smile and make small talk with the others who are standing around until Hawking has completed the composition and turned it into voice.

However, peeking over Stephen's shoulder is not cool. It's actually just impolite and annoying. One learns rapidly that there is etiquette. You might think you are assisting communication, but the effort is unwelcome. The problem is similar to what many people encounter when trying to help someone in a wheelchair to do something that they can do themselves. You are just going to speed things along. Right? Intentions being so pure, it is a bit shocking to find that they don't want your help. You are being patronizing. Nonetheless, in dealing with Hawking in social situations it is not infrequent that one of the staff peeks, but only in exceptional circumstances, dealing with the uninitiated. To make a mistake about whether Stephen is or is not responding to a question may mean that you stand there for a minute or two or three, only to realize that Stephen perhaps didn't hear you or simply chooses not to respond.

Another unexpected matter of etiquette is that the staff never answers for Stephen – even when they know the answer to your question. For instance, on one occasion, when reconnecting with the group, seeking an update on the schedule. I asked, "Stephen, have you had lunch yet?" The nurses could have spontaneously answered for Stephen: "Yes, he ate about an hour ago."

But they didn't. The question lingered. So I asked the nurses directly, "Has Stephen had lunch yet?" No response. Over time I learned that this strategy of bypassing Stephen is definitely frowned upon. First of all, you are talking about someone in the third person who is right there with you, "Has he had lunch yet?" What happens when you try this approach is that the staffer doesn't answer and looks at Stephen. The nurses interpret the question directed to them as a repetition of the original question – once again directed to Stephen. So I just waited for Stephen to answer. Finally, "Yes."

You can tell that the various difficulties in communication cause Hawking a fair amount of frustration and occasional distress. I recall being in his office in Cambridge once trying to discuss the arrangements for an upcoming lecture tour. As was typical, others were coming and going while we were talking. Stephen was multi-tasking and somewhere in the fracas the answer to my last question got lost. Stephen had composed the answer on the computer, but was interrupted by a question from his nurse, and he forgot to send his response out through the voice synthesizer. The nurses proceeded to ask Hawking several questions about what he was planning to do that afternoon.

I don't recall the particulars of our exchange at the time, but I remember feeling uncomfortable and embarrassed when I realized that Stephen had moved on to another conversation and left me sitting there – two feet away from him, waiting for an answer. I mentioned something to Pam, who was doing something behind him, and at the same moment Hawking looked at me with a sense of shock in is face, realizing what had happened. He was obviously bothered and embarrassed. Pam rescued the moment by recovering and reading Stephen's answer to me, which had scrolled down a couple of lines, off his computer screen.

Once you are around Stephen for any extended period, these miscommunications lessen and the tendency to be embarrassed or distressed, even when it happens, decreases, and the inclination is to have a little fun with it, to tease, to joke – and to laugh. It's Stephen's way.

Normal human conversation clips along at about 100-200 words a minute, even more for fast talkers. Hawking's regular pace for anything beyond the basic "yes, no, hello," and so forth is about 10-20 words per minute. One learns to develop a certain flow when interacting. The preferred strategy is to ask him something, pause a few seconds, and make unrelated small talk with the staff, or just stand silently, while you wait for his response.

Perhaps the trickiest complication is that it is not always clear that Stephen heard you or was paying attention to you, rather than to someone else, at the moment you spoke. This is doubly difficult if two people ask him a question at the same time. In normal, non-Stephen interactions, when such ambiguities arise, they are typically sorted out by eye contact – 'I am responding to you, because I am looking at you' – or by similar non-verbal body language. Since Stephen doesn't move his head easily and his eyes are usually paying attention to his computer screen or working to navigate obstacles with his wheelchair, there is often a lack of the direct eye contact that might otherwise clarify. Consequently, it is not easy to know whether he heard you and is responding. If he is looking at you, then he is not responding. If he is responding, which requires looking at his computer screen, then he is not looking at you. And there is an unavoidable uncertainty when he is hopefully composing as to whether he might be paying attention to the computer screen for some other reason.

Another complication occurs when Stephen requests something from his nurses while you are in the middle of a conversation.

For instance, I asked him one time in Seattle whether he had decided to skip the plans for lunch at the Space Needle, the schedule being rather tight and all. He said, "Lift." I wasn't sure what that meant. Having spent several years doing graduate work in London, I was aware that Brits refer to elevators as 'lifts.' I guessed that he had some concern about the elevator at the Space Needle that takes you rapidly from ground level to the rotating restaurant 500 feet up in 43 seconds. So I ask whether he has some question about the elevator. Joan, the nurse on duty, standing next to Hawking, thinks Stephen is paying attention to me, so she is mostly tuned out, taking care of some other matter. Stephen repeats, "Lift." I'm lost. I'm beginning to suspect that whatever he is on about doesn't have anything to do with what I'm talking about. These are frustrating moments for everyone. Again Stephen says "Lift." "Oh," says Joan, picking up on my distress, and hearing Stephen's request at about the same moment. "Stephen needs a lift." Here I am still lost. I am thinking, "The Space Needle has a lift, and a very nice one, so what is the problem." Fortunately, I am wise enough to keep my mouth shut at this point. Joan seems to have something in mind. She is looking around. She says something quietly to Hawking. I am wondering, "Am I part of this?" "Have I just been standing here having a conversation with myself for the last minute or two?" I finally gather enough confidence to ask Joan, who is now doing something fairly definite with Stephen, although I can't make sense of it. "Do we know what's happening?" I ask her.

Joan Godwin backstage

"Oh, yes" she says, "Stephen just needs a lift." O.K., I am down to basics now. I am prepared to make a fool of myself. "What's a lift?" I ask. I have a sense of floating. Here I am standing in the lobby of the Alexis Hotel in Seattle disconnected, floating, talking to myself. "A lift, you know, a lift," says Joan. This doesn't help me at all. I am mildly irritated with Joan. Of course I know the word 'lift' and how it is commonly used – but what does that have to do with anything else that has occurred in the last couple of minutes. I am feeling better that at least Joan seems to have some direction. I am lost, but someone knows where we are and where we are heading.

Having rearranged Stephen's grip on the chair controls, Joan now has her arms around Stephen from behind. Something coherent is happening. "Stephen needs a lift every once in a while" she says. "He slips down in the chair and can't get himself back up into a comfortable position. When that happens he asks for a lift."

I tell Joan what I had been thinking, and we have a good laugh. Stephen gives me a big smile.

When Stephen is surrounded by several staff or other experienced colleagues the problem of multiple conversations is markedly reduced by a straightforward cooperative strategy, by a different harmonic – taking turns – allowing him to answer or resolve one conversation before beginning another. Short improvisational interruptions by one of the nurses can even be accommodated. There is friendly competition for Stephen's attention, optimized through a sort of tolerant cooperation – coordination, everyone in harmony, part of the ensemble.

When the staff seriously needs to know what Stephen has in mind or needs to get his attention immediately, they just get in his face – face to face – and talk to him from 6-8 inches away. Quick questions are followed by Stephen's eye-movement answers. There is an established sequence of structured questions. "Are you OK?" If the answer is 'yes', then the next question is about his equipment, "Your computer?" If the answer is 'no', then the next question might be about his breathing. Does he need his lungs cleared? "Lungs?" If the questioner is the graduate assistant, then the next question might be, "Do you want me to get Joan (or the nurse on duty)?"

Hawking never slows down and appears not to tire. No one expects this and some find it hard to believe. I never once heard Stephen ask for a rest, or to pass some travel or cultural opportunity because he was tired. Indeed, Stephen's staff sleeps in shifts and trade-off in order to manage his pace over a period of days and weeks. A standing tease in the Hawking group is how new arrivals to the staff expect that Stephen will tire, and anticipate that taking care of him must be relatively easy. In reality, Hawking is active from fairly early in the morning until late at night, often into the early morning hours, with no rest stops. Naps are a fantasy.

One thing for sure, being a nurse for Stephen Hawking is no easy job.

The single most surprising and striking aspect of being with Stephen Hawking over a period of several days is the almost constant joking and teasing between Stephen and his entourage. It's a fun group to be with.

When people are introduced to, or greet Stephen they are usually polite and reserved. But occasionally someone comes up and gushes all over him: Hawking groupies. "Can I have your autograph?" "I can't tell you how much it means to me to be in your presence." The first time I witnessed this, on the first visit to Portland, I wasn't entirely surprised. What was unexpected was the reaction of the staff a moment or two after the groupie gushing episode. "Oh, no! Now he's going to get a big head." "Well, if they only knew the real Stephen. Boy, could we tell them some stories." "Bet they wouldn't want your autograph if they knew the real truth." Some of this was a British cultural phenomenon – British humor. In much of Britain there is a discomfort, shading into intense dislike, with the sort of hero worship expected for royalty. Hawking is not a royalist (viz. one who supports the Monarchy and its cultural accouterments). It was obvious that Stephen enjoyed the teasing more than the adulation.

Sometimes the teasing would even start in the presence of the fan. "Don't say too much, he might take you seriously. We have to live with him you know."

There were, however, a few instances where the adulation crossed the line from basic respect and admiration to what can only be described as a sort of spiritual experience – not worship, but a respect moving beyond ordinary emotional outpourings. Silence was appropriate. And there was no teasing afterwards. Stephen, his life, his struggles and his accomplishments have obviously moved a number of people deeply and personally.

I have only seen similar intense encounters with one other person – Jane Goodall. Women would approach Jane at the post-lecture receptions in tears, bowing and holding out a hand to shake hands in a submissive gesture. "You and your life have meant so much to me." "You have been the role model for my life." These are powerful experiences. One is not inclined to speak, or to move, at such moments. You are quiet for the same reason that you are quiet when people are praying. Jane didn't accept or try to deny that she was worthy or as wonderful as the person was implying. Often she wouldn't say anything, just take their hand, looking them straight in the eye, nod and say, "I understand. Thank you." Jane understood that it wasn't about her.

Hawking and Goodall had both come to understand their roles in these intense encounters – to honor the moment. They are not what it is about, only a conduit. It's about their story and what it symbolizes in the grander scheme of human affairs. These are moments whose importance we can all recognize, while their true power and significance we can hardly imagine.

Lesser spirits than Jane Goodall and Stephen Hawking might succumb to the temptation to take such celebrity adulations personally. One of the most pleasant aspects of being with Stephen Hawking was the clear and definite feeling that he didn't imagine himself to be more of a person, or a better person, than those around him. He saw himself in others and others in himself. He treated others as he would be treated. This is not to say that Hawking is saintly. But it does reveal a defining feature of his remarkable and attractive character.

After being around Hawking for a while you learn how to join the ensemble, its timing, the melody lines and the improvisations. I remember when I started helping others, new comers, to understand how things worked, reassuring them in the awkward moments. I gradually came to feel like a veteran, an insider.

5

The First Magical Meeting with Students with Disabilities

Hawking's sense of his Participant 'calling', as a role model, as a champion of people with disabilities, came upon him, I think, a little unexpectedly as his celebrity, also unexpectedly, grew. And yet, as I came to know Hawking, it was clear to me that when this calling came to him, he stepped into the role quite naturally.

As part of the Science, Technology and Society Lecture Series, I try to arrange for the guest speaker to visit one of the local public high schools. The visit takes the form either of a small classroom visit or, occasionally, a large school assembly.

Hawking wasn't technically a part of the formal Science, Technology and Society Lecture Series. His appearances in Portland and Eugene were special, one-off, unique arrangements. This had been necessitated in large measure by the fact that I only had sixty days notice of when he could come. The regular series was planned at least nine months to a year in advance. So Stephen hadn't received the standard invitation where the high school classroom visit was included as part of the package. I hadn't planned on taking Hawking to a campus or school. The only extra was that he would visit the Intel campus. That was enough.

However, Steve Carlson, the Science Specialist for the Portland Public Schools, was ecstatic from the outset about Hawking visiting one of the Portland schools. Steve had been a stellar creative and energetic science teacher in the Estacada School District, a small rural district east of Portland on the way to Mt. Hood. He specialized in geology and had initiated, on his own, a new geology curriculum for Oregon middle schools. He was a leader, a spark-plug, a master teacher, and was hired by the Portland system to head up the coordination and development of their science and mathematics offerings.

Carlson was already putting together the Hawking event in his mind. Besides the regular students, he had focused on bringing a group of students with disabilities along with their parents from all over the district to the proposed school visit. "And the students' teachers as well," he insisted. "One teacher for each student. It is so important that these teachers see Hawking, what he can do. Realistic expectations are so vital to the relationship between the Special Education teachers and their students. To have them there together..."

When Steve first started talking to me about Hawking visiting a local high school, I just sort of rolled my eyes. "O.K., but I didn't ask him about doing that," I said with a sort of pained exasperation. I was just thankful that he was coming at all. I was concerned that Carlson's extension would disrupt the delicate arrangement. Carlson just looked at me incredulously – suggesting not so subtly that I should get with it. So I, somewhat reluctantly, called Sue Masey and she said she would run it by Stephen. She wasn't optimistic. A few days later I called again and Sue told me that Stephen had said 'no' to the school visit. Not wanting to rock the boat, I just accepted that and thanked Sue for communicating the proposal.

So I told Carlson that he wouldn't do it. "Why?!" Carlson insisted, somewhere between shock and astonishment. I hadn't actually asked Sue Masey for a reason. So I just invented a reason – something about the difficulties of transportation to and from the school. Carlson was undaunted: "What if... What if we bring the students to Hawking?"

"Ask him if he would meet with a group of students with disabilities . . . at his hotel," Carlson persisted. There was a sense of triumph in his voice – he had found the solution!

I wasn't excited about going back to Hawking with this new proposal. I didn't really know why he had said no in the first place. Sue, again, wasn't optimistic, but agreed to communicate the modified proposal. A few days later Sue gave me the response: "He said, 'yes'!" She was more surprised than I was. Hawking had been immediately receptive.

At the time of these exchanges, and for several years later, I didn't know that Steve Carlson had a young, severely disabled daughter. He never mentioned it. He understood far better than I, the potential impact on the students and their parents and teachers of a visit with Stephen Hawking. And it was Carlson who insisted that their teachers be included. "It is important for the teachers to see Hawking, what he can do, despite his disabilities," Carlson repeatedly effused. "Teacher expectations are so essential to how they deal with these kids," he went on.

I wondered aloud, "Can we get these kids downtown to his hotel?" I knew nothing about transporting students with severe disabilities. I knew the Heathman Hotel, where Stephen was staying, was wheelchair accessible, having checked it out thoroughly when I made the initial reservations. The hotel touted the fact that they were ADA (Americans with Disabilities Act) certified.

Carlson had the bit in his teeth. The excitement showed all over him. He was mapping out the issues and developing practical scenarios in his mind. It was like he was planning a military campaign. There were funds for transportation – Special Ed funds. We could bus the students to the hotel. All Hawking would have to do was to come down to one of the hotel conference rooms on the elevator.

It was a couple of weeks later when I began to receive questions from Hawking, via Sue Masey, wondering just what he would do in this meeting with students with disabilities. Apparently, he hadn't ever done anything quite like this before.

Later Sue informed me that Stephen had decided on beginning with a short piece that he had previously presented to a group of medical students. He would modify and update it for this group, and then we would have a question and answer session.

That magic afternoon in the Symphony Room at the Heathman Hotel in Portland, Oregon was the somewhat serendipitous beginning of a series of powerful meetings with students with disabilities. These became a part of Hawking's standard format in the coming years, in each city where I organized a popular lecture for him. It became clear almost immediately that Stephen believed in the importance of these meetings, and, moreover, later, that he really enjoyed them.

Carlson's people began arriving promptly at 10:00 am and kept straggling in until eleven and after. There wasn't really any organized structure. There were no chairs. I guess Steve had thought about it and rejected having any in favor of space for the wheelchairs. So the parents and teachers and media all stood – behind a growing collection of wheelchairs. A few chairs were brought in later for the half-dozen students who didn't use wheelchairs.

The students were arranged in a semi-circle with about a dozen gaining front row seats – the others scattered behind. Only a couple of the wheelchairs were the same style, so the overall scene was irregular and visually a little chaotic. Parents who needed to be with their child stood or crouched next to them.

Most of us have had some passing experience with someone in a wheelchair. Fewer have had an extended experience – usually with a grandparent or an injured friend who soon recovers. Far fewer of us have spent any time with a young person with disabilities due to cerebral palsy or developmental disorders, whether confined to a wheelchair or not. The experience of being with twenty such young people gathered together accompanied by their parents is extraordinary. There was no reason to expect anything different, but I was struck by the ordinariness of the parents; what I mean is 'from-my-(any one's)-neighborhood' sort of ordinary.

The number of fathers and mothers was about equal. The fathers were businessmen. One father in particular stood out for me; he was there early and was trying to entertain his son who was about six years old, or at least he looked about six. The boy's later question sounded more like it came from a nine or ten year old. There was this curious revelation that these parents were just like me – 'there but for fortune'… You know these students with disabilities are in the community and that they have parents, but it is somehow unreal unless you know someone personally and intimately. It is just that you would never know when you meet someone whether they have a child with a disability at home. These parents don't tend to share with other parents with children without disabilities. The communication doesn't go very far. There is no connection, no real appreciation of their situation. As a result, parents with students with disabilities are isolated and largely invisible to the rest of us in the community. But here they were – en masse. A few of them knew each other but only if their children had similar disabilities or if they lived close or if their children happened to attend the same school. There was a strong sense of community here, but not a strong sense of connectedness.

Steve was a little disappointed that more hadn't shown up. But he understood. "Transportation is difficult for these people. And parents needing to accompany meant skipping work," he conceded.

Everyone waited for Hawking who arrived on time precisely at eleven.

On that first morning in Portland, Stephen gave a prepared 20 minute talk to a group of about twenty students with disabilities, who ranged in age from about 6 to 18, their accompanying parents, a collection of their teachers, a few media types and the rest of us who watched in silent awe. His talk, fed sentence by sentence into his voice synthesizer, was followed by nearly an hour of questions and slow answers. It was often awkward and disorganized. But there was magic.

I had had only cursory contact, in person, with Stephen and his entourage before the meeting with the students. It was literally the first actual event of his visit to Oregon. I had met them at the airport and made sure they were settled in their rooms at the Heathman. Pretty brief. At this early stage, I wasn't included; I wasn't one of the crew, a member of Stephen's inner-circle – as I gradually became over the years, although never completely. Pam and Diana had immediately checked out the wheelchair accessible bathroom in Stephen's room. "O.K. Not the best, but we can work with it," they said. The bathtub had the needed railing and adequate space next to the tub to allow the nurses to be able to lift Stephen in and out. These were matters of primary concern.

As Hawking had entered the room, his entourage – Jonathan, Sue and the nurses – peeled off, losing themselves in the crowd as Stephen forged ahead. I don't know why but I vaguely expected either Jonathan or one of the nurses to stay close to Stephen during the meeting with the students. I thought, 'he is disabled, doesn't someone need to be serving as backup or something?' So I was surprised and a little worried when, as Stephen wheeled into the large room with nearly 75 people, his crew just seemed to disappear. "Ah, let's see… how do we get Stephen over here – the students have formed a semi-circle?" I am talking out loud, but to no one. Now I am talking to Stephen directly, "Well, let's see… over here is where we think it would be best. Is that OK?" Hawking's wheelchair is fully powered and he is the driver. He doesn't need assistance. This 'fading into the background' by the staff wasn't coincidental or casual. It was policy. If Stephen doesn't need you, then turn invisible. One of the reasons for this is the tendency of people to talk past Stephen to his staff, talking about him in the third person: "Would Stephen like this or that? Could you have him move over here?"

I don't see Pam or Diana or Jonathan anywhere. They aren't just stepping back a little; they are making themselves deliberately, intentionally inaccessible. But what if Stephen needs something? "He can ask," Pam tells me afterwards. I learn much later that one of the nurses, in this case Pam, has one eye on Stephen from a distance. And there is a particular facial expression of Stephen's – no words – that will bring her to his side instantly. But it never happens.

It is difficult to describe but everything was magical – surreal – from the moment Stephen wheeled into the room. No one who was in that room that day will ever forget it.

"After Dr. Hawking's brief presentation, we will have a question period," I say addressing the room.

I turn and fade into the background as quickly as possible. Now there is only Stephen and the kids. Everyone else, everything else, is undefined, unfocused background. The reader can gain better sense of the moment by realizing that this presentation, and all Stephen's presentations, are not smooth and continuous as you might read them aloud. Stephen is feeding each sentence individually, in turn, to the voice synthesizer. Each sentence needs to be cycled so there is a pause of 5, or sometimes10, seconds between each sentence reaching the audience.

"I am quite often asked," Hawking began, "how I feel about having motor neuron disease, or ALS as it's known in America. (pause)

"The answer is I try to lead as normal a life as possible and not think about my condition or regret the things that it prevents me from doing. (pause) Which are not that many. It was a great shock to me to discover that I had motor neuron disease. (pause) I had never been very well coordinated physically as a child. (pause) I was not good at ballgames and my handwriting was the despair of my teachers. (pause) Maybe for this reason I didn't care much for sport or physical activities. (pause) But things seemed to change when I went to Oxford at the age of 17. (pause) I took up coxing and rowing. (pause) I was not boat-race standard but I got up to the level of intercollegiate competition.

"In my third year at Oxford, however, I noticed that I seemed to be getting more clumsy and I fell over once or twice for no apparent reason. (pause) But it was not until I was at Cambridge in the following year that my father noticed and took me to the family doctor. (pause) He referred me to a specialist and shortly after my 21st birthday I went into hospital for tests. (pause) After all that, they didn't tell me what I had except that it was not multiple sclerosis and that I was an atypical case.

"I gathered, however, that they expected my condition to get worse and that there was nothing they could do except give me vitamins. (pause) I could see that they didn't expect them to have much effect. (pause) I didn't feel like asking for more details because they were obviously bad. (pause) The realization that I had an incurable disease that was likely to kill me in a few years was a bit of a shock. (pause) How could something like that happen to me? (pause) Why should I be cut off like this?"

There is a somber mood in the room as everyone here identifies deeply with Stephen's struggle to come to grips with his prognosis. One of the cameramen from Oregon Public Broadcasting moves a few steps to find a better angle, but everyone else is motionless – all fixed on Hawking and his powerful narrative.

Hawking continues, "However, while I had been in hospital, I had seen a boy I vaguely knew die of leukemia in the bed opposite me. (pause) It had not been a pretty sight. (pause) Clearly there were people who were worse off than me. (pause) At least my condition didn't make me feel sick. (pause) Whenever I feel inclined to be sorry for myself, I remember that boy." (pause)

This moment, this image of the young boy dying in the bed opposite him is like a still point, a touch point, defining a frame of reference in Hawking's emotional universe.

Hawking continues, "Not knowing what was going to happen to me or how rapidly the disease would progress, I was at a loose end. (pause) The doctors told me to go back to Cambridge and carry on with the research I had just started in general relativity and cosmology. (pause) But I was not making much progress because I didn't have much mathematical background – and anyway, I might not live long enough to finish my Ph.D. I felt somewhat of a tragic character. (pause) I took to listening to Wagner but reports in magazine articles that say I took to drinking heavily are an exaggeration. (pause) The trouble is, once one article said it, other articles copied it because it made a good story. (pause) Anything that has appeared in print so many times must be true. (pause)

"My dreams at that time were rather disturbed. (pause) Before my condition had been diagnosed I had been very bored with life. (pause) There had not seemed to be anything worth doing. (pause) But shortly after I came out of hospital, I dreamed that I was going to be executed. (pause) I suddenly realized that there were a lot of worthwhile things I could do if I were reprieved. (pause) Another dream that I had several times was that I would sacrifice my life to save others. (pause) After all, if I were going to die anyway, it might as well do some good. (pause) But I didn't die. (pause) In fact, although there was a cloud hanging over my future I found to my surprise that I was enjoying life in the present more than before. (pause)

"I began to make progress with my research and I got engaged to a girl named Jane Wild who I had met just about the time my condition was diagnosed. (pause)

"That engagement changed my life. It gave me something to live for. (pause) But it also meant that I had to get a job if we were to get married.

(pause) I therefore applied for a fellowship at Gondolan Caius College pronounced 'Keyes College,' Cambridge. (pause) To my great surprise, I got a fellowship and we got married a few months later. (pause) My Fellowship at Caius took care of my immediate employment problem. (pause) I was lucky to have chosen to work in theoretical physics because that was one of the few areas where my condition would not be a serious handicap. (pause) And I was fortunate that my scientific reputation increased at the same time that my disability got worse. (pause) This meant that people were prepared to offer me a sequence of positions in which I only had to do research without having to lecture. (pause)

"We were also fortunate in housing. (pause) When we were married, Jane was still an undergraduate at Westfield College in London so she had to go up to London during the week. (pause) This meant we had to find somewhere I could manage on my own but was centrally located because I could not walk far. (pause) I asked the college if they could help, but was told by the then bursar that it is college policy not to help Fellows with housing. (pause)

"We therefore put our name down to rent one of a group of new flats that were being built in the marketplace. (pause) Years later, I discovered that those flats were actually owned by the College but they didn't tell me that. (pause) However, when we returned to Cambridge from a visit to America after the marriage we found that the flats were not ready. (pause)

"As a great concession the bursar said we could have a room in the hostel for graduate students. (pause) He said, "We normally charge 12 shillings and 6 pence a night for this room, however, as there will be two of you in the room, we will charge 25 shillings." (pause) We stayed there only three nights. (pause) Then we found a small house about one hundred yards from my university department. (pause) We lived there for another four years but it became too difficult for me to manage the stairs. (pause) By this time the college appreciated me rather more and there was a different bursar. (pause) They therefore offered us a ground floor flat in a house that they owned that suited me very well because it had large rooms and wide doors. (pause)

But the college now wants to redevelop the site for more student accommodation so I am now in a flat nearby. (pause)

"Up to 1974 I was able to feed myself and get in and out of bed. (pause) Jane managed to help me and bring up two children without outside help. However, things were getting more difficult so we took to having one of my research students living with us." (pause)

My eyes shift to Jonathan, the latest 'research student' in what has now become a long tradition. Jonathan catches my glance and offers a shy muted smile.

Hawking continues, "In return for free accommodation, and a lot of my attention, they helped me get up and go to bed. (pause) In 1980 we changed to a system of community and private nurses, who came in for an hour or two in the morning and evening. (pause)

"This lasted until I caught pneumonia in 1985. (pause) I had to have a tracheostomy operation. (pause) After this I had to have 24 hour nursing care. (pause) This was made possible by grants from several foundations. (pause) Before the operation my speech had been getting more slurred so only a few people who knew me well could understand me. But at least I could communicate. (pause) I wrote scientific papers by dictating to a secretary and I gave seminars through an interpreter who repeated my words more clearly. (pause) However the tracheotomy operation removed my ability to speak altogether. (pause)

"For a time, the only way I could communicate was to spell out words letter by letter by raising my eyebrows when someone pointed to the right letter on a spelling card. (pause) It is pretty difficult to carry on a conversation like that, let alone write a scientific paper." (pause)

Many of the students here haven't discovered or at least haven't yet adopted computer technology. This inspiring meeting with Stephen Hawking will change that.

"However, a computer expert in California called Walt Woltosz heard of my plight. (pause) He sent me a computer program he had written called Equalizer. (pause) This allowed me to select words from a series of menus on the screen by pressing a switch in my hand. (pause) The program could also be controlled by a switch operated by head or eye movement. (pause) When I have built up what I want to say I can send it to a speech synthesizer. (pause) At first, I just ran the Equalizer program on the desktop computer. (pause) However, David Mason, of Cambridge Adaptive Communications, fitted a small portable computer and a speech synthesizer to my wheelchair. (pause) This system allows me to communicate much better than I could before. (pause)

"I can manage up to fifteen words a minute. (pause)

"I can either speak what I have written or save it on disk. (pause) I can either print it out or call it back and speak it sentence by sentence like I am doing now. (pause) Using this system, I have written a book and a dozen scientific papers. (pause) I have also given a number of scientific and popular talks. (pause) They have been well received. (pause) I think that is in a large part due to the quality of the speech synthesizer, which was made by Speech Plus.

"One's voice is very important. (pause) If you have a slurred voice, people are likely to treat you as mentally deficient.

"The synthesizer is by far the best I have heard because it varies the intonation and doesn't speak like a psychedelic. (pause) The only trouble is that it gives me an American accent," Stephen says.

This comment receives the first laughter, hearty and sustained, reflecting in part the underlying tension arising from both the power and uniqueness of the setting. Underneath this is the importance of having a voice. Many of these students are unable to speak clearly. Hawking's demonstration of the new voice synthesizer technology is a sort of local revelation.

Having paused his remarks during the laughter, Stephen continues, "I have had motor neuron disease for practically all my adult life. (pause) Yet it has not prevented me from having a very attractive family and being successful in my work. (pause) This is thanks to the help I have received from a large number of people. (pause) I have been lucky that my condition has progressed more slowly than is often the case. (pause) But it shows that one need not lose hope, if one is disabled. (pause) That is all. (pause) Thank you for listening. (pause)

"Thank you very much," he finishes.

There is an audible rise from the audience that had been paying close attention but with no anticipation of the endpoint of the talk. This quickly develops into applause, a few cheers and other noises from those students who are not able to annunciate words. There is a lot of stirring around, uncoordinated movement.

"Questions?"

6

Candid Q&A with the Students with Disabilities

Jonathan Brenchley, who I had hoped would handle this, is feeling shy and told me earlier that he doesn't want to address the gathering. He has explained to me how Stephen will handle the questions... and now I am on.

With some trepidation, I step forward to explain the procedure for the question and answer period.

"OK, so now Dr. Hawking has said that he would like to answer your questions. With each question he will construct an answer in his computer – word-by-word, sentence-by-sentence – and then when he is ready he will send it to his voice synthesizer, and you will hear it. This process of constructing the answer takes two to three minutes, depending on how long the answer is. So there will be this delay between when you finish your question and Dr. Hawking's answer," I say.

After a pause, I conclude, "Feel free to talk amongst yourselves while he is constructing his answers."

This last point is momentarily lost on this audience, since they have no sense of what to expect. Everyone is feeling a bit insecure due to the distinctiveness of the gathering. There is nary a peep from the seventy-plus people in the room during the interval from the first question to Hawking's answer. Waiting for Stephen's response, we are all just standing there, quietly, with nothing happening – at least that we can see. And yet – with all that – there is an overwhelming sense of the specialness of the moment – 'here there be magic'.

Anita Piercy, a Madison High School student, wanted to know how he became a scientist.

After only about two minutes, Hawking answered: "I was good at science at school so it seemed the natural thing to be. Physics was the most fundamental science. I went on to do research in theoretical physics at Cambridge University."

Aaron Rutledge, pointing excitedly to words and phrases on a board he uses to communicate, wanted to know about Hawking's family. "He would like to know if your family helps you, and in what way? And tell us about your kids," said his parental companion. Aaron could not speak, although he could make sounds.

Hawking answered, "They helped me a great deal in the past. But it is now so much work looking after me that we have had to get professional help. I have three children. Robert, age 24, who writes computer programs, Lucy, 21, studies Russian and French at Oxford University, and Tim, 12, is at school in Cambridge."

Kristopher Haines, a bright 7-year old who, like Rutledge, has cerebral palsy, asked: "Have you ever been discriminated against because you are in a wheel chair?"

Aaron's moment of insight

Before Stephen could even begin to answer, Rutledge, who had asked the second question, insisted on moving up next to Hawking to be able to view how he used his computer system to construct his answer. The gathering was loosening up. There was increasing conversation during the pauses when Stephen was putting together the next answer. It took Hawking three or four minutes to formulate the next answer.

The developing interest in Stephen's computer system motivated Jonathan to step forward and explain to the group that "the computer dictionary has about 2500 words." He went on to describe other features of Hawking's cybersystem: "He also has a portable phone on the back of his chair, which was added last year which however cannot be used in the United States. He can and does use it in England for making and receiving phone calls."

Hawking answered, "Not very often, but earlier this month, two Indian restaurants in England refused to let me in because I was in a wheelchair. I believe that is against the law in this country but it is not in Britain."

Rutledge is clearly delighted watching Stephen's screen. This led to the suggestion that Stephen turn around so that the others could see his computer screen and how his system worked. Hawking immediately obliged – rotating his chair so his back is to the students, with the secondary consequence that there is a confused surge as all the students move closer. Stephen had only been about five feet from the front row. Now there wasn't any separation.

Stephen had reached out, inspiring, interacting. Now there was no space between him and his students, they became one...

"Black hole?" was the next question.

There was accommodation so that everyone had a chance to watch Stephen select a letter from his dictionary as the cursor scanned the alphabet, resulting in a screen of words starting with that letter. Stephen then selected the word he wanted as the cursor automatically scanned over the options. The selected word then appeared in the bottom half of the divided screen. The upper half of the screen returned to scanning the alphabet – letter selected, word starting with that letter selected, appearing in the bottom half of the screen to the right of the last word selected. Gradually, a sentence began to form in the bottom half of the screen. Soon it is completed. And then somehow, I think through Stephen holding down his gravity switch, the upper screen switched from the dictionary to scanning a new set of options, one of which is 'send to voice.' And voila! The sentence is spoken through the voice synthesizer.

Hawking answered, "There seem to be many black holes in the universe, not just a single one. Some years ago I discovered that black holes can't be completely black. They will give off particles and radiation at a rate that is higher the smaller the black hole."

The next student question: "I know what physics is, but what is theoretical physics?"

Hawking answered, "I'll answer this in two parts. There are experimental physicists who measure things in the laboratory, or look through telescopes. But there are theoretical physicists like me who think about what the experiments mean or have ideas about new theories which in turn suggest new experiments."

At the end, little Andrew Fustos tried to ask about Hawking's fondness for video games, and whether he would like to play sometime. He fumbles his words and become intractably shy.

Andrew's question articulated by his father, "Dr. Hawking, my son has a really important question, now that the unimportant things about physics and the origin of the universe are out of the way. He wants to know... He has heard that you are a video game player. And that you have the secrets to defeating some of the video games, can you tell him which video games that you can play, and defeat, if that is, in fact, true."

A Small fan

A big broad loving smile spreads all over Stephen's face. This was undoubtedly his favorite question.

Hawking: "I am getting a joystick made right now so I can play Mario. But I'm not much good yet."

Finally, after an hour and a half and with questions trailing off, I suggested that Stephen should be heading to a lunch meeting. As he headed out the door there was spirited applause and cheers.

The lunch meeting, in his suite was with a trio of guys who had traveled down from Seattle: Bill Gates, Paul Allen and Nathan Myhrvold. I had vaguely heard of Gates, with this startup called Microsoft, but I had never heard of Allen, the co-founder with Gates of Microsoft.

Lew Frederick, a reporter for one of the local television stations, and later to become a major ally in helping to promote public understanding of science, asked several of the more articulate students about their experience.

Lew Frederick: "What would you tell your pals at school about this?"

Student: "First of all, I'd tell them that I was very inspired by it. And that I was very interested."

Lew Frederick: "Was Dr. Hawking what you expected?"

Student: "He's more interested in what we have to say than I thought he would be."

"This is a really huge thing," said Ryan Stott, a senior at Madison High School who has spina bifida. "It's an honor and privilege to meet such a man, who has gone through so much and yet can find time to help others cope with disabilities.

"It shows me that we all can go as far as we want to."

Kristopher Haines talked about how he now feels equal to the walking world and he was amazed at how a man of Hawking's reputation would be interested in children.

Ann Blackburn, 15, was impressed with how much Hawking has done with his life.

In his article in *The Oregonian* the next day, reporter Norm Maves commented, "All through the one-hour session, Hawking clearly was transmitting enormous jolts of hope to the youngsters."

As important as the experience was for the students, one should not underestimate the impact on the parents and the teachers that deal with these students every day. We all mingled for a while. I expressed my unexpected sense of the parents and spoke briefly about what it must be like to have a severely disabled child.

One of the parents responded: "My answer to that, since I have a handicapped kid, is that you love that kid more than anybody else and it is an absolute truth. And you are concerned and you have a compassion that goes far beyond even a normal relationship between a parent and child. When you have a handicapped child it is a bond that is stronger and deeper than even with a normal child."

Another parent who overheard the conversation commented: "They don't call them 'special' for nothing."

About a week later Steve Carlson told me of two comments from parents. First parent: "This experience changed my son's life." Second parent: "Josh is now very excited about using a computer." Josh had tried using a computer system before and gave it up out of frustration.

In the Spectator representation of inquiry the objective 'out there' universe is governed by One timeless universal order. On this magical morning Stephen Hawking and a group of students with disabilities had come together in a different type of Oneness – as Participants in a common enterprise to move the universe toward a more desirable future.

7

Portland Lecture – *Black Holes and Baby Universes*

Following the late morning session with the students, Hawking was scheduled to give a public lecture presentation that evening entitled, 'Black Holes and Baby Universes.' We had sold out 2775 seats three days before the event.

That evening on the sidewalk in front of the theatre there were people scalping tickets. Carlson commented that it was a great image for America: People scalping tickets to hear a physicist talk about recent developments in cosmology.

The evening was amazing. The crowd in Portland was more responsive than any other I have seen since. Hawking's jokes all hit perfectly. The audience groaned, cheered, and laughed and, in between, were silently attentive. It was like first sex. We have never quite recaptured the enchantment of that first evening.

Fran Gardner, reporting later in *The Oregonian*, referred to Hawking as 'the Luciano Pavarotti of theoretical physics.' The presentation had, of course, been prepared in advance. Stephen controls the pace of the presentation, feeding the talk from the computer to the voice synthesizer a sentence at a time. This control is rather crucial because the text is peppered with jokes and quips that result in audience laughter and applause. If Stephen didn't wait for these reactions to abide the audience would miss the next sentence or two. Gardner commented: "He displayed a fine sense of timing, leading the audience through his ideas a phrase, a sentence, a thought at a time. In between phrases, there was room for laughter and applause and, most important, time to think."

From the very beginning Stephen was making jokes. "Falling into a rotating Black Holes may be the key to intergalactic space travel," he says, adding the caution, "however, one might reemerge anywhere in the universe. Quite how to choose your destination is not clear: You might set out for a holiday in Virgo and end up in the Crab Nebula."

When Stephen would crack one of his little jokes, the audience would laugh – leading Stephen, to conclude, in effect, 'they liked my joke.' Stephen would then break into a broad smile – his major notable movement. This further charmed the audience and they would effervesce - a combination of more laughter, mild cheers and rising sound, as when an unexpected goal occurs in a soccer match. A rise. The audience seemed to feel privileged to be in Stephen's presence. They loved him. And the fact that he had gone to the trouble to include jokes and quips in his otherwise scientific talk made them feel like he cared about them – that he had made a special effort to please them. It was a love fest.

Hawking explained that at the entry level of understanding, the most important point about black holes is that the way they have been represented historically is just wrong. The standard approach to explaining what a black hole is uses the cannonball analogy: you fire a cannonball vertically upward from the surface of the earth and as it ascends it slows down due to the pull of gravity. After rising and slowing it eventually stops and falls back to earth. However, if the cannonball reaches what is referred to as escape velocity, then it just keeps going never to fall back. The velocity required to escape the gravity of a star depends on the gravitational pull of the planet or star; total mass is important but density is equally important. You can have a black hole with relatively small mass if it is very, very dense – like our Sun packed into, perhaps, the size of a marble.

The cannon ball analogy, originally reasoned as far back as 1783, is that since light travels at 186,000 miles per second, if a star is sufficiently massive and sufficiently small that the velocity needed to escape from its surface is greater than the speed of light, then light cannot escape. And if no light can escape, then you couldn't see it, even though its gravitational presence would still be detectable.

But as Hawking pointed out, "It is not really consistent to treat light like cannonballs." Light always travels at the same velocity – no matter what. So it isn't sensible or consistent to suggest that gravity can "slow down light."

Although the correct theoretical understanding of black holes began to emerge in 1915 with Einstein's General Theory of Relativity, according to Hawking, the implications "were not generally realized until the 1960s."

Black Holes form when space, the spatial region, around a dense, massive star begins to bend like a bowling ball on a rubber sheet, until the bending is so great that you can't see the bowling ball anymore. The rubber sheet closes around the ball. In this analogy it would look like the bowling ball had been completely lost downward with the rubber sheet closing around it. I always liked the image of a droplet of water forming on a leaf, getting larger with more water, until the droplet is heavy enough to detach from the leaf and falls, no longer visible.

Returning to the possibility of intergalactic travel using black holes, Hawking again cautioned that recent calculations showed that 'wormholes' are extremely unstable, so that it is most likely that your spaceship would be "torn apart by infinitely strong forces. "So," Hawking quipped, "black holes might be useful for getting rid of garbage or even some of one's friends. But they were 'a country from which no traveler returns.'" This last remark is an allusion to a line from Shakespeare's Hamlet.

Hawking then talked about one of his research efforts involving black holes. "Black holes," he said, "are not as black as they are painted. They can grow white hot and be the proud parents of baby universes." Around 1973 he began to wonder, 'what difference the uncertainty principle would make to black holes.' In brief, he came to conclude that because of the uncertainty relationship black holes would give off some particles and radiation, gradually evaporating. But as they evaporated some other portion of the black hole must actually separate from the universe, forming what Hawking came to refer to as baby universes. These separate baby universes, in order to exist, must exist in 'imaginary time.' This means they are still attached to the normal-time universe in some sense and will eventually reconnect, re-entering through the radiation given off by other black holes.

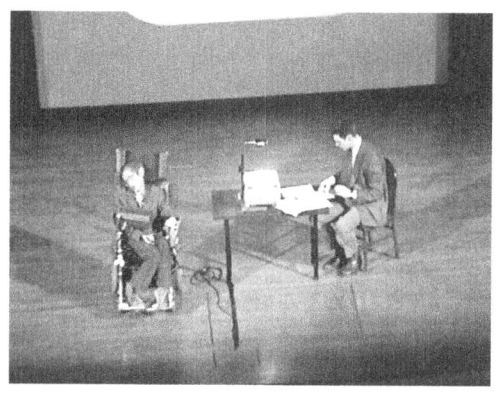

Stephen lightens this line of speculation: "The motto for anyone who falls into a black hole must be: Think imaginary."

Stephen and Jonathan

So there is good news and bad news about black holes according to Hawking. The good news is that assuming their existence makes it possible to better understand certain observations and characteristics of the universe. The bad news is that because we have no way to know how many baby universes have already formed, they represent a fundamental limitation on our ability to generate a complete Unified Theory.

His final joke, which generated yet another extraordinary response, had to do with the unlikelihood of patenting baby universes as a method of space travel. "Not an investment opportunity," he counseled. Nonetheless, they constitute an exciting, leading-edge research topic.

Stephen had specifically insisted on a live audience question and answer period. I let the audience know the procedure, which I had learned from Jonathan that morning. I told the audience that it would take Stephen three to five minutes, depending on the question, to construct an answer in his computer system. The audience was encouraged to talk quietly amongst themselves during this period. Stephen was to indicate when he was ready to answer: "I will answer now." Immediate silence ensued and it all went well.

Hawking took only a couple of minutes to answer the last question:

"What did he think was the role of God in the universe?"

"That's a good question," he responded.

"It's not for me to say."

This answer generated a long and loud applause that continued into a standing ovation as he departed the stage.

I had planned a VIP reception after the lecture to be held in the large ornate lobby of the Concert Hall. And it was a phenomenon. I honestly had no idea what to expect in planning the event. For most post-lecture receptions one could expect only about half the people with reception tickets to actually show up. They planned to come before the lecture, but after the lecture, with ample exposure, other options for the rest of the evening presented themselves. For Stephen's reception everyone showed up; partly, I think, because of the fascination that pervaded the evening. Upward of 450 people milled around, devouring nibbles from a selection of mini-desserts – drinking, talking, questioning and speculating. Stephen, for his part, was having a ball. He wheeled around the lobby freely. "Look out!"

He would stop for a while in one location, typically because someone had managed to inject a question as he was passing. Then a crowd would engulf him, peering over his shoulder as he began to construct an answer; watching him fly, letter to word, composing the sentence. Then, on his command, the miraculous voice would come forth, and he would move on.

The occasional bold attendee would ask, "Is it all right to take a picture?"

Another would push a little further, "Could I have my daughter in the picture with Dr. Hawking?" Sue Masey had briefed me on the policy: it is almost always OK, but they should ask him first. Click, click. Flash, flash. Stephen knew the routine and he knew when to move on, wheeling around, setting off in a new direction to another location. Stopped again. Click, click. Flash, flash. I am wondering whether someone is choreographing this.

"Is this all OK?" I quietly ask Pam in an aside.

"Oh, yes. Stephen is having a fine time. Can't you tell?" she laughs. I am aware and I guess a little surprised to see how much she enjoys seeing Stephen enjoying himself. She will tease him about some of this later.

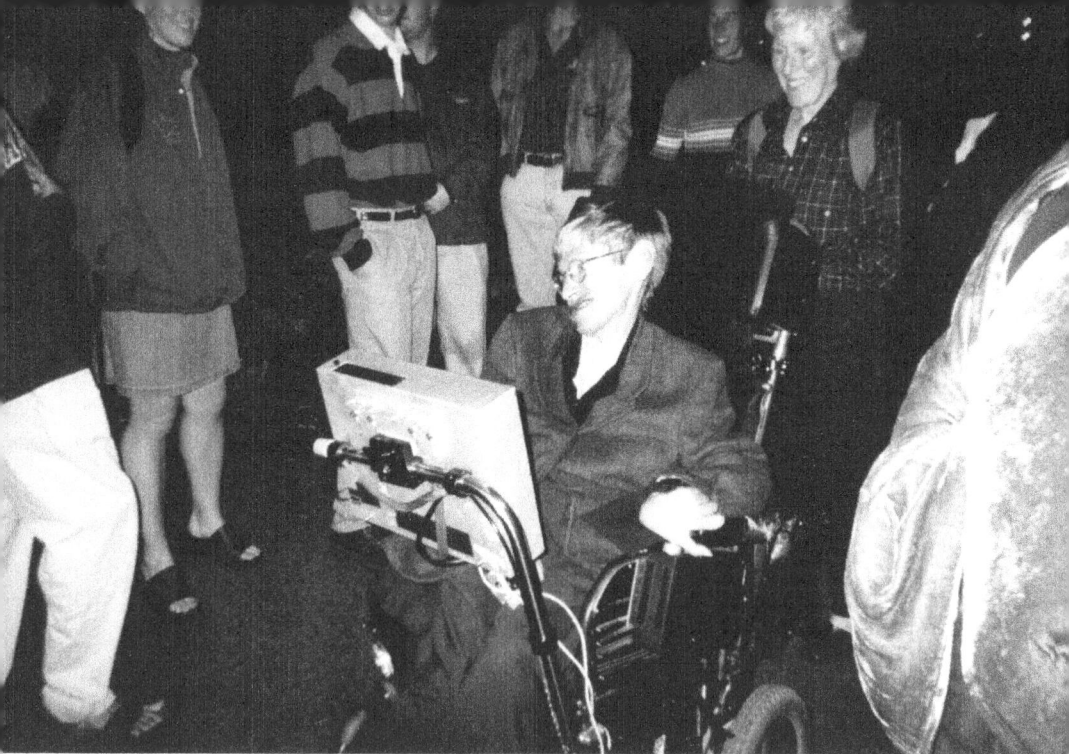

Pam and Stephen at reception

"How long should we go on?" I ask Pam.

"Stephen will let us know when he is ready to leave," she says. She is pleased. It is in her eyes as she smiles, watching Stephen from a distance.

At some point Stephen let one of the nurses know that he would like to have a couple of those chocolate-covered strawberries reserved for him for later in his hotel room. It is a good thing they grabbed these early because we ran out of all the food in about 20 minutes.

Stephen spent a few minutes with Don Hall, the Intel engineer with ALS, who attended the lecture and reception. Hall was to host Stephen the next morning when we visited Intel's Supercomputer Systems Division. Since Hall had the same computer system and voice synthesizer as Stephen, anyone listening to their conversation, except Hawking and Hall, was bound to be confused. Wondering who said what, glancing back and forth between them did not help because neither one of them displayed any revealing body-language. It almost seemed as if there were two disembodied robotic voices conversing with each other.

At some point Stephen wheeled around and looked at Pam. It was just a look, but a look with a pause.

"Time to go," she said. The crew, having scattered around the reception area, reassembled. Stephen was reassured that the nurses had secured his chocolate-covered strawberries and he was off to the hotel, located conveniently next door to the theatre.

Hawking, understood as a communicator, makes more sense in his Participant role, presupposing an evolving learning community. Learning is, at the very least, a curious phenomenon in the Spectator's mechanical worldview. If everything is governed by universal deterministic laws, learning isn't needed; it's superfluous, excludable, unnecessary. It just doesn't fit in. It really doesn't make sense. Indeed, meaningful questions and real inquiry don't seem to be necessary components of fully determined mechanical reality. What is going to happen is supposed to be pre-determined from the distant past and by the universal laws. From the point of view of the Spectator's representation of inquiry the 'knowledge' resulting from detached inquiry has no real value, can have no value. How could it?

Yet the science behind the advances in computer and voice synthesizer technology are valuable advances in the Participant enterprise, allowing Hawking, these students and pretty much everyone to live a better, more valued life, and in some higher sense a more valuable life. Advances in science are perhaps better understood as tools in advancement of technology. All this makes sense in a Participant representation, where inquiry isn't just for fun, where real inquiry begets meaningful knowledge that serves, at least potentially, to improve the world.

8

Visiting Intel's Supercomputing Systems Division

The morning was cold but bright – 'mid-winter spring is its own season.' January is typically a rather dreary month in Oregon, particularly in the wet western regions. To have a day like this was a sort of blessing, something to give thanks for. Hope reigned in the universe. The prospects for the visit to the coast were... blue sky all the way.

The admittedly ambitious plan for the day was to visit Intel's Supercomputing Systems Division (SSD) in Hillsboro in the morning, travel to the Oregon Coast for lunch, tour down the beautiful coastline for a few miles and then travel back inland to Eugene for a sold-out public lecture at the Hult Center that evening.

One thing I had learned about Hawking from Sue Masey and former graduate assistant Stuart Jamieson was that Stephen loves adventures: going somewhere exotic, doing something interesting – as long as it is a plausible adjunct to his professional responsibilities. This melding of work and pleasure is, of course, the forte of the conference business. 'How about we meet in the Bahamas to discuss the latest Cosmological Microwave Background observations?'

Stephen Hawking is definitely not your sit-around-home-and-watch-television-type of guy. Au contraire! – to put it mildly. Some outsiders reasonably imagine that anyone of Stephen's intellectual accomplishment must spend every spare moment obsessively working on his theoretical research. Stephen does work hard, but his academic commitments do not overwhelm the rest of his life. Others suppose that someone with Hawking's degree of disability would tire easily and have an extremely restricted lifestyle. I must now laugh out loud at this latter image. Reality is the opposite. Stephen is easily the most active person I have ever met – out to the best restaurants, theatre, opera, museums, partying with friends, you name it. Every couple of months it seems he is presenting a substantial scientific paper at some conference in some fascinating travel spot. One of the secrets is that he runs his medical support staff twenty-four hours a day, seven days a week. One of the little in-jokes is that because he doesn't get much exercise Stephen doesn't need much sleep. "Ha, ha – very funny," the nurses roll their eyes.

Who is this guy anyway? Really. Not what I expected.

Having been advised of his penchant for adventure in advance, in all my initial, as well as later proposals to Hawking for public lectures, I always included an adventure in the itinerary.

Over several conversations with the folks at Intel I had gathered that the rationale for having Dr. Hawking visit SSD was because they were just now bringing out their newest supercomputer, arguably the most powerful in the world at the time.

The great Justin Rattner himself would host Dr. Hawking for this tour. I had never heard of Justin Rattner. As I asked around among my Silicon Rain Forest friends and neighbors it became apparent that he was Mr. Supercomputer, and perhaps the highest-level computer hardware engineer at Intel. Rattner was one of a handful of internationally recognized supercomputer gurus, those who wrote the book on supercomputers particularly during the heyday of the Cold War. The Spooks (the folks at the Central Intelligence Agency) generously financed supercomputer research and development, as well as employing these monsters for all sorts of creative spying enterprises.

To prepare him for what to expect and what we were about to encounter at the SSD, I told Stephen, "My understanding is that part of the reason for this visit, from their point of view, is that maybe these newest computers will be able to model the behavior of black holes or simulate the development of the cosmos."

"They are not powerful enough," Stephen replied.

"Are you sure? How do you know?" I asked.

Stephen glanced at his screen line. He was about to compose, something.

"These are supposed to be a major advance," I added, waiting for whatever he was about to say. But he didn't say anything. He turned his head slightly, glancing up at me, with a sort of stare, essentially saying, "Do I need to repeat myself?"

"Right. I hear you," I said, and turned to look out the window as we swung onto Highway 26, the Sunset Highway, so named because it heads due west by the compass heading out of Portland. The traffic was light, but it still took us 25 minutes from the door of the Heathman to the door of SSD.

Gene (our driver) pulled up close to the entrance where Stephen could dismount. We are greeted at the front door by a cadre of what were, judging by appearances, public affairs staff. Well-dressed, attractive, cheery, personable, these were not engineering wonks. Just behind them, initially hidden, was someone else, an Intel engineer named Don Hall, 38 years old, who was diagnosed with ALS about a year earlier. Hall was in a motorized wheelchair much like Stephen's. His head is braced so that he can only look forward, an indication that his motor neuron degeneration had advanced to where his ability to exercise fine control of his head movements is already lost. Stephen, in a similar condition, had rejected the head restraint, preferring to be slightly reclined and willing to suffer occasional head-flops. Since Hall has the same computer system, and the same voice-synthesizer as Hawking – the same voice – there are immediate confusions. "Hello, how are you?" "Hello, how are you?" Who said that?

Hall had met Stephen briefly the night before at the reception following the Portland lecture. There had been an ineffable, yet immediately palpable, bonding between Don Hall and Stephen Hawking. Now, completely independent of the rest of us, transcending all the commotion of the surroundings, I saw them smile knowingly at each other. Recognition.

There were brief introductions all around. "Hello. How are you?" "It's a pleasure to meet you," said Stephen. Hall had been chosen, the public affairs people explained, to be Dr. Hawking's official host for the tour of the Supercomputing Systems Division. "Follow me," said Hall through his voice synthesizer. In the blink of an eye everyone else was an onlooker as Hall wheeled around a quick 180 degrees and shot forward across the lobby with Stephen in close chase. Doors opened.

From the lobby, we entered a short hallway that quickly opened into cubicle land, a large room the size of a football field divided into a labyrinth of sections and subsections separated by nothing more than five-foot high moveable walls. I am thinking that the Minotaur could be around the next corner. Hall and Hawking were trucking. The rest of us scrambled along, trying to keep up while at the same time sneaking sideways glances to take-in the strange and wondrous landscape behind the security doors of the Supercomputing Systems Division of Intel Corporation. A Dilbert cartoon is pasted on the corridor-side of one wall, but I passed too quickly to read it.

Heads turned or bobbed up over sections of the labyrinth as we passed; lots of smiles and nods. I had the feeling that everyone at SSD that day knew that Stephen Hawking was visiting and knew who Stephen was. If this had been Nike Headquarters, a couple of miles south in Beaverton, there might have been a few blank stares if we had been noticed at all and the putative significance entirely lost. I am guessing that there is a limited overlap between Michael Jordan fans and Stephen Hawking fans.

Once across the large room, Hall zipped up a ramp with Stephen in hot pursuit. The ramp is constructed over a portion of a set of four stairs – enough to otherwise block a wheelchair. I had the impression that Hall was normally the only user of this ramp. Intel had built ramps throughout the facility just for Hall. As we passed more smiling faces, I became convinced that Hall's coworkers were not only pleased, but proud, that Don Hall was leading – and was able to lead – Hawking on the tour of the facility. Intel had spared no expense to make sure Hall could keep working as long as he was able, as long as he wanted to. Hall was an engineer's engineer in an engineering company and you could tell that his wheelchair system and voice synthesizer and the completely accessible facility were the creative response of Hall and his friends to his dire diagnosis. This isn't cost effectiveness. This is heart.

Silicon Valley companies had rewritten the book on employer-employee relationships. This was not the steel mills of Pennsylvania or the car factories of Detroit. The founders of Silicon Valley, the likes of Robert Noyce and Gordon Moore, were a different cut of character. There were no unions because there was no need. Everyone was family. Intel took care of Don Hall, not because they had to, but because he was family.

Robert Noyce, nicknamed "the Mayor of Silicon Valley," who founded Fairchild Semiconductor in 1957 and co-founded Intel with Gordon Moore in 1968, seems to have provided the core vision and leadership that established the new Silicon Valley value culture of inclusionality. If you were employed by Fairchild or Intel you were family. There was no executive lunchroom, just the lunchroom. Everyone was part of the team – from janitors to CEOs. Noyce's father was the Rev. Ralph Brewster Noyce, Congregational clergyman and associate superintendent of the Iowa Conference of Congregational Churches.

Hall and Hawking stopped in two or three locations, pausing only briefly, while the rest of us were catching up, Hall explaining that this was the design section or the fabrication modeling division or whatever. The tour was so visually unique at least for those of us who had never been inside one of these technological beehives, that it was hard to concentrate on the narrative, itself in a sort of foreign language.

After 10 minutes of touring the basic work areas, Hall led us to a medium-sized conference room. A contrast to cubicle land, this was all high-tech and executive polish. Finding our way out of the facility at that point, without expert guidance, might have taken, I guessed, 45 minutes – avoiding the Minotaur. Here we are formally introduced to Justin Rattner, trim and fit, dressed immaculately, no tie but rather in a turtle-neck sweater and a blazer and unexpectedly good looking. In the world of engineering where there is an inclination to distinguish between 'the shirts and the ties' – working engineers and managers – Rattner effused management style and mannerisms. My impression was that he had designed the system and supervised its construction and troubleshooting but hadn't assembled the parts or soldered the circuits.

Rattner introduced Stephen to a half dozen other members of his Supercomputing Systems team. Of particular interest was a systems engineer flown in from the Bay Area especially for this meeting. This guy was more on the software applications side of supercomputers. Rattner's emphasis was on the hardware side. Rattner's briefing was simple – about how much faster this supercomputer could calculate the first million prime numbers or something equally straightforward: speed and power. But the software guy from Silicon Valley gave an impressive PowerPoint presentation on what they had accomplished in the development of supercomputing software and applications, and what they were projecting could be accomplished in scientific research with a new teraflop capable system.

Stephen asked a question about the current state of approaches to certain more sophisticated types of problems, indicating to me at least that he was certainly no novice here. The response was a knowing, yet obviously insecure, recognition that Stephen had put his finger on one of the Achilles' heels of the enterprise. Then Stephen complimented the team and mentioned there were a number of researchers in physics, and in the sciences more generally, who would surely find these new tools and capabilities quite valuable. There were a few knowing smiles and the supercomputing software guru said, "Of course, we don't imagine that this current system is able to compute the types of cosmological problems that you work on."

Stephen smiled and agreed and went on complimenting Rattner and his team. After Rattner and others added their comments, the conversation started to become disjointed. Before Stephen could respond to what one person said, three other comments were made. So by the time Stephen said something: "Yes" or "No" or "Except at the largest scale" or whatever, it was becoming unclear as to what or to which comment he was referring. The sense of a single conversation was lost. Where was the conversation line? Where had it been? Where was it going? In a room crammed with thirty people, this confusion degenerated further into chaos as each person made his/her best guess as to who was talking to whom and who was listening to whom. Eventually, it was just a bunch of nods and disjointed comments. People were talking to themselves – disintegrated mumbling. – This, I learned, wasn't an uncommon phenomenon in a group around Stephen.

Rattner caught Hall's eye and, speaking loudly, said, "Let's move on now to show Dr. Hawking the new Supercomputer." A sense of coherence returned to the ensemble. We had direction, a common course. More hallways and ramps led into a complex of three rooms divided by full-length, floor-to-ceiling Plexiglas walls. There was a noticeable drop in the ambient temperature – air conditioning – to keep the computer cool. In the last room was the supercomputer, a series of five or six large black sections along a wall, each about three feet wide, six feet tall and maybe five feet deep. There were lights blinking on each console: green, yellow and red. Rattner had linked 10,000 P6 chips together. I thought of HAL, the conscious and increasingly ambitious supercomputer in the movie *2001: A Space Odyssey*. We were inside, at the core – "Dave. Please don't. I can change."

Compared to the engaging movie scene, this was a bit of a letdown. One had expected to see something – dramatic. Rattner clearly sensed that for most of us, at least, this was a non-event. He had an embarrassed smile as he said, "Not much to see really – is there?" Diana joked, "Well, there are a lot of pretty flashing lights." Being the engineer of course, Rattner responded, "The lights are actually unnecessary, we just put them there so that it looks more fun. The supercomputer itself doesn't need the lights for anything – I mean, to operate," he explained.

"Very pretty none the less," added Diana, smiling with a barely discernable, politely supportive giggle in her voice. There was a flirtatious and seductive way about Diana at all times, but Rattner didn't pick up on it, except maybe for a quick smile. He was busy with Dr. Hawking.

Rattner opened the lower section of one of the consoles to reveal a entirely unedifying tangle of wires and uniform looking circuit boards. "This doesn't really show you anything but at least you can see what the inside looks like – at least this part of it," he said.

Stephen maneuvered to get a better view and seemed genuinely interested, pleasing Rattner.

"Not really representative," Rattner added with a sort of apologetic seriousness.

"Well, at least we can say that we have been in the same room with the fastest computer in the world," I offered rather lamely.

"Yes!" Rattner responded with a spark of enthusiasm – that was the point. "Of course, we are still testing and, as you heard, the software applications to take advantage of all this hardware power are still just beginning to be developed," he added.

That was the crescendo. So Hall led us back, but by a different route, through the labyrinth to the front lobby where we had entered. Hall isn't expected to live out the rest of the year. Intel kept him on at full salary and gave him meaningful work, appropriate to each stage of his decline. He remained a member of the Intel family until his death a few months later.

Our taxi van driver, Gene, who had not accompanied us on the tour, moved out to retrieve the van from the parking lot and brought it to the front security entrance. I am definitely sweating the time schedule now. With the greeting, tour, presentation and chit-chat in the conference room and the visit to the new supercomputer room, by the time we were loaded in the taxi-van we had burned a full hour. I was worried, but I kept my mouth shut about this, figuring that we'll make it. No one but I was aware of the time problem, and assuming we would make it, no one else needed to be made aware of how tight it was.

Intel is happy to support science, but they aren't primarily concerned with trying to understand 'how the world works' – right now. Intel is an engineering organization, a Participant enterprise, working to change the world, working to bring about a more desirable future. Hawking's formal relationship to Intel may have started earlier. It just wasn't overt in my dealings. For many years Intel has provided Hawking's primary hardware support and, in partnership with Microsoft his software support. The creative technological solutions developed for, and occasionally beta-tested on, Hawking naturally became creative technological solutions globally – for all people with disabilities. In subsequent years, I know that Hawking and Gordon Moore had become close friends, extending to an overt cultural camaraderie. Hawking and the Silicon Valley culture came together, as One Participant enterprise, working for social equity and inclusionality – everyone as family.

9

The Drive to the Coast – Exploring
the Limits of Science

The most significant conversational thread of Hawking's visit to Oregon that January in 1992 began on the two-hour drive to the Oregon Coast in a large taxi van. Departing Intel's Superconducting Systems Division, the morning after the successful public lecture in Portland, we headed further out the Sunset Highway, looking to connect with Highway 6, turning southwest to travel through the Tillamook State Forest. The plan was to visit the Oregon Coast for lunch on our way to Eugene, where Stephen's second public lecture was scheduled for that evening.

In addition to Hawking and me, the travelers included DAMTP administrator Sue Masey, nurses Diana Briscoe and Pam Benson, Stephen's new graduate assistant, Jonathan Brenchley and our van driver, Gene Looten.

Hawking's entourage all knew one another, so, since I was the stranger to the group, I became the focus, the target, the unexpected, novel phenomenon. I was the curiosity to be probed. Their initial impression was of a social entrepreneur with an intellectual agenda, something about the Public Understanding of Science, about placing science within the context of the broader human enterprise to better appreciate its potentials and its problems.

"How did you come to be putting on these Science, Technology and Society Lectures?" Sue asked.

"Well, I started off as a scientist, in the broadest sense. I distinctly remember having a sort of epiphany, at maybe 12 years old, about how amazingly beautiful the cosmos is. Galaxies in particular impressed me, systems of billions of stars like our Sun arranged in gorgeous spiral shapes.

"Later, in high school, I began delving into a series of books that were just coming out – the cosmology books of my, and I suppose Stephen's, generation: *Frontiers of Astronomy*; *Creation of the Universe*; *Relativity and Common Sense*; *The Universe and Dr. Einstein*; *One, Two, Three – Infinity*; and so forth. The authors were scientists and philosophers, people like George Gamow, Fred Hoyle and Bertrand Russell.

"I'm sure Stephen read most of these."

Stephen pipes in, "Gamow was at Cambridge."

"Right," I said, pausing to see if Stephen had more to say.

Born in 1904, George Gamow was a Russian born physicist, noteworthy for his pioneering work on the Big Bang. Gamow showed that the early stages were hot enough to allow formation of the light atomic nuclei such as helium. This offered an alternative to the prevalent theory that these elements originated by nuclear fusion within stars. Gamow was the first to realize and predict that there should be a background 'echo' from the Big Bang – the Cosmic Microwave Background Radiation, later confirmed by its discovery by Penzias and Wilson at Bell Labs.

Nothing more coming from Stephen so I continued, "When I went off to the University of California at Berkeley I intended to major in astronomy. It didn't take me terribly long to discover the discipline of philosophy of science. It was a new, somewhat estranged sub-discipline developing within philosophy departments, even though all the leading faculty were scientifically and mathematically trained. This was the area where the real action seemed to be, so I switched to philosophy – to philosophy of science. That is when I first met Paul Feyerabend, who eventually became one of my intellectual mentors. After Berkeley, Feyerabend pointed me to University of London, where I enrolled in a Ph.D. program, which I still haven't finished – yet.

"But you are still working on it?" Pam queried.

"Yeah. Someone said that the mark of a good philosopher is perseverance. So that's me, Mr. Perseverance. I started my Ph.D. program in 1969. This is 1992. Get the picture?"

"Where are you doing it? What college?" Sue asked.

"I moved around. Started at Bedford College, then officially half at Bedford and half at University College. But I was everywhere. I attended seminars at most of the University of London Colleges: Bedford, Birkbeck, London School of Economics, King's, University College, Imperial. Curiously, it was the American graduate students that attended most of the seminars, not the Brits. The Brits figured that if you did not have to attend, why attend."

I look to Stephen and he gives me a knowing smile. – In his *A Short History of Mine* he mentions a syndrome amongst British undergraduate students. You were branded a 'dull boy' if you attended sessions when you didn't need to and studied what wasn't necessary to pass the exit exam. In the British system, when you were working on your Ph.D. thesis, you were pretty much on you own. There were no class requirements. You just checked in with your thesis advisor periodically and arranged the occasional tutorial with a professor in a field where you might need some guidance.

"Personally, I consumed London," I said.

"So what is your Ph.D. about?" asked Pam.

This is where it all started. As Robert Pirsig pointed out in his classic *Zen and the Art of Motorcycle Maintenance*: "When you have a Chautauqua in you, it is difficult not to impose it on innocent bystanders."

"The limits of science," I answered, with a bit of trepidation, wondering if this would elicit a reaction from the great scientist. But there is nothing overt from Stephen, just an attentive expression of interest.

"What does that mean?" asked Pam.

"Lots of people talk about the limits of science but it is all pretty unclear. Peter Medawar, the Nobel Prize winner, for instance, has a full book talking about it," I point out.

I am feeling a little insecure bringing this up in the presence of Stephen Hawking. My mention of Medawar is a rather blatant rhetorical 'appeal to authority'. I just want to close off the line of thinking that only crazy nonscientists – philosophers – think about the limits of science.

"Students in the sciences are taught to stay away from philosophical questions. What's intriguing is that the question as to the nature of science is, itself, not a scientific question. The question of the nature of science, as well as the question of the limits of science, rests in a middle ground between disciplines, a gray area of inquiry between science and philosophy.

"Medawar thinks it is all rather obvious that science provides us with only a limited view of the universe. He doesn't see any difficulty in talking about science and the humanities, about facts and values, residing harmoniously in the same reality. And a lot of people agree, a lot of people think this is obvious," I add.

"But Medawar doesn't really address the controversy," I say, softening the rhetoric and trying to lay the ground for a civilized discussion.

I look to Stephen for acknowledgment. His eyes widen and eyebrows spike upwards. I don't understand this as indicating agreement, but more like, "OK, go on. Continue. I'm interested."

"So what's the issue?" asks Sue.

"One formulation of the core issue is in terms of a common belief – really the question – as to whether the scientific description of reality is the whole picture. You know, like with all the recent talk about a scientific Theory of Everything.

"It does seem reasonable, at least initially, to think of the world purely in terms of facts – scientific facts. Science gives us a testable description of the world. The scientific description is exclusively in terms of facts, a world of physical stuff, a material reality perfectly governed, apparently, by mathematically precise scientific laws. But this very reasonable, demonstrably successful perspective on the world doesn't see any values out there, just quantifiable facts and their relationships over time and space. For many people the implication that this constitutes a complete description doesn't jibe with common sense. So on one side some folks think science gives us, or at least promises to give us, a complete description of reality, timeless and universal, the only really true and correct way of looking at the world. On the other side are people like Medawar, who think that the scientific perspective must be just one limited way of looking at the world."

"Terry, you're exploring the limits of science. But what would it mean if science wasn't limited?" asks Pam.

"It would be the situation where everything in the universe is explainable by science. According to such a view, all phenomena, all relationships in the universe are governed by scientific laws," I answer.

"That's true of course," Jonathan chimes in, with what I take to be a mild condescension in his voice.

I am riding shotgun in the front seat next to Gene, both of us in individual bucket seats with a sort of console between us. Jonathan is right behind me on a four-seat bench with Sue Masey next to him, Diana next and Pam on the end, behind Gene. Stephen is behind them in his wheelchair strapped into the van's cargo space surrounded by luggage. The bed of the cargo space plus the height of the wheelchair places Stephen's head and part of his chest above the four staffers on the middle seat. Stephen is like a great eagle, silently hovering above the rest of us.

"Well, that's the question actually. How could you know for sure that everything in the universe is explainable by science? How could you know for sure that all phenomena are governed by scientific laws?" I say.

"Pretty straight-forward, I'd say," retorts Jonathan, adding, "Science would not have been successful if it weren't true. The success of science proves it."

Sensing that Jonathan's attitude is dismissive, Sue sympathetically interjects, "And what do you say, Terry?"

"Well, the real question is, how would you know for sure that science was able to give us the whole picture, a full account of all phenomena. How would you test it?

"I mean, maybe the universe is just a bunch of atoms popping around according to these mathematical laws, and that's all there is. You're suck through a funnel when you are born, and you're suck out another when you die and in between you are just doing a little dance – a little 'two-step'."

I pause to let this image sink in before I continue: "But maybe, just maybe, there is something more interesting going on – something important – something really important. Meaning and value may not be childish illusions."

"That's a novel thought where I work," remarks Sue, the friendly and ever positive Secretary in Stephen's Department of Applied Math and Theoretical Physics at Cambridge University.

"Let me start with the actual question of my graduate study: What evidence if it were to occur, would lead you to conclude that the scientific worldview is a limited, special case within a more general worldview?"

"What does that mean?" asks Pam.

Gene maneuvers the van left off Highway 26, onto the more direct, southwesterly route to the coast on Highway 6. The terrain is flat west and south of Hillsboro. We are still in farm country in the Willamette Valley, but the foothills of the Coast Range are beginning to become a more prominent part of our frame of reference to the west.

"Let me give you an example of a special case relation," I say.

"That might help," Diana adds.

"The theory, for instance, that the Earth is flat – the traditional Flat Earth Theory – is a very successful theory. I mean it works very well for most activities. But now we accept that the Earth is roughly spherical."

"Do we?" says Sue putting me on. Sue is a continuous wit, fun loving.

"Well, actually it is interesting how long the people in the Flat Earth Society were able to maintain a rational, if increasingly implausible, defense of the belief that the Earth is flat," I say, "But that's a digression."

"Sorry, I was just wondering whom you were talking about when you said 'we'," says Sue, still smiling.

I glance at Stephen, who is displaying a smiling appreciation of Sue's put on.

I continue. "Up until fairly modern times, for the vast majority of practical, day-to-day activities for just about everyone, the Flat Earth model is surely the most sensible. There is simply no practical value in changing. I mean what value does the average person gain by thinking of the Earth as spherical?"

"Well, I think that that would be rather considerable at this point," quips Jonathan.

"OK. Yeah, now that we are traveling longer distances and flying in jets," I concede, "But think of the ancient and pre-historic civilizations up until recently. You can do rather well on a day-to-day basis with the Flat Earth Theory."

"All right we accept that. Go on," says Sue.

"Then later we can see that the Flat Earth Theory is successful because the globe is extremely large in relation to our size and normal travels. The Flat Earth Theory works in special circumstances: when the ratio of the size of the globe is extremely large in relation to the size of the active observer, the experimenter, and distances traveled. So we conclude that the Flat Earth Theory is a special case within the more general, more advanced theory that the Earth is very large, relative to our size, and spherical.

"That's an example of what I mean by a 'special case' – a theory that works well, is strongly tested and confirmed, in a limited, special, set of circumstances or in terms of a particular point of view, a particular way of observing reality. The theory has validity within a limited range of applicability. So you see that a theory can work remarkably well under certain special circumstances and still be false in a more general context.

"So when Jonathan says that the scientific worldview must be true because it has worked so exceedingly well, I say that just because it has worked well, so far, doesn't mean that it might not be a special case within a more advanced, more general, more comprehensive understanding of reality."

Both Jonathan and I regularly glance up at Stephen seeking support.

"The modern textbook example for those of us raised on 20th century physics is that Newtonian physics works quite well in special circumstances, even though it is actually false. Newtonian mechanics is arguably the best-confirmed theory in the history of modern science – at least up through the third quarter of the 20th century. We even went to the Moon with the Apollo Program in 1969 on Newtonian mechanics. The navigational computers at NASA were programmed in Newtonian physics. As the story goes, Relativity gave an answer different by only nine miles over the 240,000 or so miles to the Moon. The rocket engine firings weren't that accurate anyway, requiring midcourse corrections; that is, after they figured out, roughly, where the previous rocket engine burn had taken them. So nine miles was negligible – well inside the range of normal error. And trying to program the computers of that era in relativistic physics would have been a huge difficulty. So, just as most of us navigate the small distances in our daily lives using a Flat Earth framework, NASA used the Newtonian space-time framework to travel to the Moon. In both cases we know that these frameworks are idealizations, valid only in a special limited set of circumstances, but perfectly adequate for many practical purposes.

"So, even though Newtonian physics can work extremely well, the consensus in modern physics is that it is actually wrong – false. The standard representation is that it is a special case within the more comprehensive Einsteinian Relativistic physics. Newtonian physics works just fine, at relatively low velocities and over fairly short distances and times."

"So what you are really trying to do, Terry, is not just to understand the limits of current science but to find the next, more general theory. You wanted to be the next Einstein?" interjects Pam.

"Well, I definitely started off thinking that way. I come from a generation, and so does Stephen, of science-oriented students who were fed the Newton-to-Einstein story with our mother's milk so to speak," I say, looking up at Stephen who acknowledges with a casual, quick 'eyebrows up.'

"The concept of 'an advance in science' meant, in practice, finding the limits of the current physics and transforming that current physics into a special case within a new more general and presumably more amazing and more beautiful physics.

"But there was a deeper, broader suspicion emerging, based on a dissatisfaction with the whole framework of modern science. The Newton-to-Einstein transition was confusing in many respects, even though it seemed to retain mechanics – the Mechanical Philosophy – the cornerstone of Modern Science from its beginning with Copernicus and Galileo – up through Descartes and Newton, who improved and developed it.

"With the advent of quantum theory, there were those, such as Danish physicist Niels Bohr, who thought that the entire mechanical framework itself might need to be replaced. That became a central issue in the famous Bohr-Einstein debate. Einstein was actually the conservative wanting to keep the new physics within the Mechanical Philosophy, within traditional Modern Science. Bohr was the revolutionary. To understand the depth of the crisis you need to see that this mechanical framework – The Mechanical Philosophy – seemed to literally define what we had meant by science. And Bohr was arguing that quantum theory, and perhaps even Relativity, appeared to force us beyond Mechanics. Quantum theory seemed to require us to move beyond traditional 'science' to a more general theory, to move beyond classical science to a sort of post-scientific, post-mechanical way of understanding reality. Mechanics, indeed, all possible mechanical theories, at least in the Modern Scientific tradition, would need to be newly understood as special cases with only limited validity and applicability."

"That's your limited science," says Pam.

"Right. This notion that we had found the limits of mechanics itself – really the very idea of mechanical causality – sounded like moving to a new understanding of reality more general than science itself. This was resisted by the conservatives in the science community, and still is, for that matter. It didn't make sense to them. I mean, the quandary is understandable: how can you make sense of science – mechanical causality – as a special case within a more general, post-scientific framework? For Bohr and the revolutionaries that was the fundamental challenge: to find that more general framework.

"Personally, I didn't start with the idea that I was searching for a post-scientific framework. It grew on me slowly, and it generalized. The revolution wasn't just in my personal research. Revolutionary change – paradigm change – had become an issue defining much of the 20th century intellectual milieu.

"So it seemed to some of us that if you wanted to be the next Einstein you might need to think post-scientifically, post-mechanically, outside the traditional approach of the Mechanical Philosophy.

"I think it was mathematician and philosopher Bertrand Russell who pointed many of us toward a broader context, somehow outside of 'official science,' where we could reflect on science itself, ask questions *about* science, *about* the nature of science, *about* how science actually worked. At first it was strange and curious to find that you could not ask questions *about* science in science departments. Questions about science were themselves not scientific questions. 'Science' was not one of the objects of study for the scientist. As a student in the sciences, when I raised these questions, I was told, as were virtually all science students, that these were philosophical questions, not empirically decidable, not so meaningful from a scientific point of view. This was, as it turns out, a pronouncement, an edict, a litany, not an argument and not itself based on evidence.

"It was in the philosophy of science, this relatively small subsection of academic philosophy, where I found you could pursue these questions, where such questions were taken seriously.

1 0

What is Science? – I Mean, Really?

"What's this Mechanical Philosophy?" asks Pam.

"The easiest way to get hold of it is with the image of 'billiard ball physics'. Think of how a billiards player operates. He learns the angles and just how hard to hit the ball to bring about a particular new arrangement of balls. The Mechanical Philosophy is a three-dimensional, mathematical elaboration of this image. In Rene Descartes's early formulation, the entire universe is like a giant three-dimensional billiards game – a large ensemble of atoms bouncing around according to universal mathematical laws.

"Isn't billiards what you 'yanks' call pool?" asks Sue.

"Yes, well, they are very similar. In pool the point is to put the balls in the pockets. In traditional billiards it's a different game, no pockets," I say.

"I used to be rather proficient at pool in my youth," Diana chimes in.

"What is important and common to both games – to this way of representing Mechanics – is that everything happens through contact, things banging into each other, bouncing off at mathematically predictable angles and at velocities that depend on the velocity of the impacting ball," I say.

"Anyway, the Mechanical Philosophy is commonly put forth as the cornerstone of Modern Science – the tradition starting with Galileo, followed by the formulations of Rene Descartes and Sir Isaac Newton. Descartes's theory was great for straight-line billiards-like motion but had problems explaining the force felt when things go around corners or, for instance, orbit around the Sun. Descartes's Mechanics became a limited special case within

Newton's more general theory. Newtonian Mechanics introduced two new ideas to account for phenomena associated with non-straight line motions. Inertial mass is needed to account for the force felt when you change direction, to explain the force felt when you turn a corner, like when you are driving. The introduction of the gravitational force, Newton's Theory of Gravity, is an addition, needed to account for attractive as well as curving motions like trajectories and orbits.

"Reversibility is another essential characteristic of Mechanical Systems. All processes must be able to run in both directions, forwards and backwards in time. If you run a movie of a mechanical process backwards, what you see is a mechanical phenomenon that might have occurred in the forward direction, without violating the mechanical laws. Think of running a movie of a billiards game in reverse. Everything that you see, the rearrangements of the balls, could, potentially, have happened in the forward time direction.

"Newton's Third Law expresses another essential characteristic of Mechanical Systems. Every action in a mechanical system has an equal and opposite reaction, so the net change, when you add everything up, is zero. The Newtonian universe is constant, doesn't change and, by implication, shouldn't have any 'real' net history. That is why the evidence for the Big Bang, for a beginning of the universe, conflicts with the Mechanical Philosophy. Mechanical Systems should be Steady State, always adding up to zero. Newton's Third Law is about Conservation and, even more generally, is about Symmetry.

"By the way, speaking of Sir Isaac, when is Stephen going to be knighted – Sir Stephen?" I ask.

There is an instant stirring of all four of Stephen's staff on the bench behind me. I had touched on something.

"I wouldn't hold your breath waiting for that one," says Sue.

"I think it is fair to say that Professor Hawking hasn't demonstrated that he is a great fan of the monarchy," adds Pam, teasing, looking up at Stephen, tongue in cheek, to be sure she isn't saying too much.

"He was invited to one of the Queen's Garden Parties – as the story goes," says Sue.

Stephen is laughing, his shoulders shaking up and down.

"What you are saying is that Professor Hawking is not a Royalist," I add. While studying in the United Kingdom I had learned that there are Royalists, who literally love the Monarchy (the Queen being the adored Mother figure of the country) and there are folks with the opposite opinion.

"Couldn't get the Queen, but almost ran over Charley's toes," Stephen interjects, with a large smile and still shaking with laughter. 'Charley' of course is Prince Charles, heir to the throne.

"So I am guessing that even if they offered him a knighthood... ?" I explore.

"Not likely, I think," says Sue, adding, "Professor Hawking might run Her Majesty over in the middle of the ceremony."

Stephen is still laughing, visibly but, of course, silently. His voice synthesizer doesn't have an imitation laugh.

"More about your 'limits of science', Terry," says Pam, deliberately changing the subject.

"The emergence of Modern Science is a re-emergence of the ancient attempts to understand the universe as governed by a universal, mathematical order. Galileo resurrected the ancient Pythagorean Hypothesis that the language of nature is mathematics. The Pythagorean Hypothesis was initially about a geometrical language but expanded in part on the basis of the Pythagorean's experimental observations that the notes of the harmonic musical scale had whole number ratios. The octaves are harmonic in a ratio of 1:2, the harmonic fifth has a ratio of 2:3 and the harmonic fourth a ratio of 3:4. The Pythagoreans hypothesized that all the phenomena in the universe are proportional and therefore are related to each other mathematically, as in geometry and the music example – as ratios of whole numbers. That's an early expression of the idea of law-governed phenomena, of law-governed relationships.

"Johannes Kepler, who discovered the laws of planetary motion, found that the orbits of the planets around the Sun were ellipses. A contemporary of Galileo's, in his *Harmonices Mundi* (1619), Kepler explicitly invokes the Pythagorean imagery, saying that he wishes 'to erect the magnificent edifice of the harmonic system of the musical scale... as God, the Creator Himself, has expressed it in harmonizing the heavenly motions.' He adds, 'I affirm... that the movements of the planets are modulated according to harmonic proportions.'"

"I like that," says Sue. "The universe is one great symphony."

"As the Mechanical Philosophy developed from Galileo and Kepler to Descartes and Newton the most common metaphorical image of the universe became that of a harmonic clockwork mechanism, everything causally linked by gears, by gear-trains, by 'contact causality' to everything else. The planets revolving in their different orbits, the Earth in its annual cycle, provided a persuasive clock-like image. The solar system was seen as one enormous clock. Clockworks become the popular technological metaphor, much as computers are today. With Newtonian physics the universe wasn't just analogous to a clockwork – it literally was a clockwork. If you wanted to know what time it was you looked to the heavens, to the position of the Sun, the Moon and the stars. All the parts, all the phenomena, are mechanically related to each other; everything was predictable. And what was happening in the heavens was correlated on Earth in the cycles of the seasons, with society organized around seasonal planting and reaping. Everything worked together – regularly, harmoniously.

"Newton's introduction of gravity as a force was controversial. Gravity was mysterious because it produced 'action-at-a-distance'. It altered the motion of bodies without making actual 'causal contact'. Many of Newton's contemporaries rejected Newton's introduction of gravity, charging that it was an 'occult-like' force. However, since the effects of the hypothetical gravitational force could be formulated mathematically, it was possible to accommodate it within a plausible expansion of the mathematical formulation of the Mechanical Philosophy. Newton admitted that he didn't know what gravity was, emphasizing that he was only pointing out that it was possible to describe it rather simply in mathematical terms, and to demonstrate it repeatedly, mechanically.

"The clockwork universe, once set in motion, seemed as though it must be completely predictable. French scientist-mathematician Pierre Simon

Laplace (1749-1827) famously pointed out that if a super-intelligence could determine at one particular moment 'the respective positions of the beings which compose it [the universe]' with an understanding of the forces that operated between those components, he would be able to both predict every aspect of the future as well as infer, to retrodict, every aspect of the past." ("Given for one instant an intelligence which could comprehend all the forces by which nature is animated and the respective positions of the beings which compose it, if, moreover, this intelligence were vast enough to submit these data to analysis, it would embrace in the same formula both the movements of the largest bodies in the universe and those of the lightest atom; to it nothing would be uncertain, and the future, as the past, would be present to its eyes." Laplace, Pierre Simon *Introduction to Oeuvres vol. VII, Theorie Analytique de Probabilites* (1812-1820).)

"If Laplace is correct then everything we do, even what I am saying to you now, is predetermined and entirely predictable from the very beginning of the universe. Everything that happens today was completely determined by the state of the universe yesterday or the day before or the day before that, from the very beginning of the universe, just like clockwork. This clashes fiercely with notions of freewill and moral responsibility, notions fundamental, not just in religious traditions, but in all civil societies.

"And yet, formulating the clash – understanding the problem here – isn't easy. If everything is already decided and prewritten according to deterministic laws, we don't actually have any real, meaningful choice. The moral context disappears and there is nothing to be celebrated in what we like to think of as our most heroic, generous and spirited efforts. The concern with the limits of science begins with a skepticism about these deterministic implications of the Mechanical Philosophy.

"The hope for a more general understanding is that the success of mechanics can be understood as limited within a special range of applicability – as a special case. It seems that we are able to *experience* the evidence of the limits of science while still accepting and reasoning from inside the Mechanical Philosophy. You can experience the limitation of the Flat Earth Theory without being able to understand those experiences. Ancient Science was well aware of evidence of the curvature of the Earth: the shadow of the Earth on the Moon during lunar eclipses, the gradual disappearance of ships as they sailed away, and, with travel over greater distances north and south, the positions of the stars changing in a quite regular and repeatable way. Only when you figure out that the Earth is an enormous sphere are you able to understand and explain what was experienced as phenomena that *didn't make sense* from within the Flat Earth perspective. The experiences, the evidence of the limit, are only properly, correctly understood, newly understood from within the more general Spherical Earth perspective.

"And so, by analogy, we can 'experience' evidence that *can't be made sense of* within the Scientific Research Program, can't be made sense of in scientific terms. However, we need a post-scientific theory, a different type of theory, to be able to make sense of that limiting evidence, to be able to make sense of that evidence in a new, post-scientific, way. Any More General Theory superseding all possible scientific (viz. mechanical) theories *must be able to make sense* of *both* the evidence that makes sense in, and supports, the scientific account *as well as* the evidence (viz. the 'incoherent' experiences of the limit) that don't make sense in the scientific account. We experience the limits of science without being able to make sense of those experiences, scientifically. We need a post-scientific theory that both subsumes (viz. includes the valid parts of scientific theories) and yet supersedes all possible scientific theories, understanding their old success in a new way, in terms of a new More General understanding of reality.

"So my graduate work developed into two separate questions. First: What is the evidence that, by its very nature, demonstrates the limits of science? And Second: What is the nature of the post-scientific, post-mechanical More General Theory that must be able to understand the success of science, but that can also understand the evidence that, by its very nature, demonstrates the limits of science."

11

Thinking 'Outside the Box' of the Scientific Research Program

Over a hill and then sweeping through a shallow valley we start to climb. We have passed out of the Willamette Valley into the Coast Range. You could see it coming gradually, but the experience is still a little startling. Suddenly, exiting open farmland, we are engulfed in towering trees on both sides of the road; huge trees, dark forest on either side. There is no longer any sense of the global, no distant horizon. Everything is local. The path is darker, directed toward the narrow opening ahead, toward the light. The dramatic change distracts and overwhelms the conversation and there is silence for a while.

Being the local host, I become the travel guide. "These are Douglas Firs, named after David Douglas, a Scottish botanist, who first brought them from the New World back to Scotland. They grow to around 200 feet tall and live from 600 to 1000 years. Unfortunately, the timber industry doesn't like to wait that long to harvest them. This is maybe the third or fourth growth you see here, since serious logging began about two hundred years ago. All the original, larger old growth trees, hundreds of years old, went first."

No one says anything. The visual experience is consuming: dark passages, around corners, breaking out into grand vistas of multiple shades of forest green framed by the blue sky. We enter and leave darkness and density, ascending, leveling and occasionally descending, like a gentle roller-coaster ride through a foreign, yet distantly familiar landscape of both space and time, an enchanted ecological kingdom.

"In Oregon you are hard-pressed to find any building, any structure, that is more than 150 years old. Eighty to ninety percent of what you see, what you can experience, is less than 75 years old. The Lewis and Clark Expedition only made it across the country to the Oregon Coast in late fall of 1805. Before that the limited European presence arrived by ship sailing all the way around South America. When I first arrived in England, coming from Oregon, it was an unexpected culture shock for me. The sense that people had been walking those same streets in London for over 2000 years. – Wow! The Canadian and Australian students in London had a similar experience. We just didn't have any cultural framework, no prior category, for these new types of experience."

"What about the native Indians? Don't you count them? They must have been here for hundreds or even thousands of years. I mean, I don't know much about Oregon's native tribes," says Sue.

"Most of the original tribes were quasi-nomadic and didn't leave any easily noticeable structures. Also, the vast majority of the Native population were wiped out by smallpox in the 19th century, and by now a large portion of the others are pretty thoroughly integrated through interracial marriages. A friend of mine who sits on the Warm Springs Indians Council shared with me his concern that his children wouldn't be able to sit on the Council in the future because he has only one-quarter heritage and his wife is fully Swedish. A few tribes have retained a rough identity and are linked to reservations. So the prior presence of the Native population hasn't provided much of a historical frame of reference."

"Are we feeling guilty about all that?" Sue asks.

"Welcome to the dregs of human history. European history hasn't been particularly uplifting either," I say defensively.

We pass a small mom-and-pop grocery and gas station, an outpost in the surrounding wilderness. "There are no real towns or villages along this route to the coast, not like in England, anyway." An occasional small cluster of buildings with a town name sweeps by. Invisible from the road, back in the hills, are a few homesteads.

"So, Terry, how did a nice boy like you end up organizing a Science, Technology and Society lecture series?" Pam insisted.

I glance at Stephen, seeking his approval and he gives me the 'eyebrows up'. He agrees it's a good question. Jonathan turns from taking in the view and re-engages with the conversation.

"My transition is actually a story of gradual dissatisfaction with the scientific culture, followed by the discovery that there was an intellectual niche – the philosophy of science.

"As I proceeded with my physics and math education at Berkeley I was increasingly bothered by the fact that these larger questions, raised by Hoyle, Gamow and Russell, in particular, about the nature of science and the Scientific Worldview, were not part of the science curriculum. In my junior year, I went back to New York to visit my sister, who was working at the United Nations. She had been dating a guy, Sheldon Bennett, who was doing his Ph.D. in physics at Harvard and working at Brookhaven National Laboratory. She arranged with Sheldon for me to go out to Brookhaven, meet him and talk about whether I really wanted to stay in physics and pursue a higher degree – the big issue of the day. So I received the grand tour of Brookhaven and discussed all sorts of basics. Then, as we got to know each other a little, I opened up with a couple of questions about the nature of space and time – things that I had been thinking about; the Gamow, Hoyle, Russell sorts of questions. Sheldon just sort of laughed. He told me that there were 35 Ph.D. students working on this one experiment using the Brookhaven particle accelerator. This was experimental physics, with a very specific focus. They didn't even talk about the nature of space and time. These experimentalists seemed to me to be straight along that line of narrowing specialization with which I was already feeling uncomfortable.

(Note: I recently learned that after teaching physics at Harvard for several years, the, now, Reverend Sheldon Bennett, had moved on to graduate from Harvard Divinity School, focusing the rest of his life on non-profit community service.)

"Later that fall, back in Berkeley, I took my first course in philosophy, entitled "Problems of Philosophy" from John Searle. Philosophy is concerned with these larger questions: How do we know? What do we know? Do we know anything for sure? That was interesting – and surprising. These weren't questions I had encountered in science or thought about seriously at all. I also had a provocative roommate at Berkeley, Gordon Oliver, who had read and apparently understood much more of Bertrand Russell than I had. This was physics and mathematics and yet it wasn't discussed in the physics or mathematics curriculum. I had imagined that these larger issues were to be discussed in graduate physics and mathematics, but that didn't seem to be the case either. When I would ask my physics and math professors or graduate assistants about these larger questions, they were either silent, laughed like Sheldon or just hadn't thought seriously about them.

"My sense was that just as each of the sciences branched into a narrow sub-discipline, philosophy had moved in the other direction, broadening and eventually encompassing all topics, all questions, inquiry itself. Philosophy raised questions about the physical and chemical world, but also about biology, geology, psychology, anthropology, sociology, linguistics, politics and on and on. Philosophy was the mother of all inquiry. Science in the modern sense now appeared to me to be a sort of orphan child; a spin-off, fatherless, unsupervised. One feature of the philosophical tradition that impressed me was that it was naturally self-reflexive, questioning everything, including itself and its own methods. The question of the nature of philosophy and the question of the nature of inquiry itself were natural philosophical questions. This again highlighted for me the strangeness of the modern resistance within science to question, even to think seriously about, the nature of science.

"I recall that when I took courses in psychology and political science the professors would take you through the discipline just so far. When they got to the crunch questions, to the core issues like: What is thinking? What is consciousness? What is justice or beauty? they would say, 'Well, those are philosophical questions.' The same evasion occurred in the hard sciences in response to questions like: What is truth? What is objective reality –

particles or waves? When you asked these 'deeper' questions, you were offered a quasi-religious-like litany of scientific tenets – 'official doctrine'. Then it was emphasized that these were philosophical questions, and you were warned to stay away from them.

"After researching the journals and asking practicing professionals in a wide range of fields, I began to suspect that the lack of clarity on the most basic concepts was ubiquitous in every field of inquiry. The defining concepts in every field were actually controversial, and it appeared that they always had been. Opposing concepts had opposing, competing research programs. Even in physical geography there were oppositely competing approaches. That was a shocker. One result was that all the most intriguing and fundamental questions, being controversial, were being relegated to philosophy. In scientific disciplines, where we were supposed to be 'objective', this lack of unity was enormously incongruous and 'hidden in the closet'.

"Why should the perennial issues surrounding the core defining concepts of each discipline turn out to be formally, experimentally undecidable? In some disciplines where one or the other of the perennial approaches had temporarily gained the political upper hand, the controversy wasn't as immediately obvious – yet it was there, historically. Often the opposition was just beneath the surface, carefully nurtured by the disciplinary rebels.

"As I took more philosophy courses and became acquainted with more professional philosophy I realized that most of the people in philosophy departments were as ignorant of modern science and math as the people in science and math were ignorant of professional philosophy. So I wondered, and I wondered, if someone like me with a scientific background and scientific attitude went into philosophy, resisting specialization, making an active and deliberate effort to become an 'educated generalist' competent in math and science as well as philosophy, might be able to make a contribution to both science and philosophy.

"If you had asked me when I entered Berkeley to list the most likely majors I might graduate in, philosophy would have been right near the bottom with poetry and creative dance. I had a pretty negative image of the people populating philosophy departments. John Searle had opened me up to reconsidering. Anyway, I tentatively made the switch to philosophy, continuing my science and math courses. I had discovered that you were allowed to do science, both theoretical and experimental, in philosophy departments, in stark contrast to the extreme resistance to allowing any philosophy courses in the science curricula. If I wanted to be a generalist, a true natural philosopher in the self-reflexive tradition, I couldn't do it in a science department, but I could to do it in a philosophy department.

"I remember asking one of my graduate teaching assistants in philosophy, Jesse Kalin, how you make a living in philosophy. I had always figured that I would have a substantial income working in the sciences. He said, "What do you honestly need? You can eat a long time on a bag of rice and a few cans of beans." Not an inspiring answer, but it made a point.

"When I began to ask questions about science in my philosophy courses one of my Berkeley philosophy professors, said, "You should talk to Paul Feyerabend."

"Who is that?" I had asked. I was told that he was a professor in the philosophy department specializing in philosophy of science. What I discovered in the person of Paul Feyerabend was that I wasn't the first scientific mind that had found the intellectual constraints of science departments stifling. Others had already made the same move. There was a subdivision of philosophy populated by escaped scientists and mathematicians. And they were seriously pursuing the larger questions – in the tradition of Bertrand Russell, reaching back to Newton, Leibniz, Poincare, Descartes – all the way back to the Pythagoreans and before to the ancient beginnings of the Western Scientific Worldview. I was home."

I am anxious to end my personal story, even though I think it speaks to the heart of the question of the limits of science and to our place in the universe.

Topping a hill, a broad and expansive scene opens on the left side of the road, and we can see over forested hills for several miles.

"Notice the uniformity of the trees," I say.

"Yes, why is that?" says Pam.

"This is the area of the Tillamook Burn. When I was a very young, when we drove through here on the way to the coast, my parents would tell stories of The Burn. The trees we are looking at weren't nearly so large then. What I saw in my youth was a burnt forest with many, quite small, new tree plantings. The fire my parents remembered was the one in 1945 that hit this exact area, destroying 180,000 acres of prime old growth forest. It's certainly starting to look pretty impressive now."

There is a pause for a few minutes as all attention turns outside the van to the forest scenes.

"So you were trying to be the next Einstein, and it all went bad when you switched to philosophy? Is that what you are saying?" says Diana provocatively. Jonathan smiles and glancing at Stephen for support, says, "Sounds that way to me."

Stephen doesn't respond, but his eyes show that he is clearly listening and interested.

"Well, not really. But you're not completely off track," I say, "Keep in mind that the legitimacy or status of science isn't a question that occurs *within* science. Science looks 'out at the world' – away from itself – but not at itself.

"In my formulation, I am asking about the foundations of science and yet putting it in scientific terms, insisting on evidence, insisting on a practical demonstration as to whether the scientific worldview is or is not a special case, a limited perspective.

"You want to test science scientifically? You want to test the idea of testing your theories? That's a non-starter," Jonathan quips. "The test of science is by doing science. And the proof of the legitimacy of science is the success of science," he adds again somewhat sanctimoniously.

Jonathan Brenchley is no novice in matters of science. You don't get to be a graduate student in physics at a leading British university (Cambridge University) without being uncommonly bright and well-studied in your subject.

"I agree – up to the point of the question of the limits," I respond. "We have already seen that just because something works – and works well – doesn't prove that it isn't a limited approach, a special case.

"Look, Jonathan, when you take this reflexive perspective on science you begin to ask some different and extremely intriguing new types of questions: about the nature of scientific advance, about what makes a theory 'better' and about the special case relationship.

"The standard image of science throughout most of the 20th century was that the goal of the scientific enterprise was to discover better and better theories, better and better understandings of how the universe worked, converging toward a full, complete and consistent understanding of reality. So you might start with a theory, call it Theory One, and this is followed by Theory Two, where the truth content of Theory One is included in the improved theory, Theory Two."

"But the earlier theory can't be completely included in the later theory?!" says Pam.

"Right. Only the true part – the truth content – carries forth. Theory One becomes a special case, with only limited validity. And the successes of Theory One are understood in a new way in Theory Two. The Flat Earth Theory carries forth only in part, in a modified form, its limited success, newly understood and explained within the later and better Spherical Earth Theory. The false part is supposedly left behind – for instance, the part that would lead you to false predictions, like falling off the edge of the Earth.

"It seemed reasonable in the dominant logico-mathematical representation of science that one should be able to evaluate a new theory by measuring whether it had a higher truth content than the current theory. Mathematically the natural expectation was that there should be a quantitative measure showing that the new theory is in fact better than the prior theory. The new theory should be something like more probable, or have a higher truth content. The initial strategy was to try to specify the ratio of truth content to falsity content. This was to be measured in terms of the ratio of true to false predictions. But it didn't work."

"Why not?" asks Sue.

"The problem, as it was argued at the time, was that each theory makes a potentially infinite number of predictions, both true and false. There isn't any convincing way to compare the unlimited number of possible predictions, true or false, of each theory, of any two theories. There are innumerable little predictions like 'turn left at the next corner and the Earth will still appear flat.' In the early part of the 20th century, the dominant logico-mathematically-oriented wing of the philosophy of science had talked as if you could establish meaningful ratios like this, but it just didn't get us anywhere. Counting successful and unsuccessful predictions in order to tell you which theory is 'better' just doesn't work.

"There were many examples in the history of science where, intuitively, there was a clear consensus that the later theory is better than the earlier theory. However, the 'betterness' that we were intuiting didn't seem to have a quantitative measure. Slowly it began to dawn on people that the important difference between the Flat Earth Theory and the Spherical Earth Theory had to be 'conceptual' – something qualitative, something that couldn't be expressed quantitatively, in purely logico-mathematical terms. This realization reemphasized an ongoing skepticism about the overall validity of the 'mathematics is the language of nature' presupposition, in that it neglected the qualitative conceptual aspect of 'the language of nature'.

"This problem begat a mini-crisis. Without a measure of 'betterness', how do we know we are progressing toward the hypothesized complete scientific understanding of the universe? It was a little suspicious and more than a little unsettling. What if there isn't an endpoint, a final theory? If there is no final theory would it mean that reality doesn't have a 'nature'? Could it be that we are just rambling around without actually getting anywhere, without converging on anything real, on anything timeless? How would you know if there was no endpoint, no ultimate scientific truth? If you have no way to explain 'scientific betterness', then maybe the later theory isn't 'really', objectively, better than the earlier theory – maybe it's just different.

"Rene Descartes, one of the early fathers of modern science, worried a great deal about that possibility. He reasoned that if every theory so far has been shown to be false, it was reasonable to be concerned that the current theories are false and that all future scientific theories will turn out to be false. 'Could it be then,' he asked, 'that we are just wandering from one theory to another – not actually getting any closer to 'the truth?' I like to call that worry Descartes' Nightmare."

I figure it is time for a break while all that settles in.

Everyone is now a part of the conversation except our taxi van driver, Gene. In academic terms, Gene is an outsider, a foreigner, neither physicist nor philosopher. I imagine that what we are talking about doesn't make much sense in Gene's world. Superficially, at least, Gene makes sense in our world as a simple supporter of our 'higher' enterprise. But how does Gene understand our scientific and philosophical enterprises in his worldview? "Gene, it's your turn," I say. "Who are you and how did you end up here in this van driving to the Oregon Coast with world-famous cosmologist Stephen Hawking?"

Gene starts his story. He turned out to be quite a fascinating character. I don't know that I had a stereotyped image of a taxi driver, or a taxi-van driver. Maybe I did! But whatever it was, Gene doesn't match it – isn't close. Gene looked and acted more like an executive from the corporate world, and although his dress was official taxi-company garb, he wore it... differently. The first thing that struck you about Gene was the indentation on the right side of his forehead, the size and shape of a small box of matches, mostly rectangular, and deep, maybe three-eighth of an inch in most places. He was slightly balding, maybe 45 years old, with a wide bright smile. Confident. Outgoing.

None of this fit my expectations, however vague they had been. This guy is driving a taxi-van? Gene was also articulate. Well-spoken. As we focus on Gene, my curiosity grows. He had just been 'background' before, helping us load Stephen and all the baggage, prime director in tying down Stephen's chair to the van's floor. Gene is a foreign experience. Now with more attention I also realize this guy looks decidedly out of place. He should be running a public relations firm. A secondary image intrudes: he could be successful as a used car salesman.

"What did you do before this, Gene?" Sue asks.

"I had a record company," he says. Then after a brief pause, he adds, "But it went under."

Our reaction is to be mildly stunned, and there is a notable pause. But the pieces of the experiential puzzle that is Gene begin to come together, in a new way.

"I used to manage rock and roll bands in the late 60s and 70s," says Gene.

"Here in Oregon?" Diana asks. Unlike philosophy of science, this is a subject that interests her.

"Here in Oregon and in the San Francisco Bay Area," Gene answers, adding, "We played the Fillmore a few times. Then I moved into managing recording contracts and record publishing. After a few years I figured I understood the business well enough to start my own record company."

No one asked about the indentation on his forehead; undoubtedly the result of a major injury, perhaps a blow to the head from a drugged-out lead guitarist? Perhaps a bullet wound? It was a strange and prominent mark. But no one asked about it and Gene didn't volunteer any stories or explanation.

Gene isn't formally trained in the sciences but he is curious, he asks questions, and he learns. Gene is a real world problem solver – a Participant – a team member in the composition and performance of music. I imagine he has been more mainstream than we are – part of the developing symphony of Kepler's harmonic universe.

There is a silence for a few moments.

While we are still moving Pam and Diana stand up, facing the rear of the van, to check on Stephen's comfort level.

As they sit down again we reengage the previous conversational thread. "In the philosophy curriculum one learns that Descartes, one of the greatest scientists and mathematicians, also worried that the universe might be deceptive; indeed that the 'creator' might be a malevolent deceiver – perhaps having a good laugh – deliberately trying to deceive us by making it look like there were real questions, meaningful questions, questions with practical beneficial answers.

"Real meaningful questions don't make sense in a fully deterministic mechanical world. Questions are superfluous, extraneous – have no essential function. The ideological determinist might respond that it just happens that in this universe it only 'appears' that there are meaningful questions and that successful inquiry is beneficial – as imagined in a Participant perspective. However, reality, according to the Spectator's determinist perspective, could have done without these 'appearances'. It is just a coincidence that we happen to be in a deterministic universe where it looks, deceptively, like there are meaningful questions, where it looks like it isn't deterministic, where it looks like we make real value choices.

"When I entered philosophy of science, like many others, I imagined that one practical goal of our studies would be to better understand and improve scientific method. We imagined that practicing, experimental scientists would be quite interested in our work. We were disturbed to find that they had not the slightest interest and they never read our best work. Since their inquiry was about the 'out there', it was non-self-reflexive. So they categorized our self-reflexive work as irrelevant. We were raising questions in a foreign territory.

The tension between the Spectator and the Participant representations of inquiry is over the proper understanding, over the proper representation of, our place in the universe. What is a question, anyway? What are we doing asking questions? How does all this fit into the nature of reality? What does it indicate about our place, perhaps our role, in reality?

12

Auxiliary Hypotheses and
The Promissory Note

We've topped the last elevations of the Coast Range and start to descend into the Wilson River Valley. The rivers in the Coast Range now flow to the west, towards the ocean. The highway hugs the river, winding, falling. The river itself appears and disappears. This route along the river was the early wagon trail from Portland and the country to its west, to Tillamook on the coast. The highway cuts into the hillsides near and along the river. Initially, we just have glimpses of the river. As we come more and more out of the mountains, descending, we come closer to the river. We pass the turnoff to Lee's Camp, a locally famous rallying point for serious river fishermen. The Wilson River is one of Oregon's top sport fishing rivers. It hosts runs of salmon and steelhead in different seasons – regularly, annually, predictably, correlated with the positions of the stars in the great clockwork of our universe, by the Earth's current clock anyway.

After a period of silence, as everyone takes in the new surroundings, I continue my Chautauqua. "That we could not measure what we intuitively accepted as 'better' from one scientific theory to the next was just a clue that actual inquiry and real advances in our understanding of nature were more sophisticated than any simple quantitative, logico-mathematical measure could capture.

"One thing that really bothered me, in searching for the limit of science, was that whenever you criticized a currently accepted scientific theory, pointing out some failing, there was always this natural expectation within the scientific community, that some time later, a better scientific theory would account for the failings of this current scientific theory," I said.

"It always has," remarks Jonathan.

"Well, actually, that is precisely what is not entirely clear," I say.

Jonathan just rolls his eyes and smiles at Stephen, not so subtly appealing for support. I also give Stephen a similar glance, appealing for support. Stephen, listening intently, is remaining neutral.

"That is what I came to refer to as the Promissory Note Argument – an unrestricted promise of future success of the scientific approach, of the Scientific Research Program," I continue.

"What's wrong with that?" insists Jonathan. "Science has always succeeded. The proof is in the success, like I said. It is perfectly reasonable to believe that it will succeed in the future."

"Yes, but notice that this sequence of better and better scientific theories is always *within* the scientific worldview. Failed scientific theories are always replaced by better *scientific* theories. What I am trying to get at …," I say, pausing, feeling frustrated.

"Consider the possibility that a whole sequence of successful scientific theories – Theory One, Theory Two, Theory Three and so forth – might just be selecting out only a certain range, or a certain type, of phenomena," I say.

"For instance, just to give you an image, these theories might only be selecting out low velocity or left-handed phenomena – whatever the selective characteristic might turn out to be. I want to know whether the scientific sequence of successes, by its very nature, might be limited to this certain subclass of the totality of phenomena.

"Is it possible that the sequence of successful scientific theories could go on and on but never be comprehensive, never cover all types of phenomena, never be able to claim that it is giving a full and complete scientific account of all the phenomena of the universe, never arriving at a scientific Theory of Everything?"

"So how are you going to decide that?" asks Sue.

"Maybe..." says Jonathan. "Maybe that's an issue sometime in the far off distant future. That's all."

"But that's part of my problem, Jonathan. I don't want to wait until the far off distant future," I say. "That waiting attitude is part of a Promissory Note Syndrome. I want to know *now*."

I make a long pause to emphasize the 'now' point, making deliberate eye contact with Jonathan and then with Stephen to mark my new point of departure, my revolutionary move.

I catch Stephen's eye. He sees that I want an acknowledgement, a 'go on' – and he gives me an 'eyebrows up' with a sort of casual air indicating what I interpret to be a neutral position.

"To answer your question, Sue," I offer, "the strategy that evolved in my thinking was to ask a scientific question about science itself: What evidence, if it were to occur would force you to conclude that the scientific approach gives us only a limited view of reality? What I am looking for is the evidence that would logically force you to conclude that the scientific worldview is inherently incomplete. If there were such evidence, it would have to be – by its very nature – such that no scientific theory, come what may, no matter how long science goes on, could ever explain it."

"I don't get it," says Diana.

"I want to know what evidence, if it were to occur would refute all possible scientific theories come what may," I say. "I want to know if there is some phenomenon, some *type* of phenomenon, that I can point to *now* that – by its very nature – could never be explained by any possible future scientific theory."

"Why do you ask it that way?" asks Sue.

"Good question. I am trying to address the Promissory Note Argument by asking for a type of evidence, that if demonstrated, would force one to conclude – *now* – that the scientific approach gives us an inherently limited understanding of reality, so that scientific successes are all special cases. The formulation of the question evolved. I came to call it Popper's Question. To explain, I need to give you folks a little Popper 101," I say.

"Popcorn? Do we get popcorn with this entertainment?" Diana laughs.

"No, really. That's his name. Sir Karl Popper. He is one of the dozen most important philosophers of science in the 20th century," I say.

"Well! *Sir* Karl. That makes all the difference doesn't it. I mean if the Queen likes him..." says Sue in an exaggerated tone.

I feel I am on firm ground here with Stephen since I know he is familiar with Popper's work. He refers to him favorably in some of his lectures.

"Here's some easy background," I started.

"In his student days in Vienna, Karl Popper (1902-1994) was a Young Marxist. Marxists have their supposedly scientific theory of reality, about how the world works. Their focus was of course particularly on the socioeconomic world but nonetheless they took it to be a scientific theory, within the modern scientific tradition. These students would formulate various activities – natural experiments – based on their analysis. For instance, they might challenge the newspaper barons on this, or try to enlighten the workers on that.

Sir Karl Popper

"Anyway, frequently things didn't work out the way they had predicted. The other players in these real world experiments often didn't act the way they were supposed to according to Marxist theory. So after a number of failures, Popper asked whether the group should rethink the fundamental theory, the Marxist fundamentals. He was told, 'No, there was no need for that since the failures were easily explained' because of this or that special circumstance. The failures did not constitute 'real' counter-evidence against the core Marxist fundamentals.

"Popper was not happy with this answer. He had begun to suspect that maybe Marxism wasn't completely correct. I mean, perhaps you could understand a lot of what was happening in terms of Marxism but there seemed to be a lot of excuses for things that didn't quite fit the theory so easily. Maybe Marxism was able to make sense of only a select subcategory of socioeconomic reality, while also tacitly incorporating some fundamental misunderstanding.

"Then, on one occasion, there was a street demonstration by some Young Marxists, leading to an unexpected and violent confrontation with police. Clubs swung, heads cracked and one of Popper's friends was actually killed.

"When the group reconvened later to consider what had happened, Popper argued that surely with the death of his friend, their comrade, as a result of this completely unexpected, unpredicted violent encounter, they should reevaluate the Marxist fundamentals. Instead, Popper was told, there was just a minor misunderstanding and that the core Marxist framework was not to blame.

"Popper then began to wonder whether there was anything that might have happened or could happen in the future – any evidence – that would lead his fellow Young Marxists to seriously question their Marxist framework, their supposedly scientific Marxist worldview.

"Popper was studying physics at the university and he speculated that, although the Marxists considered themselves to be scientists, they were actually pseudo-scientists. Marxism, Popper hypothesized, was not real science, because unlike physics, it was not sensitive to counter-evidence, it was not responsive to experimental failures. Marxism, he concluded, was a pseudo-scientific ideology.

"Later in the philosophy of science literature these after-the-experiment defenses of a theory came to be called 'auxiliary hypotheses'. These were excuses when the evidence appeared, at first glance, to count against a theory. In other words, pseudo-scientists would offer these after-the-experiment auxiliary hypotheses to excuse the failings of their theory.

"Based, at least in part, on these early experiences, Popper later, in his work in philosophy of science, formulated a Demarcation Criterion, designed to distinguish – to demarcate – real science from pseudo-science.

Popper wanted to find a way to separate the bad guys of pseudo-scientific ideologies from the good guys, the real scientists.

"Popper proposed that a claim about the world was meaningful only if it was possible to specify how one might potentially refute it. Sounds simple.

"What Popper believed he had identified as differentiating pseudo-science from real science was this practice of 'after-the-experiment defenses' – despite the failure of a prediction. After-the-experiment defenses of a theory were a potentially endless series of excuses designed for no other purpose than to preserve and perpetuate the core ideological belief.

"Popper imagined that real science didn't make after-the-experiment excuses to cover up the failure of its theories. For Popper, then, to be scientific meant, at least in part, that you were empirically critical. – It all seemed perfectly reasonable.

"One of the initial, and still common interpretations, at least for many people, has been that as soon as there is evidence against a theory, the theory should be rejected as false. This oversimplified version of Popper's theme eventually came to be known as 'naïve falsificationism.' Whether Popper ever believed this is arguable, but it certainly wasn't what he was after, it wasn't a solution to the problem that was motivating him.

"Popper's more sophisticated formulation of his falsificationism arises from the insistence on the 'prior specification' of the evidence that would lead you to abandon your core hypothesis. I came to formulate and formalize this as, what I call, Popper's Question: what evidence if it were to occur – not saying that it has or will – would force you to conclude that your current theory must be limited, only partially true, a special case within some more general understanding of reality?

"Popper reasoned that if you were credibly scientific you could answer this question, specify – *ahead of time*, before you actually performed any experiments – what evidence, if it were to occur, would falsify your theory, what evidence, if it were to occur, would lead you to give up your core theory.

"This 'prior specification' strategy was precisely what I needed in my quest to understand the limits of science. What I needed was a 'prior specification' – *now* – of the evidence that would falsify science, that would falsify the Scientific Hypothesis. The Promissory Note Argument constituted a 'faith' that each failure of a current scientific theory would *eventually* be explained by a later better scientific theory. Popper insisted, 'Tell me now!'

"I reasoned that by Popper's standard, if the Scientific Hypothesis itself was scientifically credible then it must be possible to specify the evidence, some phenomenon, some type of phenomenon, that – by its very nature – would count decisively against the core hypothesis.

"Tell me now – what evidence would convince you that the scientific approach to understanding the universe is inherently limited?

"A lot of people liked Popper's falsificationism – this idea of a Demarcation Criterion between science and pseudo-science, based on whether the core hypothesis was falsifiable. It looked like it captured a defining aspect of science in a fairly common sense way. The essence of science, many imagined, was this critical responsiveness to evidence – to the facts. And hadn't we all experienced some friend or colleague inappropriately trying to defend a favorite pet theory against what we saw as overwhelming evidence against it?

"Another group of philosophers of science, the conservative Logical Positivists, also based in Vienna, like Popper, had also offered a Demarcation Criterion. They maintained that a theory was meaningful only if you could specify how to verify it – how to demonstrate its truth, how to prove it. This was called the Verificationist Criterion of Meaning. On the surface their intention was the same as Popper's: to separate the scientific theories from pseudo-scientific theories. However, their original problem and consequent strategy was quite different from Popper's. In particular they were concerned about distinguishing or demarcating science from religion. They were opposed to believing anything based on what some authority figure, like the Pope, claimed. So if you believe in gods or angels or some supernatural force controlling events, the Positivists would say that these were meaningful claims if, and only if, you could specify how to verify them experimentally;

how to demonstrate the existence of these actors or forces. This applied quite generally to beliefs in unicorns or leprechauns or aliens or any sort of 'invisible' or non-scientific force controlling events. The Positivists argued that unverifiable beliefs were simply meaningless nonsense. The Positivists argued that a belief, or causal explanation, was meaningful if and only if you could specify how to verify it experimentally, how to demonstrate, in particular, the existence of these actors or forces. If you wanted to label such beliefs as 'unfounded speculations' or a 'favorite fantasy', that was fine – just don't pretend that they are scientific beliefs about reality.

"Both Popper and the Positivists were trying to distinguish science from what they took to be non-scientific beliefs and reasoning.

"The common ground between Popper's Falsificationism and the Positivist's Verificationist Criterion was the sensitivity to evidence. Both Popper and the Positivists were saying that a theory or truth claim is meaningful if and only if it is empirically testable. But there are two sides to the testability coin: one is verification, the Positivist insistence on specification of possible evidence in favor, and the other is falsification (or refutation), insisting on specification of possible evidence against.

"Noting the asymmetry in the two aspects of testability, Popper argued that falsification was a much more important criterion, since no matter how much (quantitative) evidence you have confirming – verifying – a theory, it is still possible that it is false. On the other hand, it takes only one instance of counter-evidence to show that a theory is false – not completely true. It was this 'one instance of counter-evidence' argument that led some to think that Popper was, at least initially, a naïve falsificationist.

"Applying Popper's Question to science itself – to the Scientific Hypothesis – I could acknowledge all the successes, all the verifications, of scientific theories. What I wanted to know was what type of evidence, if it were to occur, would lead any proponent of the Scientific Hypothesis to conclude that the scientific approach could provide only a limited understanding of reality.

"I realized that I was asking for 'a prior specification' of the type of evidence that would demonstrate the limits of science, 'a prior specification' of the evidence that would demonstrate the inherent incompleteness of the Scientific Research Program. More to the point of the Promissory Note Argument, I was asking for 'a prior specification' of the *type* of evidence that would demonstrate the incompleteness of the Scientific Hypothesis in such a way that no possible future scientific theory, 'come what may', could rescue the core Scientific Hypothesis.

"Ready for the punch line?" I said.

13

The Dark Matter Promissory
Note and Popper's Question

We break out of the confines of the narrow winding passage of the Wilson River Valley into a spectacular view of the broad, flat and open Tillamook Valley: open vistas on either side; expanses of luscious green grassy meadows, dappled with black and white spotted milk cows. This is dairy country. We descend into the territory of world-famous Tillamook Cheese.

"Historically, what happened to Popper's Demarcation Criterion in 20th century dialogue about science and pseudo-science – to his falsificationism – was bizarre," I say.

"Science, it turned out, by Popper's Demarcation Criterion, wasn't scientific.

There was a problem in providing 'a prior specification' of what evidence would lead one to abandon any specific scientific theory, as well as for the Scientific Hypothesis in general.

"What was really disorienting was that it turned out that scientists do the same thing as the Marxists. They make excuses to save their theories in the face of counter-evidence. Using 'auxiliary hypotheses' to defend one's theory against counter-evidence is actually a very common practice in science."

"Imre Lakatos (1922-1974), another of the 20th century's leading philosophers of science and mathematics, pointed this out in an extremely straightforward thought experiment. Lakatos, who had agreed to be my

Ph.D. advisor just before his early, untimely death, was one of Popper's colleagues at the University of London. Born to a Jewish family in Hungary, his mother and grandmother had died in Auschwitz. When the Soviets crushed the Hungarian Revolution in 1956, Lakatos fled to England. He received his Ph.D. from the University of Cambridge, applying a version of falsificationism to generate a new understanding of the supposed 'proofs' in the official history of mathematics. Lakatos, the escaped Jew, and Feyerabend, the former lieutenant in Hitler's army, became the best of friends and famously collaborative colleagues.

Imre Lakatos

"Lakatos's thought experiment goes like this. Say you have a theory of planetary motion. And it works quite well for the seven planets you know of in the solar system. Then one day you observe that the outer planet isn't behaving according to predictions. Should you abandon the theory? Does this counter-evidence falsify your theory? Well, in this thought experiment, Lakatos's scientist introduces an auxiliary hypothesis. He hypothesizes that there is another, previously undiscovered planet, still further out, that is disturbing the motion of the known outer planet. So he calculates the position of the newly hypothesized planet, turns his telescope to that position – and voila! – oops! – he doesn't see anything. No new planet.

"Undaunted, the scientist introduces a second auxiliary hypothesis, namely, that there must be a dust cloud between the Earth and the new, hypothesized planet that is obscuring the telescopic view from the Earth. So, being a prestigious and influential scientist, he convinces NASA to send a space probe out to observe the new planet, positioning the spacecraft so as to avoid the hypothesized dust cloud. This takes a few years, but finally the spacecraft arrives and sends back its data. It doesn't see anything. No new planet.

"Still undaunted, the scientist introduces a third auxiliary hypothesis – that there must have been some sort of electromagnetic radiation, in the area from which the spacecraft was observing, that interfered with the space probe's instruments. He calls for a second electromagnetically shielded spacecraft.

"I could go on, but I think you must begin to see the point. The scientist here is doing exactly the same thing that Popper's Young Marxists in Vienna were doing: introducing auxiliary hypotheses to explain away the 'apparent' counter-evidence – thereby defending the core theory."

"But what if they had found the planet?" asks Diana.

"Well, exactly. You're way ahead of me," I said.

"As usual," she says.

"The point is that although the scientist is introducing a series of, what are from a critical perspective, suspicious-sounding auxiliary hypotheses, defending his original theory of planetary motions, he might well have been correct at any point along the way. These are all quite reasonable, plausible auxiliary hypotheses. He might have seen the new planet in the telescope right away. Or the NASA probe might have seen the planet by avoiding the dust cloud. Or maybe a new NASA probe with more sophisticated electromagnetic shielding would see the new planet.

"This illustrates that the simplistic interpretation of falsificationism – naïve falsificationism – where you just reject a theory as soon as it experiences a failure, doesn't correspond to actual scientific practice.

"Lakatos's thought experiment brings to our attention that defending a theory with auxiliary hypotheses can be an entirely reasonable and scientifically credible practice. And once you start looking at what scientists really do, reviewing their publications, you find that this is not just occasional but quite common practice. You can find examples, literally, in almost every issue of every major scientific journal."

"But not forever," interjects Jonathan. "At some point you know that there is no planet. At some point you know that the Earth isn't flat."

"Yeah, well, that certainly seems right," I say. "But how long is it reasonable to hang on to a theory – say Newton's enormously successful theory?"

"Jonathan, for a long time I pursued your point. At some stage of inquiry you must know. I looked for what I called 'a forcing argument', some evidence that, by its very nature, would logically force any scientifically reasonable person to accept or abandon a particular theory. I mean, at some point we do move forward. We are convinced that the Earth is spherical. Newtonian mechanics, the clockwork universe, is now commonly accepted to be a special case within the more general Relativistic Theory.

"But I couldn't find the forcing argument. It just isn't clear how we end the sequence of potential auxiliary hypotheses. Finding a final refutation, conclusive falsifying evidence, is as elusive as finding evidence that would provide a final proof of the truth of a core hypothesis. Naïve falsificationism, where any counter-evidence is supposed to refute a hypothesis, is undermined by the viability and credibility of auxiliary hypotheses in a progressively developing research program.

"So, if there are forcing arguments – they must be based on a *type* of evidence, a type of phenomenon, that precludes further auxiliary hypotheses – come what may.

"By the way, how much longer until we arrive at the Inn at Spanish Head? And where is the ocean?" asks Diana.

"Not far, we are more than half way," I say.

Conversation pauses as we pass through the city of Tillamook, everyone taking in the sights and main street scenes of small town America. "If we had more time we could stop at the cheese factory," I say.

"How exciting! Sorry we're going to miss that. People must come from all over the world to visit the cheese factory," says Diana.

"Actually they do," I say.

"What's the name from, Terry?" asks Pam.

"The Tillamook were the native tribe in this area when Lewis and Clark arrived in 1805. Archeology indicates the tribe had been here since the 1400s with a maximum population of just a few thousand. Best guess is that they were wiped out by the small pox introduced by the European settlers. My understanding is that the tribe is no longer even registered as an official tribe by the U.S. government," I say.

We head south-southwest out of the city of Tillamook, rolling hills, grass-fields full of milk cows. After a few minutes of silence I continue.

"Let me give you a simple, real life example of what Lakatos is pointing at," I say.

"Yes, that would be helpful. Thought experiments can be clever and cleverly misleading. Not saying that yours was, Terry, but...," says Jonathan smiling, wryly.

"Here is an example that begins to overlap with Stephen's research. It's about gravity," I say.

I look to Stephen for reaction. He's listening – engaged – but nothing beyond that at the moment.

"Although there is a prequel, this scientific odyssey really takes off with the research of American astronomer, Vera Rubin," I say.

"A woman scientist? All right, I like this story already," says Sue.

(The rest of this chapter is an expanded version of the account I presented in the van.)

Vera Rubin's initial research project in the late 1960s was designed to determine the mass of our sister spiral galaxy, the Andromeda Galaxy. The strategy was to measure the rotational velocities of the stars in the spiral – starting near the center and moving out toward the edges. We know a lot about the masses of stars in our galaxy, so, if one could measure the speed of rotation of similar stars in the Andromeda Galaxy one could calculate the total mass of the galaxy.

Rubin, who was at Georgetown University, teamed up with Kent Ford, a physicist, also at Georgetown. Ford had just developed an advanced spectrograph, using the then-novel photomultiplier tube technology. Using this device on a telescope reduced the exposure time to 1/10th that of a regular photographic plate – a considerable step forward in instrumentation for astronomical research.

Vera Rubin

Rubin and Ford set out to determine the velocity of the orbiting segments as far out as they could make measurements, as far out as they could find stars or gas clouds rotating in conjunction with the galaxy as a whole.

The expectation – the prediction – is quite simple. According to standard Theory of Gravity, formulated in either Newtonian or Einstein versions, stars orbiting in galaxies should behave very much the same as planets orbiting our Sun. The closer a planet or star is to the central gravitational source, the Sun, or the center of the galaxy, the more gravity it feels and the faster it has to orbit to resist falling into the central mass. Conversely, the farther the orbiting object is from the central mass, the slower its orbital velocity needs to be to maintain its orbit. The planet Mercury travels around our Sun in 88 days, Venus in 225 days and the Earth, of course, in 365 days, one year. As you move further out from the Sun, the periods of rotation get longer and longer – Jupiter takes about 12 years, Saturn nearly 30 years and good old Pluto takes a whopping 250 years to complete one orbit. This is referred to as 'Kepler's fall-off' since Kepler, researching the Copernican model of the solar system, and using Tycho Brahae's observations, calculated this regular 'fall-off' – this orbital slowing – with increasing distance from the Sun.

There are actually two effects that result in the longer orbital periods: one is that the outer planets are actually traveling slower, and the second is that they have much farther to travel in order to complete one cycle of their larger orbits.

What Rubin found was, as she put it, "puzzling." The stars farthest out from the center of the galaxy traveled almost as fast as those much closer to the center. Rubin's work on Andromeda was quickly confirmed and extended to other galaxies.

The observations of Rubin and her colleagues 'appear' to be counter-evidence to the universality of the Theory of Gravity – both the Newtonian and Einsteinian. The advocate of naïve falsificationism would say that we should just reject the Theory of Gravity and start afresh.

Instead, astrophysicists put forth an auxiliary hypothesis, analogous to the move in Lakatos's thought experiment, postulating that there was some unobserved matter (viz. currently referred to as Dark Matter) distributed in giant halos around spiral galaxies. If correct, it would explain the observed rotational behavior, thereby preserving the core faith in the universal applicability of the Theory of Gravity. In the case of galaxy rotation, Rubin's evidence counted pretty nearly the same against the Newtonian and the Einsteinian treatments of gravitational phenomena. Moving to Einstein's General Theory of Relativity didn't offer any sort of resolution.

Astrophysicists had a choice. Either discard the Theory of Gravity and come up with a new theory to explain 'gravitational phenomena', or propose an auxiliary hypothesis to 'explain away' the counter-evidence leaving the current Theory of Gravity (either Newtonian or Einsteinian version) unchanged. Just like the Marxists in Vienna that Popper had criticized, the astrophysics community chose to defend their initial interpretive framework – the Universal Theory of Gravity. They tried to explain – really, to 'explain away' – all this counter-evidence in terms of some 'extenuating circumstance'.

This defensive, 'explaining away' approach has turned out to be rather difficult. What Vera Rubin had documented was not a minor discrepancy. To account for the orbital behavior of the Andromeda Galaxy they needed to postulate several hundred times as much mass outside the outermost, observable stars. This would be a dramatic inversion of the mass distribution in the solar system, where over 90 percent of the mass of the system is centralized, concentrated, in the Sun. Nonetheless the initial auxiliary hypothesis was that there must be a 'halo of unseen matter' surrounding, outside, the galaxy.

Historically, when astronomers turned their telescopes out to look for this halo of matter, they didn't find it. Following the naïve falsificationist model of science, this failure to observe any surrounding matter refutes the first auxiliary hypothesis – the halo of matter – and removes this first defense of the Theory of Gravity.

'But wait,' the astrophysicists said. 'Just because we can't see the halo of mass by normal telescopic observation, doesn't mean that it isn't there.' And they proposed a second auxiliary hypothesis – to defend the first auxiliary hypothesis, designed to defend the Theory of Gravity.

New more sophisticated observational strategies were developed. 'Don't be too quick to conclude that the mass isn't there just because you can't see it by standard telescopic techniques. 'Absence of evidence isn't evidence of absence,' as the aphorism goes.

The next auxiliary hypothesis was actually a series of auxiliary hypotheses, each new one arising from the failure of the last, but all with the common research agenda of finding the missing halo of gravitational mass. The question morphed again and again, generating new hypotheses as to the form or nature of its presence. The 'hot gas halo' auxiliary hypothesis was quickly abandoned, since, if present in that form, it would have been easily detectable at different electromagnetic wavelengths – in the ultraviolet or x-ray, or maybe infrared or microwave. The halo of matter just wasn't there in a gaseous form of matter understood in terms of the well-established Standard Model of Particle Physics.

Another line of auxiliary hypotheses considered whether the missing matter might be collected into clumps, formally designated as MACHOs (Massive Compact Halo Objects), and there were numerous investigations to discover them. Such highly massive objects should leave distinct gravitational footprints that astrophysicists could observe within our own galaxy. The individual clumps couldn't be too large or they would collapse, go nuclear and become stars, luminous, easily observable from the beginning. The search for brown dwarfs – what you might think of as 'almost stars' – developed and many were detected, but way too few and in the wrong locations.

As in Lakatos's thought experiment, NASA, and its Congressional funders, have been tremendously supportive of the astrophysical research efforts designed to defend the Theory of Gravity – against the growing counter-evidence.

Vera Rubin's story has a fascinating pre-history, a prequel. Rubin's research wasn't the first to show the failure of the Theory of Gravity at the galactic level. In the 1930s Fritz Zwicky, at Cal Tech, examined thousands of galaxies looking for supernovae. Zwicky's research had an unexpected, serendipitous spinoff. He found that galaxies occurred in clusters. According to Zwicky's calculations – technically similar to Rubin's – the clusters shouldn't exist. The individual galaxies were moving so fast in relation to each other the clusters should have flown apart long ago. There wasn't enough matter in the total galaxy cluster to gravitationally hold the cluster together. In his 1937 research publication, Zwicky introduced an auxiliary hypothesis: there must be enough undetected – what he dubbed – 'dark matter' supplying the extra mass needed to hold the clusters together – saving the Theory of Gravity. Zwicky's research was at the leading edge of innovations in observational astronomy and, although noted, it was only recalled and its importance recognized after Rubin's seminal work in the 1960s.

These attempts to defend the Theory of Gravity against growing counter-evidence date then from 1937 at least, and it is still ongoing. One of the latest auxiliary hypotheses as to the nature of this Dark Matter invokes some perplexing particles called WIMPs (Weakly Interacting Massive Particles). The reasoning is straightforward. What is needed to defend the Theory of Gravity are 'Dark Matter' particles that are pretty much invisible – therefore 'dark' and 'weakly interacting' – and yet somehow producing a great deal of gravity, therefore 'massive'. There is no evidence of WIMPs in the laboratory, and no natural expectation for such particles in terms of Standard Particle Theory. The Dark Matter Hypothesis may well be just a place-holder, pointing to phenomena that, by their very nature, *cannot be made sense of* within the Mechanical framework defined by the presuppositions of the Scientific Research Program.

Pulling this all together, I said, "Just as the success of the Flat Earth Theory didn't prove that the Earth is flat, the fantastic success of the Theory of Gravity (viz. Newtonian and Einsteinian) within our solar system and, indeed, in recent laboratory measurements, doesn't prove that it is completely true, rather than a limited idealization, a special case success, applicable only within certain circumstances. Similarly, the success of science so far – which Jonathan finds so impressive – doesn't prove the Scientific

Hypothesis: that all the phenomena in the universe are governed by universal scientific laws.

"Popper had criticized the Verificationists arguing that no theory could be proved true – once and for all – since each theory has an unlimited number of consequences and possible tests. No finite set of confirming evidence can ever prove a theory true beyond all reasonable doubt. There are no final proofs – at least short of some fanciful final Theory of Everything. However, disproving a theory is not as easy as Popper initially imagined. The Promissory Note Argument is bolstered by the use of auxiliary hypotheses. Lakatos's thought experiment and the Dark Matter example demonstrate that it is not easy to falsify a theory once and for all – beyond all possible auxiliary hypothesis defenses.

"The forcing argument – the evidence that would falsify a theory once and for all – is strangely illusive. And yet as Jonathan emphasized and I agree, at some point, somehow, we learn and move on.

"Lakatos went further, arguing to me one day that even when we have moved on, no theory that has had *some* success has ever been *completely* refuted – *completely* falsified – and so should never be *completely* rejected. A theory that has succeeded in some limited range of applicability might still return and be valuable in future research, might make a come-back, finding novel applications, reemerging as the basis of the new leading-edge research effort. I thought of this in terms of the perennial disciplinary oppositions found in virtually all academic research disciplines.

(Note: In an article [*Scientific American* (June, 1992, page 80)] entitled "Single Electronics," Konstantin Likharev and Tord Claeson consider the question of whether electric current is the motion of individual electrons or the continuous flow of a fluid of charge. They argue that recent experiments confirm that both ideas may lead to novel electronic devices. The sub-theme of their discussion is that on a simple objectivist interpretation of the research enterprise, as something that is to be understood in the classical scientific framework, one or the other of these exclusive models might have been favored, based on crucial experiments. Yet each of the two distinct lines of research (viz. based on these incommensurable models) has yielded important new technologies. To paraphrase, there are technologies that can be developed on the assumption that electrons are particles, and other quite different technologies that can be developed on the assumption that electrons are waves. If we had decided that electrons were one and not the other, one of these engineering research programs would have been lost. In short there is a research policy implication (for both science and engineering) deriving from the recognition of the reality of complementarity.)

"Despite the occasional claims of final victory by one side or the other, the perennially opposite research programs re-emerge, sometimes after a couple of generations, sometimes in different cultures. Lakatos argued that in the history of both science and mathematics supposedly refuted and defeated research programs just keep coming back, often to reclaim the high-ground making new leading-edge advances.

"Pursuing Popper's Question as applied to the Scientific Research Program meant that I was searching for the 'forcing evidence' that would demonstrate the limit of the core Scientific Hypothesis. The forcing evidence had to be such as to preclude further auxiliary hypotheses. The forcing evidence had to preclude any future 'scientific' theory that would be able to give a 'scientific explanation' of the forcing evidence – come what may. Only such 'forcing evidence' would answer the Promissory Note Argument.

"Just what would evidence look like that could demonstrate the limits of science, evidence that could never be explained by any future scientific theory?" asked Jonathan.

14

The Quantum 'Situation' and Mysticism

"The answer to the forcing argument, the answer to the question of the nature of the forcing evidence, really began to emerge, paradoxically, with the rise of the new physics in the early 20th century.

"There were two major advances in 20th century physics – quantum theory and relativity – that led many scientists to question whether these were actually advances *within* the Scientific Research Program, consistent with the Scientific Hypothesis, or whether making sense of them required a post-scientific understanding of reality.

"Quantum theory was, and still is, for that matter, unexpectedly weird. I always loved Richard Feynman's remark that 'if you think that you understand quantum theory you simply haven't studied it long enough.'

"Quantum theory involved such a strange conceptual departure from the traditional bedrock scientific view of reality that leading physicists such as Niels Bohr began to argue that it required a new post-scientific theory – a superseding of all possible scientific theories, come what may. Bohr suggested that no possible future scientific theory could account for this new way of understanding reality.

"The paradoxical revelation was that the evidence suggesting an inherent limit to the Scientific Hypothesis didn't arise from any failure to predict. On the contrary, it arose from the enormous success of two, conceptually distinct, scientific research programs: the Newtonian particle program and the Maxwellian electromagnetic wave program. It arose from the success of two theoretical frameworks that – unexpectedly, by their very nature – didn't fit together. It arose from the success of two conceptual frameworks that *don't make sense* in terms of each other.

"To gain a sense of what happened, start with one of the defining maxims of traditional scientific reasoning: that we are investigating an objective world. Objectivity insists that in the competition of theories there is 'only one right answer', and that's the truth that science seeks – the objective truth. What is true is true and its opposite is not true, and there is no quibbling on this principle.

"The final objective truth that science seeks is that which resolves all competing opinions into 'one right answer', maybe not now, but eventually, at least in principle.

"This maxim of objectivity is closely associated with the logician's Law of Excluded Middle – something is either true or it is not true. Period! There is no middle ground, no little bit true, or approximately true, or sometimes true, or true from one point of view, but not from another. There may be uncertainty for us, for human observers, here and now, but there is no uncertainty 'out there'. In objective reality the world is definite, one way. Our final real true, objective scientific description of the universe and how it works must be one thing, one description, unambiguous.

"The full objective description and explanation, the Scientific Theory of Everything, must be complete and consistent. 'Consistency' means that objective truth is logico-mathematically continuous – one coherent system where everything fits together consistently. Excluding the possibility of inconsistencies also means that there must be a conceptual homogeneity – only one conceptual type of phenomenon, only one conceptual type of law governing all phenomena. 'Completeness' means that objective truth must include all truths and the final theory must be able to account for all phenomena. At least in principle there will be no unanswered questions, no unanswerable questions, no 'real' uncertainty about 'objective reality in itself' – 'out there'.

"Much of this is captured in the Newtonian image of a clockwork universe – everything causally connected to everything else in both space and time. The clockwork is meant to represent the One coherent reality, governed by One order – both consistent and complete. Nothing is allowed in the clockwork that doesn't belong there, and everything that is there must be there for it to be the One universal clockwork.

"Historically, scientists have always accepted that they were currently in a position of uncertainty regarding the complete truth, both about many specific matters as well as about the universe as a whole. However, adherents of the Scientific Hypothesis expected that the Scientific Research Program was advancing, converging, on the 'objective answer' – something that at least in principle we could arrive at – eventually. If, as mere humans, we couldn't literally arrive at certainty, the Scientific Hypothesis was that there still must be, in reality, One definite, complete and consistent objective causal order.

"For the scientific mind, for those committed to the Scientific Hypothesis, even to contemplate the opposite of One objective truth would be irrational, would be to embrace inconsistency, violating the Law of Excluded Middle, like saying that what is true about the universe is also what is not true. It would be, for the scientific mind, to enter a foreign world of inherently inconsistent, incoherent nonsense.

"That the universe had to be objective, that there must be One true description, is a presupposition of the Scientific Research Program so deep, so fundamental, that no one who challenged it was taken seriously in the scientific community. No one could oppose it *within* the mode of thought and reasoning defined by the Scientific Hypothesis. No one could practically oppose it *within* the Scientific Research Program.

"Another way to define and clarify what we mean by the 'objective' is to contrast it with the 'subjective'. Objective truth is separated from the subjective by saying that what is objectively true is what is true independent of whether anyone is aware of it, independent of whether anyone believes it, independent of whether anyone observes it and independent of how anyone observes it. Scientific, objective reality just is what it is, independent of any observer, including the scientific inquirer. In the scientific perspective this is commonly represented as the 'detached observer' formulation of objectivity. The 'detached observer' entailment of 'objective' scientific inquiry is what I have characterized in terms of the Spectator representation of inquiry.

"The presumption of objectivity, reasoned in keeping with the Law of Excluded Middle, means that there must always be a way, at least in principle, to decide between alternative theories. Either the disruptive planet, or the Dark Matter, is definitely there or it isn't. Accordingly, if two conceptually distinct theories both seem to account for most all of the phenomena in some area, there still must be, at least in principle, some way to decide between them. A common example used in history of science to illustrate this 'only one right answer' maxim involves the competing chemical theories of oxygen and phlogiston, each able to account for a large portion of the evidence at the time. They can't both be true. These competitive situations arise frequently enough so that the decision strategy has a somewhat official name. The decision strategy is to design a 'crucial experiment'. In the case of oxygen and phlogiston, the crucial experiment strategy worked and the oxygen theory won.

"What happened in quantum theory challenged the fundamental tenets of objectivity and the Law of Excluded Middle. The challenge didn't arise in an obvious and straightforward way, but, nonetheless, in an unavoidable and undeniable way.

"The prelude to the official formulation of quantum theory arose in that there were two highly successful competing research programs, the Newtonian program, primarily successful in explaining the motion of objects (particles) and the newer Maxwellian program, primarily successful in explaining electromagnetic wave phenomena. One particularly important research question that had emerged in the investigation of atomic structure and function was whether the electron was to be understood as a particle or as a wave. What was happening was that if you observed the electron in one way it behaved in a wave-like manner and if you observed it in another way it behaved in a particle-like manner. Crucial experiments were designed to decide whether the electron was to be understood, in the final analysis, as a particle or as a wave.

"In such a crucial experiment you are supposed to demonstrate something that the electron could do if it is a particle (or alternatively a wave) but, at the same time, could not do if it was a wave (or alternatively a particle). So these crucial experiments would decide once and for all whether the electron was to be understood *exclusively* – objectively – as a particle or *exclusively* – objectively – as a wave. There were at least a dozen different crucial experiments devised. The problem was that both types of crucial experiments succeeded. One set of crucial experiments demonstrated that the electron must be a particle and could not be a wave. Another set of crucial experiments demonstrated that the electron must be a wave and could not be a particle.

"The unexpected results of these experiments, at the heart of the new 20th century physics, were the first clue of the nature of the 'forcing' evidence demonstrating the limit of the Scientific Hypothesis, the first clue to the demonstration of the limit of the hypothesis that all phenomena were governed by One universal *type* of order. The nature of this evidence of the limit was not anything that could have been expected by naïve falsificationism. The forcing evidence was not encountered as a failure of prediction, but rather as a sort of logical 'confusion'. It was weird.

"How the electron behaves is now accepted to be observer-dependent, depending on how you look at it, depending on what type of experimental setup you use to observe it. The famous, popular illustration is the two-slit experiment. When you fire a beam of electrons at a plate pierced with only one slit, the pattern on a photographic film on the other side is a characteristic clustering, just what you would predict if the electron is a particle. Changing the experiment only slightly so that the plate is pierced by two parallel slits, the pattern on the photographic plate shows a characteristic interference pattern, just what you would expect if the electron were a wave. The novel problem here becomes apparent only with the realization that particles and waves, are different *types* of phenomena – not just a little bit different, but fundamentally different. Particles were the furniture of the Newtonian universe – always having completely definite positions and definite motions. Electromagnetic waves, on the other hand, were the furniture of the Maxwellian universe – with distributed positions and distributed motions allowing them to overlap and interfere.

The natural expectation, based on the Scientific Hypothesis, had been that either particles must be a type of wave or waves must be a type of particle. And yet particles, by their very nature, are local and definite, while waves, by their very nature, are non-local and indefinite. The concept of a particle and the concept of a wave seemed to be conceptual opposites. And yet, enigmatically, the electron observed one way displayed particle behavior and observed another way displayed wave behavior.

"I want to emphasize that it wasn't that any experiments failed to demonstrate what they were designed to demonstrate. The problem was that too many experiments succeeded – exclusively different *types* of experiments. We had too much success – 'an embarrassment of riches.' The diverse successes supported too many objectivities. The crisis was that there was apparently no way to decide between the particle objectivity of the Newtonian physics and the wave objectivity of the Maxwellian program. The success of numerous oppositely oriented crucial experiments to decide whether particle physics or wave physics was more fundamental, representing the 'real' objective world generated a crisis. There seemed to be no way to settle on the 'one right observer-independent answer'. There seemed to be no way to settle on the 'one right experimental-setup-independent answer'.

"Out of the confusion one insight was that the competing particle and wave observations, the evidence for each alternative conception of the electron, weren't actually, formally, inconsistent. The successful experiments in Maxwellian physics didn't seem to conflict with the successful experiments of Newtonian physics. They were different types of experimental setups, and the resulting observations were of different types.

"The evidence demonstrating the limit of the Scientific Hypothesis, per hypothesis, didn't arise in a direct counter-evidential manner to a scientific theory. It couldn't be represented as a simple conflict, as an inconsistency, what one would expect in a strictly mechanical, logico-mathematical model. One of the founders of quantum theory, Nobel Laureate, Louis de Broglie, later articulated the situation: "Two seemingly incompatible conceptions can each represent an aspect of the truth... They may serve in turn to represent the facts without ever entering into direct conflict." (de Broglie, Louis, *Dialectica* I, page 326)

"De Broglie pointed out that we seemed to be forced to the conclusion that every electron initially conceived as a particle has an irreducible wave aspect and every electron initially conceived as a wave has a particle aspect. The depth of the enigma becomes clear only when you realize that particles and waves are, conceptually speaking, opposites, mutually exclusive alternatives, mutually exclusive types of phenomena. Particle behavior and wave behavior, particle phenomena and wave phenomena are 'irreconcilably different *types* of phenomena.' The 'quantum principle', as it came to be expressed, affirmed that as in the case of the individual electron, every observational situation, every observable system had an irreducible aspect of particle-ness and an irreducible aspect of wave-ness.

"De Broglie's insights were fundamental to the formulation of quantum theory: "The entity "electron," as well as the other elementary entities of physics, thus has two irreconcilable aspects, which, however, must be invoked in turn in order to explain all of its properties. They are like the two faces of an object that never can be seen at the same time but which must be visualized in turn, in order to describe the object completely. Bohr calls these two aspects "complementary", meaning that, on the one hand they 'contradict' each other [although not in the traditional sense], and, on the other hand, they complement each other. And, in its essence, this notion of complementarity seems to have taken on the importance of a true philosophical doctrine." (de Broglie, Louis, *The Revolution in Physics*, page 218, The Noonday Press, 1953, Third Impression 1958)

"Particles are 'things' localized in both space and time. Waves, on the other hand, are spread out, non-localized in both space and time. Particle phenomena and wave phenomena are qualitatively, conceptually, incompatible in essential, demonstrable ways. As Niels Bohr expressed it, like left and right, up and down, constancy and change, rest and motion – they are complementary.

"If the quantum principle actually calls for a post-scientific theory, as many had begun to suspect, the fact that the evidence for the limit of the Scientific Hypothesis arose in this scientifically confusing way shouldn't be entirely unexpected. You can't make sense of 'multiple objectivities' as forcing evidence in the old objectivist way, by reference to the old objectivist interpretations of experimental results. The evidence is demonstrably repeatable 'scientific-like' evidence as I had insisted it must be in my research strategy. Quite unexpectedly, the forcing evidence comes in the form of two complementary types of phenomena associated with two, highly successful, complementary types of research programs.

"Those intent on defending and preserving the Scientific Research Program and the traditional One order, 'one right answer', sense of objectivity are faced with three bad choices.

"First, they could stick with the classical, Newtonian particle physics and search for auxiliary hypotheses to explain the wave phenomena in particle terms – in effect proposing a Promissory Note. Second, they could convert to Maxwell's wave physics and search for auxiliary hypotheses to explain the particle phenomena in wave terms – in effect proposing the opposite Promissory Note. The third choice is to accept this quantum weirdness and conclude that there are at least two distinct types of order in the nature of reality and that there are at least two different ways to experience 'reality'. The novel characteristic of this third view of the new 'reality' is that how you experience 'reality' depends on *how* you chose. The type of phenomena, and the type of order that you experience depends on how you chose. And, as de Broglie points out, each type of observation is conceptually incomplete, each way of observing is incomplete, is not conceptually representative of the whole. Each physics gives an incomplete account of reality. Every 'situation' has an irreducible aspect of each of the complementary types of physics.

"The third choice needs to find a post-scientific, post-objectivist More General Theory that can understand both Newtonian and Maxwellian physics as limited special cases.

"In 1927, in what is sometimes referred to as the mature phase of quantum physics, Erwin Schrodinger offered a Wave Mechanics formulation – apparently favoring the second option. Einstein liked Schrodinger's approach in that it seemed to retain a single mechanics and thereby to reestablish One 'visualizable reality composed of waves.' However, colleagues soon pointed out that Schrodinger's wave-function description of reality was incomplete. To Einstein's consternation, the 'waves' described by Schrodinger's wave-function were not Maxwellian waves. As Max Born was one of the first to realize, they were a curious new species of 'probability waves' specifying a non-local, distributed range of potentially observable realities. The good news seemed to be that the 'potential realities' evolved mechanically and deterministically when not being observed, as one would expect from a singular, objective Maxwellian physics. The bad news was that the 'potential realities' weren't *actual*. In order to observe the 'potential realities' you had to choose to observe in one way or another. The observer's choice is often characterized as a 'collapse' of the 'distributed non-local potential realities' into one observed actuality. The 'collapse of the wave function' actualizes one localized observational reality from the initial, otherwise inaccessible, potential field.

"The third choice embracing the wave-particle complementarity and the observer-dependence (or experimental-setup-dependence) was developed and championed by Werner Heisenberg and Niels Bohr and is commonly, although loosely, referred to as the Copenhagen Interpretation of Quantum Theory. The main difficulty for traditional scientists with this new observer-dependent alternative is that it forces us out of the Scientific Research Program. It forces us out of the Spectator's observer-independent conception of One coherent objective reality. In the Copenhagen Interpretation it is not possible to refer to or characterize 'reality' without reference to how it is observed. It is not possible to make sense of any observation without making reference to the observer's mode of observation, without making reference to the observer's frame of reference. It becomes impossible to characterize reality without making reference to the observer's method as part of the observation. In the final analysis, according to Bohr, it becomes impossible to characterize reality without making reference to localized conscious observation.

"Walter Heitler, who, together with Linus Pauling, led the scientific community in the hugely successful application of quantum theory to chemistry, put it this way: "The separation of the world into an isolated 'objective outside reality' and 'us', the self-conscious onlookers, can no longer be maintained." (Heitler, Walter, *Albert Einstein: Philosopher-Scientist*, ed. P.A. Schilpp vol. 1, page 218, Harper (1959))

"Heitler went on to specifically suggest that classical science – with its Spectator notion of an 'observer-independent reality' – was an idealization to be superseded by a more advanced worldview: "Object and subject have become inseparable from each other. Their separation is an idealization which holds, approximately, where classical physics holds. ... No sharp line can be drawn between an outside world and the self-conscious observer who plays a vital role in the whole structure and cannot be separated from it."

(Heitler, Walter, ibid, page 194-196)

"This is analogous to the Spherical Earth Model superseding the idealized Flat Earth Model – but the transition is far more radical; more fundamental. The move to quantum theory is a move away from simple classical objectivity toward something conceptually much broader, something post-objective, post-scientific.

"What was most important for my research was the recognition that the transition from the classical to the new quantum physics was a transition from a 'detached' Spectator representation of inquiry to an 'included-as-part-of-reality' Participant representation of inquiry.

"But here is where the third option is stuck. All attempts to provide a coherent representation of Bohr's Copenhagen Interpretation have failed to gain a comfortable consensus in the scientific community. It is a little like the situation with the Dark Matter Hypothesis, in that no one is really comfortable with the situation, but it is what we have to work with at the moment.

"In the philosophy of science, the community is similarly split. The conservatives associate with the Promissory Note search for a new 'objective' reality. The rebels tend to favor Bohr's insistence that there is no going back to traditional, classical scientific objectivity. The search now, among the rebels like myself, is for a coherent post-scientific Participant worldview that can make sense of the previous success of both the Newtonian and Maxwellian physics as based on limiting special case idealizations.

"Unfortunately, just to accept that we have been forced beyond the Scientific Hypothesis, that we need a new worldview to supersede the classical scientific worldview – as Heitler and Bohr and many others have argued – does not actually produce the new more general worldview."

I thought of Eugene Wigner, Nobel Prize winner in Physics in 1963, and I paraphrased him for the group: 'It is too early to say whether the current formulation of quantum theory will continue, but it will always remain amazing that, in our search for the determinants of the phenomena we thought of as a purely material reality, we were forced to the conclusion that consciousness is an ultimate constituent of reality.'

"Mysticism!!" booms Stephen from the back of the van. He had obviously been listening intently. The great eagle that has been hovering above my extended Chautauqua has just swooped in, both talons flaring.

As I glance at his face it is clear that his expression has changed, he is intense, with a look that I find a little intimidating, and his eyes are focused right on me. I had to turn away. "Mysticism?" I muttered rather softly with a sense of bafflement, searching to make sense of this comment.

"Mysticism?" I repeat quizzically a little louder.

I was noticeably shocked by Stephen's interjection. He hadn't said anything for the last half hour. I see Diana smile. She senses a lively fight. She is wondering where the popcorn is.

"Heitler helped establish quantum physics, and Wigner won the Nobel Prize. How can you call that mysticism?" I counter rhetorically.

My first guess was that Stephen wasn't making a pronouncement in favor of the objectivists. He was just aggressively rejecting the last step, the extension to the mention of consciousness. The quantum principle seemed to imply that since 'reality' was not 'objective', is was no longer correct to think of it as 'out there'. For many, the 'natural' implication was that 'reality' was 'our localized observations themselves', and, so the reasoning goes, these 'observations' are 'inside our consciousness'. This led some to suggest a connection between the implications of the quantum principle and mystical traditions. Over time, mainstream physicists, like Stephen, just rejected this entire line of reasoning – with increasing vehemence.

I could tell that all this talk of quantum physics had already begun to lose the nurses.

"But look," I say with a pause, "all this is getting a bit technical and obscure."

"You noticed!" says Diana, laughing.

"Sue, did you notice that?" Diana continues.

"I haven't really decided in my own mind about the electron, you know," says Sue.

"I thought it was becoming clearer when Stephen said 'mysticism,'" says Pam.

"Oh you're always taking his side," laughs Sue.

"Perhaps so," says Pam, "just my bias, I guess."

"Welcome to the world of confusions about quantum theory, about the limit of science, about the limit of objectivity, about the limit of the Law of Excluded Middle."

"In any case, what is important is that the evidence of the limit of the Scientific Hypothesis arose, not as the failure of experiments, where something was predicted, and the experiment failed. It arose *paradoxically* through the demonstration of two, highly successful, 'irreconcilably different' scientific research programs: the Newtonian and the Maxwellian – each limited, each a special case, to be understood, perhaps somewhat differently, in terms of a hoped-for More General Theory."

The scenery along the coast highway is spectacular. But because of the time constraint I discouraged interest in a walk on the beach and dipping toes into the Great Pacific Ocean. The nurses are disappointed, but business is business at this stage. We are well past lunchtime for Stephen.

Lincoln City is a seven-mile long sprawl of what were once five smaller individual cities that gradually grew together. Stoplights, motel and hotel signs and shops of all kinds now temporarily replace the sweeping vistas of the Coast Highway. My suggestion for a quick stop at the local McDonald's is unceremoniously rebuffed. "Stephen needs a proper meal." Nurses' Rule! The ocean comes into and out of view as we meander slowly through the long, merged-city that is only about half a mile deep, away from the ocean. As we arrive on a high bluff with a spectacular view south along the coast, everyone agrees that lunch at the Inn at Spanish Head is a most appropriate choice.

15

Common Sense Version – The Limits of All Ideologies

After our meal at the Inn and Stephen's admonishment not to chastise the waiter, I grow nervous. By now I'm not hiding my concern about the time. "We're running a little behind schedule here," I say to no one and everyone. They appear to be a bit surprised. Not that they had any clear, detailed expectation; it was simply that they had been relying on me. I fumble around, blaming it on everything but my miscalculation. I expedite our departure from the restaurant. "You go ahead and get loaded in the taxi-van and I will take care of the check," I say. By the time we are pulling away from The Inn at Spanish Head it is nearly 3:30 pm.

Heading south down Coast Highway 101, we need to reach Florence, a two-hour drive, before turning east to reach Eugene, another 90-minute segment. This doesn't give us much time to register at the Eugene Hilton, have dinner and be on-time at the Hult Center for the 8:00 pm lecture – to a sold out crowd of 2500.

I think I have made it clear to Gene that we are no longer in easy-going tourist mode. Smiling, I tell him that we are a little tight on time – and stare straight into his eyes for an extra couple of seconds. We are moving, and there is nothing more I can do. There is a knot of tension in my abdomen making me a little breathless, but I don't mention it to anyone.

There is silence for a few minutes as everyone settles in. "So where were we?" says Sue, I am still trying to decide if the electron is a particle or a wave?" As the highway passes close to the shore, Pam remarks, "I am rather favoring waves at the moment – on purely aesthetic grounds, I confess."

Everyone pauses to take in the ocean, sunlit in January, white turbulence rolling in from a blue expanse. The highway is cut through one of the foothills of the Coast Range, the time-weathered mountains coming down to the shore. As we top the next hill the long view south down the rugged coastline is breathtaking. Gradually, on the right, the ocean horizon reemerges.

"Look, I know that it is easy to get lost in the technicalities. Once you start talking about particles and waves in quantum theory it is tough to find your way back to common sense," I say.

"You noticed, Terry. How sensitive of you!" Sue repeats.

"I am sympathetic, really! This always bothered me as I pursued my studies in philosophy of science. My strategy had been to become 'an educated generalist' so I could explore all the sciences and talk competently to anyone working in any of the sciences. This strategy turned out to be even more valuable than I had anticipated.

"When people ask me what a philosopher of science does, I sometimes say that he learns all the sciences and then looks for common themes and common issues. You look for patterns of common problems that arise in more than one or perhaps in all the sciences. Over time this led me to the suspicion – to the hypothesis, I guess, that most of the truly deep philosophical problems of the individual sciences are actually common to all the sciences. They manifest themselves in slightly different terminologies, but the logic and structure reveal common underlying issues.

"For instance, I came to the rather bold hypothesis that the particle-wave dilemma in physics is structurally the same as the nature-nurture and heredity-environment dilemmas in biology and the human sciences. Another similar guess, or hypothesis, is that the common political divide – political left from political right – is analogous to similar conceptual and methodological divides within each of the sciences. These perennial dualities are often hidden in the science communities, in part because of the emphasis on objectivity, because of the insistence that there must be just 'one right answer.'

"So are you suggesting that the political right is like a bunch of particles?" asks Jonathan in a less than sympathetic tone.

"Not so simple, I think," I say, ignoring his overt dismissal. "But there is an interesting grain of truth in the analogy. The political right tends to think in terms of the nature of individuals, that is, free-market individualism, whereas the political left tends to think more holistically in terms of a distributed, yet coherent, social ensemble.

"Anyway, when I start to get lost or really confused in quantum theory or evolutionary biology, I look for analogies in chemistry, geology, psychology and politics. Moving by analogies among the different sciences turns out to be quite illuminating.

"As an element of my overall research strategy, I have tried to develop 'a common sense formulation', where my natural intuitions aren't so easily confused and undermined by abstract terminology."

"That sounds rather sensible," says Pam.

"Let me give you a simple example that everyone can relate to. Think of this as a sort of thought experiment. Everyone is familiar with the phenomenon of First Impressions. If someone makes a really pleasant and good-natured first impression, you are inclined to continue with that initial impression. Even when the person does a couple of things that 'seem' not so nice, you make excuses. You say, "Well, he is certainly a good person but he was just having a bad day." Or you might say, "He is really very polite but just not so much when he has been drinking." You might say, "He didn't really mean that, he was just trying to impress his girl friend, or his boss, or his friends – or whatever."

"You are offering the equivalent of auxiliary hypotheses to defend your initial impression, your initial theory, based on your original positive impression, based on your original observation.

"Similarly, if someone makes a bad first impression, and you distrust him, for instance, then despite his doing a number of positive, favorable things, you still don't immediately abandon your negative theory, formed at the first impression.

"If you expect people to act badly, then when they 'seem' to act good, it is reasonable to suspect extenuating circumstances. Auxiliary hypotheses again! The 'apparent' counter-evidence is dismissed as illusory or misleading evidence, not representative. You argue that the good behavior isn't properly understood as counter-evidence to your bad impression theory. Another common consideration is to suspect their motives. His good behavior, you might reason, is deliberately deceptive. Perhaps he is trying to give a false positive impression to gain your trust before taking advantage of you.

"Stephen's like that," interjects Sue. "He tries to give a good first impression when he advertises for nurses."

There seems to be a good-humored recognition and consensus among Stephen's crew about this. I gather it has been a topic on earlier occasions. Stephen pipes in, "We advertise for a nurse to give light assistance to an elderly professor. We make it sound really easy." Stephen is shaking, laughing almost uncontrollably – no sound, no voice, shaking in silent laughter – exuberantly.

I glance around at Sue and the nurses to try to understand from their expressions and body language what this is all about. The nurses are smiling, but not laughing. Clearly this is an insider's joke that I am not quite grasping.

"Once they have signed on, I work them hard for slave wages," says Stephen, who is obviously enjoying this humor immensely. On the nurses faces I am reading, 'Very funny. Ha, ha, ha.'

I am beginning to understand but Pam sees that I am still out of the loop. "They don't get the real picture until, Stephen, up past midnight writing a scientific paper, wants a bath at 2 am," says Pam. Everyone agreed that if Stephen told the truth in his advertisement, no one would show up for the interviews. It was necessary to pretend that taking care of Stephen was easy.

It's early January, and as the afternoon progresses the Sun has dipped. There is a lull in the conversation for a few minutes, and then we pick it up again.

"Instead of talking about Marxism or scientific objectivity, let me give you a really straight-forward, common sense version of an ideology and how the reasoning works. The question of the limits of science and objectivity is the same as the question of the limit of any ideological thinking generally. It is about the method of reasoning more than about the content of the specific theories.

"For instance, consider the following: There is a fairly common view or line of thinking that says that all people are really selfish," I say.

"Of course, they are," says Diana.

I am rather taken aback by her remark but I ignore it for the moment.

"So my question is: How would you know if it weren't true?" I say.

"To put it in terms of Popper's Question: What evidence, if it were to occur, would lead you to conclude that people weren't really *completely and consistently* selfish?"

"There isn't any such evidence, because it is true," says Jonathan.

Again, I am taken aback by the affirmation. "You're kidding me. You don't really believe that do you?" I ask with a sense of shocked disbelief.

"I believe it because it is true," says Diana.

"Well, look, my point is that, according to Popper's Question, you aren't really making a meaningful claim unless you can give 'a prior specification' of what a 'non-selfish act' would look like. To say that all possible actions are selfish doesn't say anything. Claims aren't meaningful unless they are potentially falsifiable. Popper is saying that if you claim that there is a property, or nature, that literally everything has, then the claim doesn't make a distinction, precluding the possibility of counter-evidence. If it doesn't make a distinction, then it isn't falsifiable and it doesn't say anything. You aren't saying anything.

"In other words, if you are saying something meaningful, making a meaningful claim about the world, then you must also, at the same time, be denying something – something that you should be able to specify, something that you should be able to describe. For instance, to say that everything is white (or constant, or local, or particle-like) isn't a meaningful claim unless you can tell me what it would be like for something to be non-white (or changing, or non-local, or wave-like). You need to be able to 'make sense of' the possibility of something being non-white (or changing, or non-local, or wave-like).

"Some people would say that giving someone else a gift is a selfless act. But a proponent of the Universal Selfishness Theory can counter that it isn't a selfless act because the giver either receives or expects to receive a net benefit, to get something out of it – more back in return somehow.

"This is the sort of question they hit you with in a Philosophy 101 curriculum – Theory of Knowledge: 'What do you know?' 'How do you know it?' 'How can you be sure?'

"But isn't it possible to give something away without expecting to get something out of it?" says Pam.

"It may seem possible, but it isn't real," says Diana, adding, "Even if they don't expect to get a benefit right away, people expect to be rewarded in heaven or something."

Jonathan and Diana have a well-considered impression, similar to a First Impression, that everyone is selfish and that we live in a selfish, competitive universe.

"Yes, well, there are these extreme cases that always come up in these discussions, for instance: Is sacrificing your life for someone else, or for your country, a selfless act?" I ask. "Say you dive in front of a bus to save your child or your neighbor's child."

"No, because people who do those things believe that they will be rewarded in the after life or they expect to survive and to be a hero and get a medal or something," responds Jonathan.

I am feeling rather uncomfortable at this point. I am talking to two of Stephen's caregivers, and they are telling me that they believe that everything they do is for selfish reasons. This strikes me as peculiar. How can this be? I mean, are they saying that they don't really care what happens to Stephen? That's not my idea of a caregiver. How can you be a nurse, a professional caregiver and believe that everyone is really selfish?

I taught at Linfield College for several years in the early 1980s. One of my main courses was entitled "The Political Environment of Health", attended primarily by nursing students, seeking the new Bachelor of Science in Nursing degree. Most of the women entering nursing said that one of their main motivations had been to help other people. But here was one of Stephen's nurses telling me that this was never their real motivation.

I mention this to Diana. "Those sorts only want to help others because it makes them feel good. They enjoy it. That's why they do it. The motive is always selfish," she responds.

Jonathan points out the evolutionary argument, saying, "It's biologically natural to act to advance your genes – even if it costs you your life."

Pam remarks, "It is rather strange they can't even identify a possible exception. Aren't people sometimes selfish and sometimes selfless?"

The difficulty is perennial. For any example that is put forth as a non-selfish act, as a counterexample to the theory that everyone is completely selfish, one can add plausible sounding auxiliary hypotheses that reinterpret, re-represent, re-conceive the act as selfish.

I think about saying that the theory of universal selfishness tells you more about the people who believe it that than about human beings in general. But it is confrontational, a sort of 'cheap shot', and I don't really believe it anyway, so I decide to let it pass.

I am happy to have Pam in the conversation, who is beginning to sound like an ally. Pam is 'middle ground', seeing people acting both ways, as having a sort of dual nature.

Gene, the taxi-driver, chimes-in supporting Pam. "It just can't be necessarily that everyone acts selfishly in principle," says Gene.

"But look, this is the whole point," I continue. "If there is no action that is, by its very nature, selfless, somehow non-selfish, then it doesn't mean anything to say that all actions are selfish. According to Popper's Question, understood as a criterion of meaning, such universal claims just aren't meaningful. One *type* of action or observation can't be meaningfully representative of the whole."

Addressing Jonathan and Diana, I say, "Popper would argue that you are being ideologues – just like Popper's characterization of the Marxists in Vienna. Whatever counterexample is proposed, you try to 'explain it away', you try to find a way to reinterpret it the other way – as selfish."

"That's because people are selfish!" asserts Diana.

"But how would you know if you were wrong?" I insist.

"I'm not wrong. I've had a lot of experience in the matter," says Diana.

"But look, some people believe just the opposite – that everyone is really selfless – cooperative, community-oriented," I say.

"They're naïve," says Diana.

"Confused," adds Jonathan.

"Marx actually argued that people are deep-down, community-oriented. People who act selfishly are really trying to gain acceptance in the group. They might have felt rejected, or that there was an injustice and they were trying to reestablish a moral balance. Marx felt that people who resisted the communist revolution shouldn't be killed or punished but en-lightened, re-educated, convinced to see that what they really wanted was to be part of a just, cooperative community. For Marx humanity is one, unified family.

"I have a friend who worked with the Hopi Indians. The Hopi believe people, children in particular, are naturally selfless and identify their in-terests with the family and tribe. They must learn – if they do so – to be selfish."

"That's so ridiculous. Look at society. People take whatever they can get away with," says Diana.

"OK, maybe you're right and maybe not, but at least now we can see that there are two different theories of human nature. One says that people are 'really' completely selfish and competitive and the other that people are 'really' completely selfless and cooperative.

"The famous Irish author James Joyce once commented, 'there are two types of people in the world: those who expect others to act well and those who expect people to act badly. The former are regularly disappointed and the latter occasionally surprised.'

"What Joyce is expressing is a scaled-up social version of the two sides of the First Impressions phenomenon.

"Here's another thought experiment. Imagine someone landing on a dis-tant planet and after some observation, forming an initial, first impression – that everyone on that planet acts selfishly.

Laughing, I say, "It looks to this traveler like it's a right-wing, individualist paradise.

"Now consider another traveler landing on a different planet and after some observation, forming an initial, first impression that everyone on that planet acts cooperatively for the welfare of others, for the common good. This second traveler believes he sees a politically left-wing paradise, perhaps Marx's ideal socialist paradise.

"Each has formed an initial theory – a first impression – of human nature, and what I am driving at is an analogy with the question of the nature of the electron – particle or wave. Historically, one group of scientists was convinced the electron was a particle while another group was convinced it was a wave. Evidence to the contrary to each position – 'apparent' counter-evidence – was explained away, or set aside with a Promissory Note to handle it later. Each side was quite reasonably more excited about advancing their research program – mapping out the wider area of applicability, where the particle model or the wave model worked – than they were about looking for, or trying to explain, where their research program didn't work.

"The reasonableness of this sort of 'positive bias' helps us to understand the tendency to ideological reasoning. If you believe reality must be objectively 'One', one way – mathematically and conceptually consistent – and if you have clear and demonstrable evidence that your one type of phenomenon (say particle) is a highly successful way of understanding much of the world, then there is a natural tendency to extrapolate, to hypothesize – perhaps to conclude – that your type of phenomenon, your type of knowledge, is representative of the whole. Consequently, any 'proposed counter-evidence' of a different type of phenomenon can't be representative of the whole. Such a proposed alterative is 'somehow' misleading, so it is legitimate to ignore it or to explain it away.

"Now consider a third situation, where both travelers arrive on the same planet – the planet Earth – and, after some observation, they form these same, precisely opposite first impressions. Each traveler can cite many examples, positive evidence – verifications, of behaviors that they see, clearly and distinctly, as either selfish or selfless, competitive or cooperative. When confronted with 'apparently' critical counter-evidence from the opposite traveler, each traveler offers auxiliary hypotheses to defend his theory. For each traveler there is evidence confirming his theory, so the attempts by one traveler to explain away the other traveler's evidence seem implausible and distorting, involving some sort of misunderstanding.

"It is essential for the analogy and for what I am trying to get at that each traveler is arguing for the universality, the completeness, of his theory of human nature – no exceptions. Each is arguing for how people actually are, universally, objectively, by virtue of human nature, perhaps by virtue of the nature of all of reality, despite any proposed 'appearances' to the contrary.

"How one tries to understand how society works often differs on precisely which of these two opposite impressions – perennially opposite theories – one accepts. The theory we accept is the basis of how we understand the actions of other people in our life – and ourselves, for that matter. Furthermore, the theory you accept is likely to influence your beliefs about how you *should* act – your ethics, your sense of right and wrong.

"And here is a clue as to how ideological reasoning links to moral reasoning. If, in the final analysis, I can't tell the difference between a selfish action and a non-selfish action, one way or the other then, Popper's Question insists that the sense, the meaning, of right and wrong actions disappears. It is with the universality of the claim that human nature, or the nature of the universe, is universally – objectively – One, one type, conceptually uniform, that all *qualitative* distinctions disappear, all moral distinctions disappear. To say that someone is being selfish is no longer a value statement, no longer evaluative, no longer making a distinction. Actions are just actions, not good or bad. It is with the extrapolation to universality, the objective claim to have the 'one right answer,' associated with ideologies, that meaning – and the right-wrong, good-bad moral context disappears. (The more subtle, self-referentially paradoxical implication is that with the embrace of an ideological, one-type 'objective' reality, any meaningful distinction between true and false also disappears.)

"If you can't distinguish, then there is no real difference," I say, thinking that I am winning them over with deft reason.

"That's because there is no difference," responds Jonathan.

"But notice the key point is that if you can't tell the difference then the moral sense of the term 'selfish' is lost. To say that everyone is selfish all the time doesn't tell me anything. The moral order, which presupposes real choice between real moral differences, disappears. Justice is whatever happens.

"That is close to the argument of Thrasymachus in Plato's *Republic*: might makes right. The only real or relevant sense of right and wrong in the world is, in practice, whatever the current ruler or ruling group says. The powerful define what is right. And it changes arbitrarily, depending on who is in power. Everyone is selfish, and the most powerful of the selfish rule and exploit the others.

"Sounds like normal politics and economics to me. Another confirmation of the selfishness theory," adds Jonathan.

"And those who are currently being exploited would do the very same thing if they had the opportunity, if they were the rulers," I suggest. "Right?"

No one answers me. Jonathan and Diana just sort of smile.

I'm frustrated. What evidence, if it were to occur would convince them that this ideology of universal selfishness isn't reality, isn't objectively true, isn't even a meaningful claim?

"You actually experience the world differently if you believe people are always selfish," says Pam reflectively. "They live in different worlds."

Pam's simple reflective insight is reasoned from her sense of the middle ground. Reasoning from the framework defined by the Law of Excluded Middle one must insist that people are exclusively either selfish or selfless. If people have an objective 'nature' then they must be one or the other. There can be no conceptually continuous, consistent middle ground. Selfishness and selflessness are co-defined opposites – formal contraries.

The middle-ground position is that people are both: selfish in some circumstances, in some experimental settings, if you like, and selfless and cooperative in others. Maybe the right way to say it is that there is a selfish and a selfless component to every action.

This middle-ground position in the new physics is Niels Bohr's complementarity solution to the wave-particle question. Somehow the electron is both, even though particle and wave are two irreconcilably different types.

In the case of quantum theory, despite the verifications and crucial experiments supporting each research program, there is a realization in both opposite research programs that the electron is somehow oxymoronically both a wave and particle. So somehow some people are capable of seeing past their dominant theory, beyond their dominant approach to understanding the world – beyond their first impression.

16

Time to Panic and the Origin
of the Universe

A sign whooshes past indicating that we have just entered, Florence Oregon.

"Gene, watch for the turn off to Eugene," I stress.

"I'm familiar with it," he says.

It's 5:30 pm, past dusk, starting to get quite dark. We are right on schedule – the revised schedule that is. If it is really 90 minutes to Eugene from here, we will arrive at the Hilton at 7:00 pm. Plenty of margin to make the 8:00 pm performance time at the Hult Center. Conveniently the Hult Center is next door to the Eugene Hilton.

Gene makes the turn east, inland, back into the Coast Range towards Eugene. The road narrows to two lanes, forest on either side of the road – leaving behind into the past, into memory, the ocean waves pounding the sandy, particulate beach.

"How are we doing on time Terry?" Pam asks.

"It's a little tight, but we should arrive by 7:00 for the 8:00 start time," I say with a tactically manufactured air of confidence.

"That doesn't sound like much time to me," she says. "What about Stephen's dinner?"

I just stare at Pam in mild disbelief as this sinks in. I am of course by now acutely aware of the 'Stephen will eat' policy of the Nurses' Rule Principle. Judging by the time it took for lunch…

Pam has now fully grasped the problem. And she comes at me.

"And you don't imagine we can get Stephen ready in that time do you? Besides dinner, he needs to wash-up and a change of clothes," she says quite forcefully.

"We will not rush his meal," she adds.

His meal. His clothes. Arriving. Moving all the luggage from the van to the lobby. Checking in. As I glance forward out the windshield the road is dark and this all comes rushing at me, matched by the headlights of the oncoming traffic, intermittently blinding.

A mild sense of fear begins to grow as I again imagine 2500 people sitting in the Hult Center, wondering where we are. How long would they wait? This takes over my consciousness, demanding assessment, evaluation, problem solving – trying to devise a strategy.

"There is a restaurant in the Hilton," I add. "But if dinner takes anywhere near as long as lunch, that won't work."

"What kind of food do they serve?" asks Pam. The undertone is an aggressive reaffirmation of the 'no compromises' on Stephen's meal policy.

I get a look from Pam that I don't understand – an idea!

"Stephen has a cell phone," she says. "Perhaps we could call ahead and order the meal so that it would be ready when we arrive."

My God! She is on my side! This isn't intransigent Nurses Rule. This is cooperative problem solving. Jonathan scrambles over the middle seat to gain access to the back of Stephen's chair, where the cell phone is stored. Stephen's cell phone – circa 1992 – is the size and weight of a brick. But it works.

"What's the number? Terry do you have the number?" she asks.

"Yes. Wait a second; I have it right here," I say sorting through my itinerary folder.

Pam calls the Hilton. After a confusing attempt to explain the situation to the person at the front desk, she is forwarded to the restaurant – The Big River Grill. The maître d' is clueless: "You want reservations for how many?" Pam asks to speak to the manager. She explains the situation: "Professor Hawking is arriving a bit behind schedule for his public lecture at the Hult Center."

"Thank you. Could you read me the menu?" she says.

"Yes. OK. Just a moment," she says, turning to Stephen and repeating several of the menu items, asking for his preferences. Pam enters into a rapid-fire exchange with Stephen, "The Chef says he has a lovely New York steak. Or would you prefer the cordon bleu? Yes? Was that a yes for the cordon bleu? No? For the steak? Yes. How would you like it cooked? Rare? Of course."

Stephen is responding to Pam, sometimes quickly, sometimes only after a little deliberation, in a well-practiced eyes-up, eyes to the left-side communication routine. The nurses don't decide what Stephen eats. To an outsider it might seem sensible that the nurses might act as dieticians, selecting Stephen's meals for him with an emphasis on nutritional balance. But that isn't happening here. Stephen is in charge. Stephen makes all the decisions. The nurses don't typically even voice an opinion, except as a familiar friend might point out that one of Stephen's favorite dishes is on the menu at this particular restaurant.

Pam is talking to Stephen and the maître d,' "What kind of vegetables are you serving this evening? Stephen, would you like the carrots or broccoli? I think the carrots sound best. Yes? How are the carrots prepared? Could we have them without the sauce? Yes? No? Stephen, the carrots are already in a sauce. I think maybe the broccoli would be best. No?"

Stephen makes a face of disgust.

"I know you don't care for broccoli. But..." she says, moving on.

"Do you have a soup this evening? French onion? No? Another? How is it thickened? With wheat flour? Because Professor Hawking has a slight allergy to wheat gluten. Oh, lovely. Stephen I think they have a very nice tomato bisque that would be alright. Yes? OK?

"We'll have that," she says to the manager.

"What do you think? What would you like, Stephen?" Pam asks again.

Pam is efficient but she isn't rushing him. "Hang on. We'll be right with you in a moment," she says into the phone.

While Stephen is thinking, Pam consults Diana. "What do you think? I mean given the time question?" Diana defers, "What does Stephen think?"

"So Stephen... Steak?... Lamb?... Salmon?..." Pam asks, pausing after each suggestion reading Stephen's eye signals. There seems to be ambiguity in Stephen's response.

Pam reengages on the phone: "How are the steak and the salmon prepared?"

After a moment, she comments, "Stephen, I think the filet mignon sounds rather nice."

She waits for Stephen's response and then glances at Diana looking for consensus.

"And how would you like that cooked, Stephen?

"Rare, I suppose. Of course," she says.

"No? Not rare? Yes? Then what do you mean?" says Pam with a momentary spike in exasperated frustration.

"You don't want the steak? Yes? Then what?" says Pam in calm pursuit.

"Another type of steak? Yes?" Pam asks.

Pam queries the manager as to other types of steak. Several other options are contemplated. Pam begins to show more than a little impatience with Stephen. I don't recall the details of the final choice, but it was beef – not the filet – and it was to be rare. I am largely incapable of reading the communication between Stephen and Pam. It is rich and subtle, reflecting their history together.

Pam continues on, asking about details. Eventually, the manager connects Pam directly with the chef. This makes Pam very happy – a personal connection. She explains the situation once again and goes over all the details. The chef has heard of Stephen Hawking. That helps. Complete cooperation. Teamwork.

Despite all the complicated decisions, it all happens rather quickly. Pam, the neurological surgical nurse, knows how to work fast in critical situations. Her personable, polite, diplomatic style is naturally winning.

"Terry, when will we arrive at the Hilton?" Pam asks.

"Gene, what do you think?" I say.

"I think we are about half way, so in about 45 minutes," says Gene.

So I imagine the meal will be ready to eat when we arrive. Cell phones!

"How delightful to work with such people?" says Pam. Needing a little reassurance at this point, I take it that this somehow reflects on me – my choice of hotels I guess – since Pam is smiling and looking at me as she says it.

What I just witnessed was the performance of a real professional at a moment of crisis. I have a brief flight of imagination: Pam taking charge in an emergency in the surgical suite. Something unexpected has happened: "Yes, doctor. And should I increase the plasma flow." Pam quickly slaps the perfect size of hemostat into the surgeon's palm. Crisis averted. Now I understand the smile differently. The operation isn't over yet, but the patient – like our plans for the evening – didn't die during that crisis.

Diana, who is also fully cognizant of our time crunch, addressing Pam, says, "I wonder if we could have them deliver the meal to Stephen's room? That would certainly save time."

"Yes. Brilliant. What do you think, Stephen?" Pam asks. "Yes?"

Pam calls the hotel again and speaks to the restaurant manager, "And could we have that as room service to Dr. Hawking's room?"

Pam takes this opportunity to review a horde of detailed instructions to the folks at the Hilton.

"And could we have a pot of hot water and a pot of cold water. For tea, of course. The cold water is used to mix, allowing us to precisely control the temperature of the tea," she explains.

Stephen drinks a lot of tea, part of keeping him hydrated, so the temperature question is rather crucial. Stephen isn't able to test the temperature before it reaches his mouth. So the nurses take extraordinary care to make sure the temperature is right. No scalding, please: no tepid tea, please.

Everyone begins to contribute to the arrival strategy. Stephen will go straight to his room with Pam and Diana. "We better call ahead to make sure we can do this?"

Sue Masey and I are to take on the task of checking everyone in – room assignments. Thinking ahead, to make room assignments before we arrive, we make another cell phone call to the Hilton's front desk.

Jonathan, along with Gene, will be in charge of unloading the luggage and making sure it is delivered to the right rooms. Most salient is Stephen's medical luggage and the extra batteries for his chair and his computer. "No batteries, no presentation," Jonathan emphasizes.

It doesn't all happen smoothly. But we are unloaded and Stephen is in his room having dinner and preparing for the presentation, which is supposed to be about 45 minutes from now.

I rush next door to the Hult Center to find the University of Oregon folks who are coordinating Stephen's introduction. I am met on the way by Hult Center staff, with whom, fortunately, I have worked before. I am barraged with ticketing and logistical questions, topped by: "Where is Dr. Hawking? Can we do a sound check from his voice synthesizer?"

I work through the setup with the stage crew. "A standing microphone. Forget plugging into his computer for a direct feed. There is no time for a sound check. Just adjust it on the fly," I say. That's not their first choice, but these are professionals and they are 'OK' with it.

The main squeeze now is to free up Jonathan from check-in and luggage duty so that he can bring over the other laptop computer. Do I need to go get him? Jonathan arrives guided backstage by one of the Hult staff. He hooks his laptop to the projector behind the screen for rear-projection of Stephen's PowerPoint slides. Jonathan has a script printout of Stephen's talk, annotated as to when to make a slide change. He adjusts the connection to the projector – and it works.

"This is why Stephen pays you the big bucks," I joke.

"Yeah, right. Don't I wish," says Jonathan.

People are everywhere. The audience is packing in.

Dr. John Moseley arrives. He is a Professor of Physics, former head of the Department of Physics at the University of Oregon and now Vice President of Research at the University. He is advised of the time crunch. "We might have to start late – I mean fifteen minutes, half an hour at the outside," I say. Moseley is composed. He isn't critical, which is rather reassuring.

It's about ten minutes to eight. The place is packed and no Stephen. Time for me to head next door to the Hilton to see how things are progressing. I just hit the sidewalk between the buildings and there are Pam and Sue and Diana with Stephen, wondering which way they are supposed to enter the Hult Center. They have gone the wrong way. The way they were headed, following signs, was for wheelchair seating in the audience. "We have to go through the lobby, but it's OK. Around this way," I say, leading, blazing the path. Wheel-chair access to the stage, it turns out, is much more complicated. Once in the lobby, we take the audience elevator to the basement. Then, moving down a dreary, musty smelling hallway, we find the freight elevator. This is the elevator used by production companies to bring their staging, their sets, to the stage-level. We arrive stage left, greeted formally and politely by a gracious John Moseley.

We actually start at 8:10 pm. No one complains – Stephen is here. The title of the Eugene Lecture is *The Origin of the Universe*. I greet the audience, say thank you to all the cosponsors and introduce Moseley who introduces Stephen.

"The most dramatic moment of the evening is when Stephen moves out onto the nearly bare stage in his motorized wheelchair and is met with loud applause – a sustained standing ovation. This ovation is followed by an equally long silence as Hawking, motionless, but for the unseen manipulations of his fingers, boots up the presentation on his computer.

Finally: "Can you hear me?" Again, cheering affirmations and applause. Another short pause and he begins.

In the lecture Hawking covers the galactic gamut from small white dwarf stars to monstrous black holes, from the theory of chaos to speculations on mysterious 'dark matter'. He expounds on current theories about the birth and death of the universe.

"The universe is 'on the knife edge' between expanding forever or collapsing back to its origin as a single, infinitely dense particle of matter, says Hawking. The ultimate fate probably won't be known until we find a better way to measure the distance between galaxies. In the meantime, not to worry. The universe started with a Big Bang. If a collapse, or Big Crunch, is in the cards, the serious action won't begin for another 100 billion years or so. That should give us time to sort out the Middle East and one or two other problems," he says.

The audience loves the injected humor, and Hawking loves their response.

"Physicists, he says, can now estimate with a fair degree of accuracy the total mass, or 'weight', of all the observable matter in the universe. But the figure is only a small percentage of the mass needed to hold galaxies together, to keep them from flying apart.

"Something else, something unseen, must be out there, forming the 'critical mass' that allows the universe to function. The unknown is 'Dark Matter', something with physical mass and gravity, but not emitting or reflecting light, as stars and planets do.

"Understanding Dark Matter is the key to understanding the workings of the universe," Hawking says.

People come to popular science lectures to learn. However, you can only communicate so much in an hour. The real strategy is not to educate completely but to stimulate curiosity – to encourage the audience to learn more on their own. "It was a little like going back to college," said Nancy Dunn, while Hawking was preparing to answer a question from the audience. "I feel like I went to a lecture and I hadn't done the reading."

Hawking offered an argument about wandering backward through time. Then he added, "But the best evidence we have that time travel is not possible is that we haven't been invaded by hordes of tourists from the future."

Stephen's PowerPoint slides are all black text on a white background. I remark to him later that he is behind the times. His PowerPoints were also designed for a smaller venue – like a seminar room or a small lecture hall, seating perhaps 300-400. This is a symphony hall with 2500, and I am sure that the folks in the back of the main floor and especially in the upper balcony, must have had trouble reading them. At future public lectures Stephen's PowerPoint presentations were much improved – leading-edge quality – even including video clips.

For the audience question and answer portion we had set up several microphones in the audience, as well as inserting question cards in the program-magazines. The ushers collected the cards and brought them backstage, while Stephen was answering the first questions. Stephen wasn't keen on it but we convinced him to include the inevitable God question at the end. "It's not for me to say," he responded, to a huge affirmative applause.

Since there was no time for an earlier separate meeting with the local Eugene students with disabilities, we arranged for them to come to the reception after the lecture at the Hilton. There were about a dozen, and Stephen spent a disproportionate amount of the reception time with them. They are learning from Stephen. However, what they are learning here is more likely Participant knowledge about how to work in the world, about how to live. Stephen's lessons to the students with disabilities are different from the Spectator-like subject matter of the main lecture at the Hult Center. Everyone is a natural inquirer, a natural problem solver. But there appears to be more than one type of question, more than one type of answer, more than one type of solution.

17

Aporia (Puzzlement) at the Eugene Airport

The possibility of flying out of Eugene had been a selling point for the Eugene lecture. No need to return to Portland. However, the flight options out of Eugene, heading south to California, were limited. There was nothing direct to Santa Barbara. The departure to Los Angeles International – LAX – was scheduled for 11:00 am. From there, Hawking and crew would drive north to Santa Barbara to meet and stay with Hawking's colleague, Jim Hartle, for a few days.

Sue Masey had disappeared earlier in the morning, taking a cab to the Eugene airport to catch a flight connecting, eventually, back to Cambridge, England.

Loading up all the baggage in the morning is comparatively easy. Everyone is focused on his or her specific task. Gene, our driver for all day Saturday, had gone back to Portland, after dropping us off at the Hilton the evening before. So, for the short trip to the Eugene airport, I engage a local taxi-van. Everything is business – the logistics of packing up, checking out, loading the van ... "Is Stephen's chair securely tied down?"

Jonathan is responsible for double-checking Stephen's considerable luggage. The batteries, backup computer and medical paraphernalia are all accounted for.

Only as we reach the airport with plenty of time to spare does a calm emerge. There is a sense that we have all been through something together – a voyage, an odyssey – and, at least, this portion of it is about to end.

"The drive to the Coast was supposed to be an alluring adventure," I say.

"Certainly was an adventure," says Diana. We all laugh and smile in a good-natured manner.

There is a round of compliments and thank yous, recognizing the particular contributions of each person in the party. "Thank you." "Nice working with you." "You were great!"

"Nothing really went wrong, did it?" adds Pam, and she smiles.

As we are disembarking from the van, bag by bag, Pam asks, "So what happens to you now, Terry? I mean, what do you do now?"

I pause, wondering how this will go down, then say, "Actually, I am going to meet my mystic."

That turns heads, and I add, "The fact is that I am off to see Fritjof Capra in Berkeley.

"Oh really, that is interesting," says Pam. Stephen perks up. He is listening.

"And who is that? I've heard the name. But just who is that?" asks Pam.

"Capra is a physicist. He was in the first lecture series we did in Portland 1989-1990. Capra is most famous for his first book, *The Tao of Physics: An Exploration of the Parallels between Modern Physics and Eastern Mysticism*. He is definitely into mysticism by Stephen's standards," I say.

"Definitely mystical," quips Jonathan.

"So you know of this guy? You've read this book?" Pam asks Jonathan.

"Yep. Well, most of it," says Jonathan.

"Anyway, when we were together, while he was here in Portland, we developed a pretty good intellectual rapport. I had used his second book, *The Turning Point: Science, Society and the Rising Culture*, as a required text in the main, regular, five-unit course I taught for several years at Linfield College.

"So what are you doing with him now?" Pam pursues.

"One of the local high schools wants to establish itself as, what's called, a magnet school – a high school with a specialty in a certain subject area. They want to create an environmental magnet school. Central to Capra's writing is his multidisciplinary systems approach. He sees all the different sciences, as well as the humanities, as equally valid perspectives on the same reality. And that 'web of perspectives' is essential to thinking about environmental issues. So the high school faculty asked me to work with them to engage Capra as a sort of consultant. So Capra and I are working together on this. I am mostly just a catalyst. We visited with the key faculty at the school a couple of months ago and now Capra has invited me down to Berkeley – to his Elmwood Institute – to meet with other folks he has been working with on developing multidisciplinary curricula for high schools," I say.

"Why is he a mystic?" asks Pam.

"I'm joking. He isn't really – except perhaps to people who think the universe is purely material – just dead stuff that moves around mechanically according to mathematical laws.

"Capra recounts a turning point in his thinking about physics. He was at a friend's house high on a cliff near Big Sur, one of the most beautiful parts of the California coast. It was dusk, just at sunset, and he was meditating, facing out over the Pacific. Facing the afterglow, he began to imagine the particles, cosmic rays, entering the atmosphere, colliding with the atmosphere. The incoming particles collide and are destroyed, while creating new particles, and these, in their turn are destroyed, creating other new particles. There is a sort of continuous shower, a cascade of creation and destruction and re-creation. Then, at some point, imaging all that, he had a revelation about Shiva, the Hindu god of creation and destruction, the Hindu god of transformation. And he began to experience, to realize, to embody, a conversion. In his book he offers at least a partial interpretation of his intellectual insight, proposing that the Eastern Philosophies – the *Tao Te Ching* in particular – had arrived at a more general framework for understanding the universe millennia earlier. This ancient Eastern framework – the sort of yin-yang stuff – is what we in the West have just begun to rediscover in our encounter with complementarity.

"In terms of our earlier discussion, Capra proposed that quantum theory does represent – does demonstrate – the limits of classical mechanics, the limits of 'one-right-answer' objectivity. For Capra, classical science is an idealization, a workable perspective only within a limited range of applicability; a special case within the broader wisdom realized much earlier by the ancient Hindu and Chinese mystics – ancient natural philosophers.

"Capra wasn't the only one, and certainly not the first, to have been thinking in this direction. Niels Bohr, one of the main characters in the early development of quantum theory and usually identified as the author of the dominant Copenhagen Interpretation of quantum theory, chose as the central image of his coat of arms the yin-yang diagram – you know, the circle that is half black and half white, divided by a sort of s-shape swirl. And each half contains a small element of the opposite – small black circle in the white, small white circle in the black. Both Bohr and Capra saw in this ancient symbol the theme of a dynamic unity of contraries, of opposites, of essential complements. The motto on Bohr's coat of arms is "contraries are complementary."

"Why did you use his book in your course at Linfield College?" asks Pam.

"Well…" I pause, wondering whether I should get into this.

As we are talking, Stephen has dismounted from the taxi-van and everyone is moving from the parking lot into the terminal. The Eugene International Airport, despite the impressive name, is about what one would expect in a small university town. But it does receive medium-large jets and has flights south into the major California airports. At the moment the terminal building is largely empty.

"OK. Yes, you nurses should appreciate this. The Portland campus of Linfield College, where I was teaching, was primarily serving nursing students and was connected to one of the finest local hospitals. Capra's conversion, thought of in the medical context, was from seeing the world as composed of inanimate objects, to seeing the world as composed of subjects – as composed generally of living processes. In his book, *The Turning Point*, he argues for a conversion, actually suggesting that western medicine is already undergoing this fundamental paradigm shift. The old way, associated with the dominant medical culture over the last 500 years, is 'the physician's view' of patients – as objects, as sophisticated material mechanisms, like smaller Newtonian clockworks. According to Capra, the 'new' (or perennially oppositional) enlightened view that is being rediscovered is historically associated with the nursing tradition. Nurses treat the patient as a person, as a subject – as a conscious, feeling, intelligent subject. I used the book because Capra was, in effect, encouraging the nurses to see themselves and the nursing tradition as the new, more general paradigm and nurses as representing the new leadership in healthcare.

"That'll be the day," says Diana.

"Capra's theme, applied to nursing, captures the shift away from the overarching Mechanical Philosophy. Physicians today, particularly surgeons, are taught to see people as objects, as complex pieces of living meat. Rene Descartes, one of the primary modern formulators of the Mechanical Philosophy, is famous for suggesting that the screams of pain from the animals he experimented on were analogous to the screeching sound one might elicit from tightening a bolt on a machine. He didn't believe they were 'really' conscious subjects. Everything 'out there' from the Spectator perspective of human inquiry was just a mechanical, material clockwork.

"Capra's theme was that the nursing tradition was all about dealing with the patient as a person, as a subject. And Capra at least pointed toward looking at the whole interconnected planet and, perhaps, even the universe, as 'alive' – as some sort of sensitive, feeling subject," I say.

"I like him. I think I'll read that book," says Pam. "Can you write in down for me, Terry?"

Jonathan has all the tickets and confirms reservations with the airline, while the rest of us stand and wait and chat. Gradually all the baggage gets checked-in.

It is agreed that Stephen will board last. Boarding first had seemed like a natural strategy, since the airline standard is that anyone needing extra time or assistance, anyone with young children, boards first. But on several occasions Stephen's boarding had taken enough extra time to aggravate passengers that had to wait inside the terminal. By now Jonathan and the nurses had developed a sophisticated set of procedures. After Stephen drives right to the door of the aircraft, two nurses, one on each side, carry Stephen from his chair to his own seat in the plane – typically in first class, where he might be more easily tended to if the need were to arise. The greatest time challenge is in dismantling and storing Stephen's chair – the computer, huge heavy batteries, voice synthesizer and the complex superstructure of the chair.

We had started to the airport early so as to provide extra time in case there were any complications. There weren't any. So we have about an hour to kill before starting the actual boarding process.

"Is there somewhere to have a cup of tea?" Pam asks. Whenever there was uncommitted time, pouring tea down Stephen to keep him constantly hydrated was a favorite activity for the nurses. More than in any other way, we all give off water from our lungs, as we exhale moist air, and I learn that Stephen tends to lose water off his lungs even more easily. This would be of particular concern on a long flight. I noticed that the tea was often diluted, in part, to maximize the number of cups consumed per tea bag. It wasn't about the caffeine. It was about the liquid.

We all gather around one rather small table in an otherwise empty corner café in the outer terminal area.

"English Breakfast or Earl Grey?" Pam asks. This morning Stephen indicates 'Earl Grey.'

Almost immediately, Jonathan, Diana and I start up again on the "Is everyone selfish?" question. They have been thinking about it. No longer dismissive, they're genuinely interested, and yet none of us is quite sure where to go with it.

"If you can't distinguish selfish behavior then you aren't saying anything," I reiterate.

Stephen is also clearly interested in the dialogue. He could be reading off his computer or daydreaming, diverted, but it is clear from his focus and attention, that he is right with us, intensely. It's a very small table, and no one is more than two feet from anyone else. Through his facial expressions and eye moments, Stephen is part of the conversation. Each of us looks at him after making a point, looking for a sign of approval – or disapproval. We guess at his expressions, but he isn't saying anything at the moment – just listening, taking it in.

Judging from his attitude toward the Queen and the monarchy, I was guessing that Stephen's politics are leftish, and I guessed that selfless behavior was real for him, as it was for Pam. It was just so ironic that these two people, Jonathan and Diana, who were charged with taking care of Hawking's needs, didn't believe that 'apparently' selfless acts – truly caring acts – were real.

"Consider Freud's theory. Just as in the Universal Selfishness Theory, the 'real person' – everyone's real nature – is characterized as the selfish Id. But the Id finds itself in a world where other people – other Ids – are acting and reacting to what the Id does. Your Superego is a negative feedback, your understanding of how you are supposed to act according to the dominant socio-political powers. If you want to get along, you need to work within societal rules and constraints. Your Ego is the negotiated middle ground. In a way it is your selfish Id's adjustment to the rules of the current social powers. Your core nature is still selfish, but it has developed a more informed and clever strategy for how to best get its way, as much as is possible," I say.

"I remember that I never liked the Freudian theory – just intuitively, my gut reaction. So over the years, and as part of my philosophy of science research, I kept looking for definitive scientific evidence against it – for a 'forcing argument' that would show that it was incorrect, limited – that human nature wasn't actually, completely, consistently, objectively – Id-like – selfish.

"I came to suspect that I must be looking for evidence that somehow I already had. I just needed to articulate it in modern scientific and philosophy of science terms. I kept looking for the crucial evidence – you know, along the lines of Popper's Question – 'the evidence if it were to occur...' that would conclusively demonstrate the limits of the Freudian theory. Then, one day, I realized that I didn't have any 'objective' evidence that people weren't actually always selfish – no evidence that couldn't be interpreted the other way. Freud was just giving us a First Impression argument that humanity was selfish, and then making excuses, explaining away counter-evidence by the standard reinterpretations – such as that 'apparently' selfless actors 'really' believe they are being as selfish as possible at the moment.

"So I wondered why I was so clear intuitively that the theory must be wrong, even though I couldn't specify, couldn't define the evidence against it. Finally, I realized that the reason I didn't like the theory was because it implied that my friends weren't actually my friends. I mean, not just whether they accepted me as their friend, but also whether I was fooling myself as to whether I was actually their friend. I took it – intuitively I guess – that we honestly were friends, and that therefore Freud must be wrong. I had always liked Aristotle's characterization of a friend as being like 'a second self.' For example, if you found your friend's lost wallet or purse somewhere you wouldn't think twice about returning it perfectly intact. It is, perhaps, as if friends have a unity of purpose. You are on the same team, in the same family.

"Freud's idea that my real motivations were those of a selfish Id implied that friendships were simply alliances of convenience: we are friends just until I have a chance to stab you in the back and take advantage of you. Real friendship, real loyalty, in Aristotle's sense, was an illusion for Freud," I said.

"And you don't think that's the way people behave?" says Diana.

"I am not suggesting that people never take advantage of others – even their friends, occasionally. The question is whether that's our ultimate nature – and that it is true, whether we like it or not – objectively true, the one correct understanding of human nature, whether we believe it or not. And the question here is how we decide," I say.

There is a silence. Attention shifts to our tea. Pam is on duty with Stephen and is mixing the hot and cold water that the waitress has provided to reach the optimum temperature – middle-ground – between too hot and too cold.

"What troubles me most," Diana interjects, "about not being able to distinguish selfishness in terms of specific evidence is that what you believe really does affect the way you think of yourself and how you treat others."

"I agree – totally," I respond with some enthusiasm. Diana is now thinking about the question personally, no longer just defending the ideological position.

"This question isn't a matter of idle curiosity, some sort of abstract philosophical question. What you believe about yourself and others alters how you experience the world – qualitatively," I add, building on Diana's insight.

"I'd like to see people as being good, but I just don't see it. I guess you could say I have trouble trusting," Diana says.

"We need a crucial experiment," says Jonathan, who, to my surprise, is also now taking the question seriously.

"Shouldn't we be able to decide the question? It seems clearly to affect how we understand each other, ourselves and how the world works," Pam offers, continuing, "The experience of dealing with others when you believe that they are selfish or trying to take advantage of you, is different than when you trust them."

"Same person, different experiences," Diana comments.

"In my reflections on Freud and Marx, like Popper, my scientific attitude insisted that there must be a testable resolution. There must be an answer to Popper's Question. But testing whether people are really selfish or selfless turns out to be rather tricky. As you suggest, Jonathan, I kept looking for a crucial experiment," I say.

"One of my roommates, in a house a group of us students had rented in the Berkeley Hills, had an intriguing approach to this crucial experiment question. He was rather successful at the initial steps in attracting girlfriends. But he had what I came to see as an unusually deep and abiding personal insecurity, a problem trusting, as Diana put it. He wanted to be sure that his girlfriends truly loved him or, at least, in these early stages, cared for him and the relationship. So he would test them. At first he was exceptionally personable and complimentary and attentive. He managed to convey a brilliant first impression. Then, later, he would throw in some minor dissonance – like being late for a date. If they were still with him after that, then the tests would escalate. His logic, his reasoning, seemed scientifically impeccable: if they truly cared for him, they would forgive him each insult. But the more tests they passed, and the more he would let himself trust them and care for them, the more he needed to test. Well, I think common sense would tell you the outcome. He left a trail of destruction. And with the eventual failure of each relationship, his need to test the trust and caring of the next relationship grew. This guy was extreme, but I think we can all sympathize with the dilemma.

"The opposite approach to testing whether people are selfish is to treat them always as though they are selfless, as one of the family, on the same team, inclined to reciprocate and work together – by nature," I say.

"The downside of this test is that the person who constantly and uncritically trusts, someone who helps too much, too easily, almost invites abuse. In a different world, where no one was at least occasionally selfish, this might work. But once such persons meet someone who is less than morally perfect, or entirely appreciative, they are likely to be taken advantage of. No matter how poorly you treat them, they never complain, they always forgive.

"Freud is strangely, rationally, hypocritical on this count, since, if everyone is really selfish, then it stands to reason that for those who realize this, it would be wise 'to shut up.' The best policy would be to try to convince others that everyone isn't selfish – thereby making it easier to take advantage of them. So I always wondered why Freud, if he believed everyone was competitively selfish, was letting 'the cat out of the bag', so to speak.

"The Marxists argued that the Church maybe formalized religion in general, since it encouraged people to think about others in the other extreme, to be uncritically forgiving – as in 'to turn the other cheek' – inadvertently, or perhaps deliberately, played into the hands of those who believed in being selfish and taking advantage of others," I say.

As we sat huddled around that small table in the drafty, nearly empty Eugene International Airport, I sensed that an all-inclusive sympathetic feeling had developed. Everyone recognized the contrariness and the complementarity of selfishness and selflessness, and yet, at the same time, intuitively, these opposites seemed to be harmonious, a set of balancing processes, fitting together, somehow, like the pieces of a larger puzzle. Jonathan and Diana were no longer dismissive or combative. The question was real, and everyone recognized its importance.

With this emerging concord, my role began to change. I, who had been the questioner, challenging them to come up with the answer to Popper's Question, now was equally challenged by the question.

Pam is intermittently holding a cup of tea up to Stephen's mouth, tipping it slightly. Stephen manages to take in better than half of the liquid offering, the rest spilling down into the waiting pocket of the bib.

"So, Terry. What is the answer?" Pam interjects.

I felt a little embarrassed by the question, since I had just recounted, in my dealing with the Freudian ideology, that I had been unable to explain – in terms of scientific evidence, actual or potential – why I took the Freudian theory to be false. Again it wasn't that people weren't selfish – sometimes. The issue was whether they were universally selfish, whether the

Universal Selfishness Theory even made sense, whether it was a scientific claim, whether it passed the meaningfulness test of Popper's Question.

So now the challenge applied to all of us. Surely there was an answer to this simple question: what evidence, if it were to occur – not saying it has or even will – would convince you that people are not universally selfish, that human nature is not objectively selfish?

I felt like I knew. I must know. I knew it intuitively, as I had expressed in my Freudian experience. But every attempt to formulate the evidence failed. All suggestions could be reasonably questioned, reinterpreted, reconceived.

"So Diana, Jonathan, what would convince you?" I ask, again, but in a much more sympathetic tone. Gently, I kept pushing. I guess I had this feeling that I wanted to save them from this belief – from seeing others and themselves as 'really' selfish. And I had the sense that they were willing partners, collaborators.

I kept asking and probing, not because I knew how to answer the question. It wasn't like a test or an exam. It had become a common sympathetic dialogue. Sometimes that is how you find answers, by asking people to share their intuitions. When you think about these things too much by yourself, the confusion can be blinding.

Finally, it was time to board the plane. We all moved through the main airline gate and up the entrance to the gangway – to the point where I wasn't allowed go any farther. The conversation didn't pause. One more thought. One more exchange. One last chance.

The final scene is etched in my memory. Pam is already inside the plane, preparing Stephen's seat. Diana and Jonathan are standing in front of me, only a step or two away from entering the plane's gangway. We are waiting for the go ahead from inside. Stephen is next to me on my left still in his chair.

As I glanced at Stephen, an image suddenly rushed into my mind. I turned and asked him, "Why did you stop us from trashing that waiter at the Inn at Spanish Head?" It took him just a few seconds to answer.

"I felt for him," he said.

I look at Diana and Jonathan. They look me. We look at Stephen. Stephen looks at us. Eyes meet. Nothing is said, but we all take it in. There was no rejoinder. That was the unsettling end of the conversation. It was time for Stephen to drive his chair down the gangway to the door of the plane. I don't know that Diana, Jonathan or I – or Stephen, for that matter – were instantly enlightened or convinced of anything, or converted. But what Stephen had just said touched on something we had all been missing – a contribution to the nature of the larger puzzle... obvious, and yet, in terms of how we had been formulating the question, not easily interpreted.

We all heard it, and it penetrated – deeply.

Something had happened. Once again the universe had turned slightly, heading in a new, unexpected direction.

The 'word' at the Inn had been an act 'out of' a feeling – out of empathy – not selfishly 'for' anything. But it wasn't simply selfless either, there was no cost. The 'word' was about taking a different path into the future.

Stephen's comment at the door to the gangway pointed to some larger, more general, more comprehensive framework. The lesson wasn't something we were able to articulate, but, nonetheless, there was a sort of deep recognition. In some more general understanding of reality, these contraries would be special cases, each with limited applicability. The contrary alternatives could be locally competitive in specific situations and yet globally compatible, possibly balancing, perspectives.

Perhaps, the difficulty of deciding the complementary nature of humans and the difficulty of deciding the complementary nature of the electron were the same difficulty. Perhaps, there was just one difficulty, with many diverse manifestations.

Part Two

The Two Paths to Complementarity
in the 20th Century

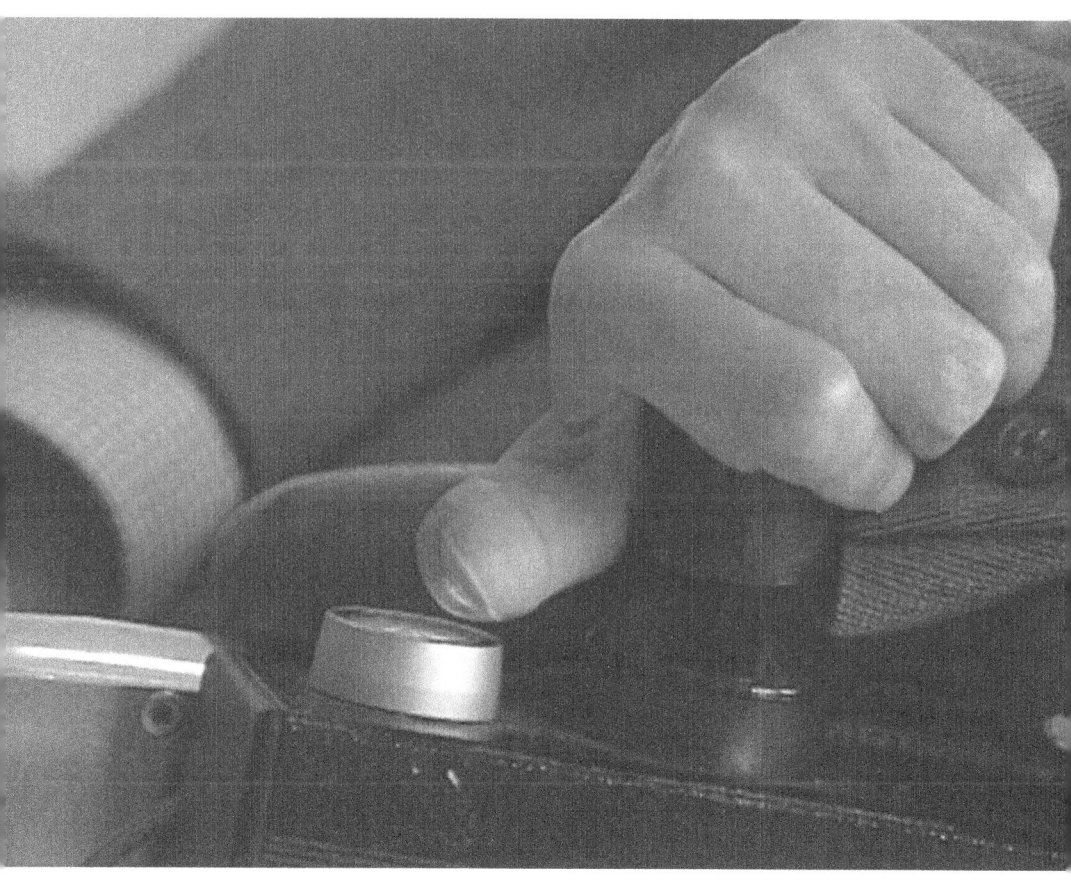

18

The Surprising Answer to
Popper's Question

The original invitation had asked Stephen to give four lectures in four different cities in the Pacific Northwest: Portland, Eugene, Vancouver, BC and Seattle. The first two lectures, having been quite successful, gaining his acceptance for the second leg of the original invitation was straightforward.

With the inclusion of the nurses, through the discussion of selfishness, my initial presentation of the question of the limits of science had expanded into something much more general, into the consideration of the hypothesis that all meaningful beliefs or belief systems might be inherently limited. Meaningful beliefs could not be 'universally true' – true from all points of view, everywhere, for all time. Meaningful beliefs, in other words, were not 'objectively true' in the classical scientific sense – fully independent of the observer's perspective, of the way of observing, of the who, what, when and where of observation and belief.

In the case of the Scientific Research Program, the plurality of successful scientific theories should be understood as idealizing special cases within some, yet to be articulated, More General Theory.

I had begun to speculate that the traditional representation of scientific beliefs, as observer-independent, as if the scientist were a detached Spectator, was an idealization. If successful scientific theories always involve idealizations, then there must be a More General Theory, in terms of which these idealizing scientific theories can be understood as special cases. My working hypothesis was that the More General Theory might well be found in the Participant framework, where the observer of the universe, the inquirer, is also an embodied. As active inquirer, the Participant necessarily chooses between alternative approaches, alternative ways of questioning and acting. In making these choices the Participant-inquirer selects, at least in part, the future course of events.

If correct, then the two seemingly incompatible worlds of Stephen Hawking, as the detached Spectator-scientist searching for the unlimited, timeless laws that determine the inevitable course of events in the universe, and, Stephen Hawking, as the embodied Participant-activist in a developing universe, working to steer the universe toward a more desirable future, could be resolved into one coherent whole.

One central motivation in asking about the limits of science has always been to make sense of the relation between scientific facts and human values. The question of the relation between science and values had now for me merged with this attempt to understand Stephen Hawking. On the one hand, he seemed to be pursuing a strictly deterministic, law-governed understanding of reality, while, on the other hand, he was obviously demonstrating his social and moral leadership, promoting values, pointing toward – and selecting through his actions – what he understood to be the better future.

Stephen's support for students with disabilities, and the rights of persons' with disabilities in general, is a deliberate attempt to alter the universe, to influence the course of events, as if he were a Participant in the development of a better universe, working toward a more desirable future.

I surmised that Stephen had been sympathetic to my line of thinking on the issue of the uncritical, ideological thinking of the Universal Selfishness Hypothesis. In my mind, this issue had now become inseparable from the question of the limits of science. The traditional, universal, objective representation of scientific knowledge now began to appear, at least potentially, as an idealization, and, consequently, as a reflexively uncritical, ideological approach to understanding reality. Science itself, so represented, had become, in practice, just another 'one right answer', non-self-critical ideology.

What had most intrigued me about the parallel nurse-focused dialogue about the selfish-selfless and competitive-cooperative dichotomies, particularly when each member of the pair was expressed as an ideological position, was their apparent complementarity. The oppositeness of their understandings of reality paralleled the wave-particle complementarity that had emerged in 20th century quantum physics. In both cases, the concepts are co-defined or inter-defined, so as to make them mutually exclusive, so as to make them qualitative, conceptual, opposites – what the ancients had called 'contraries.'

Whereas Popper had been critical of the ideological character of cooperative Marxism as a socialist ideology, its proponents considered it to be a scientific theory – scientific socialism. I was impressed by the fact that there was an opposite competitive free-market ideology, whose proponents likewise considered their theory to be scientific. Much of 20th century Western Economic Science is based on free-market presuppositions about an 'objectively' competitive reality. (Actual realization of the perfect free-market reality might require either convincing or eliminating everyone who either thought or acted otherwise.) Much of scientific socialist economics is based on the opposite presuppositions about reality. Proponents of both approaches have displayed this sort of ideological defensiveness, using auxiliary hypotheses and Promissory Notes to defend their core hypotheses – their core presuppositions about reality. By Popper's initial reasoning, both Marxism and free-market promoters are being un-self-critical and perhaps pseudoscientific.

Could it be, I wondered, that these sorts of conceptual oppositions, these complementarities, were quite widespread, even perhaps a general characteristic of how we understand reality, and perhaps, as the *Tao* suggested, of the nature of reality itself? Reflecting, I imagined linking complementarity back to my experience in Berkeley of the perennial controversies that seemed to populate all academic disciplines.

I had secretly been embarrassed in my challenge to Jonathan and Diana, pressing them to answer Popper's Question, applied to their theory that everyone was completely and consistently selfish – deep down selfish, despite any appearances to the contrary. I kept asking, "What evidence, if it were to occur – not saying that it has or will occur – would lead you to conclude that people were not completely and consistently selfish? Just tell me what the evidence would look like."

The argument I confronted them with was that if they couldn't answer that question then they weren't making a real distinction between selfishness and non-selfishness; indeed, between selfishness and any other behavior. And if they weren't making a real distinction, then the statement – the claim – wasn't meaningful, didn't say anything. Their position, at least in the beginning, had been that selfless and altruistic behaviors were 'mere appearances' because 'real' human motives and 'real' human actions are always selfish. For them, human nature was objectively, universally selfish – making reality objectively competitive. Any interpretations to the contrary were misunderstandings.

What was embarrassing, and hidden behind my slight pretense of intellectual self-confidence – as if I knew – was that I didn't know how to answer the question any more than they did. To give myself a little credit here, what I imagined I was doing was exploring their relatively unexamined, unbiased intuitions, looking for at least a clue. My own intuitions were by now hopelessly suspect, since I had been thinking about this for several years now and had read far too many confusing philosophical commentaries.

How to proceed? Perhaps putting this common sense question to these philosophical innocents, I thought, might produce a clue.

It seemed so obvious that there should be an evidential – empirical – difference between selfishness and selflessness, between a world of universal competition and a world of universal cooperation. I had been obsessing about this for some time now. I recall a comment attributed to Isaac Newton, when he was asked how he had figured out his famous theory. He answered, "I thought on it," and then added, "Constantly." Perhaps obsessive perseverance is a plausible strategy.

I have a recollection of sitting in a tavern just a block or two up the hill from the Oregon Health Sciences University, where I had been working. Having a beer, I was vaguely watching some associates playing pool. Others were socializing. But I was thinking about selfishness and selflessness in my own little personally obsessive self-dialogue. I remember that moment because it was then that I saw a new approach. It struck me that the term 'selfish' was a negative attribution, a pejorative, only from a cooperative perspective, if you believed the world should be cooperative. And the term 'selfless', sometimes thought of as laudatory, was a negative attribution, if you believed that the world was or should be selfishly competitive.

It was as if there were two ways of looking at the world – the competitive and cooperative – and that each of these established its own standard of normalcy. Selfishness is abnormal in the cooperative world and selflessness is abnormal in the competitive world. Say you are with an associate, and you believe that the world is competitive, and your associate comments that your recent behavior was rather selfish. Your reasonable response might be 'Duh! Of course. What else? What did you expect?'

In the middle-ground world 'selfish' and 'selfless' would be outliers – both referring to abnormal or critically inappropriate behaviors. The 'normal' middle ground is preserved and enhanced by the oppositely balancing critiques. This middle-ground image suggested that selfish and selfless (competitive and cooperative) interpretations must actually each be limited special cases, special ways of understanding and doing, with limited applicability within a more general context, a more general middle-ground reality. I suspected that it was only from that more general middle-ground perspective that I would be able to make sense of Stephen's 'word' – not simply selfish, not simply selfless – clarified somewhat by his later reflection, 'I felt for him.'

The following middle-ground image had come to mind at that time – appropriate to Fritjof Capra's Taoist line of thinking. If life is a developing path and the problem is to stay on the path, then ideologies offer 'speakable' – time-space invariant conceivable – strategies. One tells you, when in doubt, always turn to the right. The opposite tells you, when in doubt, always turn to the left. Pursued rigorously and exclusively, each of these uniform 'paths' will curl in on itself – self-destructing. To stay on the real, 'eternal path', the non-pre-conceivable, unspeakable, eternal Tao, involves an

ongoing, irreducible 'interplay of opposites'. The eternal path is the under-
determined 'middle-way'. The notion of a self-critically balancing middle-
way is one of the great wisdom offered by the ancients. The Greeks carved
in the stone at Delphi their ultimate wisdom: 'Everything in its measure.
Nothing in excess.'

If you believe the world is completely, selfishly competitive, then you
see examples of that everywhere. In the terminology of modern philosophy of science, the observations by people holding such a view are said to
be 'theory-laden.' You can cite volumes of confirming observations and
even experimental demonstrations. On the other hand, if you believe that
the world is completely cooperative, then you see examples of cooperation
everywhere. You can cite volumes of confirming observations and even experimental evidence. It now seemed entirely possible that both were right –
and yet neither one completely.

When one of these theory-laden, ideological research programs is confronted with apparent counter-evidence – examples of the other type of behavior –
the response is simply to 'explain away' such observations with auxiliary
hypotheses or, viewing such 'apparent counter-evidence' as due to a misunderstandings, to make corrections by reinterpreting them, for instance, by
suggesting extenuating circumstances or hidden ulterior motives.

The reason the advocate of one of these universalist – all one way or all
the other – positions needs to 'explain away' the 'apparent' opposite type of
behavior is that there is simply no possibility of *explaining* it as part of the
rational structure of their reality. Each theory and its associated research
program presupposes that the world has a universal, objective – complete
and consistent rational order – what, in the scientific tradition dating back
to the Greek scientists, has been called the *logos*. For the selfish, competitive model, selfish, competitive behavior is the rational order. And that's the
way it is, whether you see it or not. It is the objective, plausibly scientific,
order of nature. Now, if this is true, then there simply are no selfless behaviors, no 'real' selfless behaviors, no real cooperation. There can't be. 'Real'
selfless behaviors would be 'incompatible' (viz. incommensurable) with the
rational order of things. Cooperative behavior, in a necessarily, objectively

competitive world, would be incoherent – would be irrational, could not be made sense of, could not be translated into competitive terms, could not be understood in competitive terms.

From an objectivist perspective, if one member of a pair of contraries is completely, objectively, universally true then the other member must be completely, objectively, universally false. When dealing with conceptual contraries it is not possible for one to be a limited special case within the other.

'Apparently cooperative behavior' could be explained away by the competitive ideology as being part of a competitive strategy, as possibly a deceptive ploy to deceive and out-maneuver a naive competitor. But this is different from saying that there is 'real' cooperative behavior – it's not. In this case, it is 'really' competitive behavior.

Likewise – let me spell it out – for the cooperative ideology the rational order is cooperative. That is the way it is, whether you see it or not. We are all parts of one big system, one big family – one team – life, the universe. There is no 'real' selfish behavior, no 'real,' 'zero-sum – me versus you' competition between members of this One universal team – there can't be. If we are on 'essentially' different teams then we are 'really' competitive. But if we are actually on 'essentially' the same team, even though our efforts may not be tightly coordinated, our overall efforts are necessarily cooperative.

People who hold these extreme ideological positions tend to be less than diplomatic toward the contrary position. One tendency is to suggest that people who act selfishly in a cooperative world or selflessly in a competitive world are either stupid or irrational. Such attitudes are self-reflexively inconsistent in that they seem to be admitting that the other type of behavior, contrary to their theory of the universality of their logos, can 'really' exist. The least sympathetic of such ideological attitudes is that the irrational opposition (viz. to the extent that it is allowed and admitted to actually exist) should simply be eliminated, intellectually cleansing reality. (Jonathan Haidt, Professor of Ethical Leadership at New York University's Stern School of Business, in his recent book, *The Righteous Mind: Why Good People are Divided by Politics and Religion*, provides an excellent characterization of how such conceptual contraries naturally divide societies and cultures, and he describes the moral psychology of the occasional (but not necessary) tendency to vilify the opposition and, in the extreme cases, to either convert or eliminate the opposition.)

When the question of whether the world is competitive or cooperative is formulated as a factual question, asking which of these theories describes objective reality – a scientific formulation – then we have a fundamental problem. If these two apparently opposite views are formally complementary, and their separate verifications are theory-laden, then the opposite ideologues can debate the issue until the cows come home and never resolve it. To a very large extent, the advocates of each perspective will just 'talk past each other', because the two types of evidence don't actually, scientifically, logically, conflict. They don't actually contradict each other in the formal, logical sense because you can't make sense of one in terms of the other. They are irreconcilably different types of behavior, irreconcilably different types of phenomena – conceptually different, qualitatively different, incommensurable. Each corresponding research program just sees reality – literally experiences it – differently. You can't generate one of these types of phenomena from the other type. They don't translate. They are irreconcilably, irreducibly contrary.

It is important for the larger issues that each of these contrary worldviews – the selfish and the selfless, the competitive and the cooperative – sees the other's 'normal' behavior as irrational. To act selflessly in a selfish world would be irrational. It simply doesn't make any sense, which is another way to say that you can't make sense of it in terms of selfishness. Likewise, if the world is truly a cooperative venture, then acting selfishly doesn't make sense, is irrational, can't be made sense of in terms of selflessness.

Seeing other people's beliefs and behaviors as irrational is quite a common occurrence. A simple expression of the experience is the typical advice in terms of behaviors: 'if I were you, I wouldn't do that.' In other words, in terms of my rational understanding of the world, what you are doing, or about to do, doesn't make any sense – is irrational.

On just a moment's reflection, the more common experience seems to be middle-ground. I feel sure that I have never met anyone with whom I agreed on everything about the world or about how best to act in all specific instances in the world. You meet someone with whom you agree on very fundamental beliefs, only to discover later that the two of you seem to disagree, to have opposites views, on other fundamental issues of the day.

In an era when it has been common for anthropologists to talk about what people in this or that culture believe, I think the broad middle-ground distribution of different types of beliefs in every society is underappreciated. Beliefs, what one accepts as the rational view of the world, have something more like the variation of traits within a biological population, perhaps analogous to the genetic variation within a species. And they change dynamically. Rationality, what makes sense, is not experienced as a continuity, as a logico-mathematically generated uniformity – the same for everyone, everywhere for all times. This is, at least should be, particularly troubling for those who want to maintain that there is one objective, right, correct and *rational* way of understanding, observing and acting the world.

The middle-ground perspective might be that the rational structure and function of society is not a single uniform logico-mathematical clockwork, but a dynamic, conceptually diverse, distributed rationality. The middle-ground perspective might anticipate that if you placed a group of Republicans on an island they would soon divide into Republicans and Democrats. And likewise if you placed a group of Democrats on an island they would soon divide into Republicans and Democrats. Each group, by some 'higher, more general rationality,' would tend to a balanced middle-ground, through a dynamically balancing middle-way.

The Surprising Answer to Popper's Question

Although I still wasn't confident that I could properly answer Popper's Question as applied to the selfishness, competitive ideology – the question I had pressed on Jonathan and Diana – that dialogue with them had led me to a tentative, abstract, in-principle type of generalized answer to Popper's Question. It was a sort of non-answer answer.

What I came to refer to as 'The Surprising Answer to Popper's Question' was that you couldn't actually answer it. Since it had seemed like such a reasonable question, my conclusion that it couldn't be answered was – to put it mildly – *surprising*!

The key insight was that you couldn't answer Popper's Question from *within* an ideological position. You simply don't have the language to describe 'the evidence,' you don't have the conceptual machinery to generate a description of 'the evidence'– the contrary evidence – because the contrary evidence is conceptually outside of your ideology's core defining concepts. You can't even describe a selfless act – that is, 'really' selfless, by its very nature – in terms of the concept of selfishness. You can't even describe a cooperative act – that is, 'really' cooperative, by its very nature – in terms of the concept of competition. The inadequacy is not just in the concepts alone, but also in the associated logic, in the structure and reasoning of the ideology, that tells you how things fit together, how the world works.

On the drive to Eugene, Pam had candidly interjected the common sense, balanced, middle-ground answer to Popper's Question, observing that there were some selfish behaviors and some selfless behaviors. Rationalistically, in logico-mathematical terms, Pam's middle-ground position is conceptually paradoxical – does not form one consistent whole – because what is rational in a world of selfishness is irrational, doesn't make sense, can't be expressed coherently, conceptually, in the language of a world of selflessness.

The tricky part here is that evidence for selfish behaviors and evidence for selfless behaviors are not formally 'contradictory', logically inconsistent. They don't 'conflict' in the standard logical sense anticipated by the Law of Excluded Middle. The two types of evidence are contraries, complements – qualitatively opposite. This is why, when two people holding these opposite hypotheses try to resolve the question by citing evidence, they talk past each other. The evidence that either cites in support of their hypothesis simply doesn't make sense in terms of the contrary, complementary hypothesis; in terms of the contrary, complementary conceptual language. They listen...

but they can't hear, can't understand, at least in terms of their conceptual language. What each one is saying simply doesn't translate conceptually into the other's way of understanding.

Jonathan and Diana couldn't answer the question the way Pam did, because for them there simply were no 'real' selfless behaviors, there couldn't be. Properly speaking, such proposed 'evidence' must be considered a 'confused illusion' at best. The suggestion that there were selfless behaviors was rejected and couldn't qualify as an answer when reasoning from *within* the core Universal Selfishness Theory. They couldn't express the answer to Popper's Question in terms of their core theory, from *within* their ideological position, from *within* their claim to have 'the one right', objectively true, rational way of understanding. And that impossibility was a surprise. That was unexpected.

The irrational cannot be either expressed or explained in terms of the rational. That was it! That was the Surprising insight!

At least that was one way of putting it. If your theory is that all 'rational' behaviors are selfish, then proposed examples of selfless behaviors are incoherent, irrational – such 'observations' are indescribable, inconceivable, in rational terms.

Moving from a balanced, common sense middle-ground to a universal, objectivist ideology, one direction or the other, is like transforming a black and white yin-yang diagram into a uniform circle – either all black or all white.

If you are always out looking for more examples of selfishness to confirm that hypothesis, it is unclear whether you could find, or even notice, examples of the conceptually contrary phenomenon. There would be a 'failure to appreciate' the value and the reality of – by their very nature – truly cooperative behaviors and relationships. And the reverse is equally true: ideologues of the selfless research program would fail to appreciate the value and reality of – by their very nature – truly competitive behaviors and relationships.

With the insight of the Surprising Answer one can see that Popper's Question is a rather crazy question, asking you: Tell me something that you can't possibly make sense of. Tell me something that you can't possibly tell me in the terms of, in the language of, your way of understanding the world. As long as the responders are restricted to answering from within the conceptual framework defined by their core ideology, they can't possibly give an answer. You are asking them to specify something that they can't make sense of in their way of understanding the world, in terms of their way of making sense – come what may, in the long-run, no matter what.

On the other hand, as Pam demonstrated, from a non-ideological point of view, it isn't typically a tough question; that is, if you are a middle grounder. But if you are a serious ideologue the question makes no sense – you can't answer it. People who are trying to think that 'other way' – the contrary way, the foreign way – appear to you as misguided, uninformed, conceptually confused or simply irrational.

When the advocate of one ideology tries to give a 'forcing argument' to the advocate of the contrary ideology, it doesn't work, because he can't understand the argument, because it doesn't make sense in his way of understanding. Haven't we all experienced this? Doesn't this sound rather familiar?

19

The Avinash Dialogues and
Complementary Conversions

My consideration of the complementarity of the selfish-selfless and competitive-cooperative belief systems and how each side experienced the other as irrational had brought to mind already the socialist versus free-enterprise economic polarization, that had seemed to provide some insight into the perennial 'talking-past-each-other' phenomenon in normal political dialogue. A couple of experiences during my graduate study in London helped to clarify. I had offered these stories and images in brief versions to Hawking and the crew on the van ride between Portland and Eugene on the first visit.

"During my second stint of graduate work at the University of London there was a Saturday morning gathering of graduate students in the History and Philosophy of Science Department at University College. The organizer was Nicholas Maxwell, Lecturer in the Department and Ph.D. thesis advisor for almost everyone in the group. Nicholas had become my thesis advisor as a sort of default, following Imre Lakatos's unexpected and untimely death. Lakatos had died, at age 51, a few months before I returned to London. Lakatos had never actually been my advisor, but had agreed in a letter, based on our interactions during my previous period in London, to be my advisor when I returned to London for this second stint of graduate work.

"One of my fellow graduate students in Maxwell's group was one Avinash Puri. He was from India, a quite loveable and jovial character – and not inclined to shy away from a good-natured argument on pretty much any issue. A few off-hand and entirely innocent remarks about some political issues of the day gradually morphed into an extended dialogue – what I have come to refer to as The Avinash Dialogues.

"What happened was that Avi would provoke me with some critical remark about the foreign policy of the United States, and I would counterpunch with a spirited defense, typically, because I thought his analysis and conclusions were superficial, impugning motives that I thought were laudable. I am not an ideological defender of capitalism, but I can see its good side. Indeed, I tended to see the good and the bad of both ends of the political spectrum. My wife, Suzanne, was my regular defender at parties where I would be accused of being a capitalist when the group was politically left and of being a socialist when the group was politically right. My natural inclination was to 'enlighten' each group by pointing out the limitations of their general, ideological position. I countered ideologies – whether from the left or the right. It is not that I just like to argue, at least that's my story, but, rather, that I feel a moral obligation to guide people to a balanced perspective – toward a middle-ground. My efforts to bring people to a balanced view were not at all conscious or deliberate, at least initially – just my natural way. A former girlfriend in London during my first stay, Lynn Lindholm, recently Professor of Philosophy in North Dakota, put it nicely that I had 'a delightfully innocent, and entirely well-intentioned way of insulting people.'

"Avi had many of the same characteristics. Avi was no more an ideological defender of socialism than I was of free-market capitalism, but he could see its good side. So we engaged. The memorable core of the dialogue centered on the correct understanding and explanation of U.S. foreign aid to South America. These exchanges were never formal, often reemerging week after week at lunch after Maxwell's seminar had ended. But they gained in intensity and sophistication over the period of a couple of months. We were philosophers of science so the real issue that overlay the specifics and began to emerge as the most interesting, was whether we could settle our differences – scientifically – on the evidence. This turned into an incarnation of Popper's concern about whether political ideologies were scientific or pseudo-scientific.

"We went over a number of specific issues, such as the intent and eventual status of U.S. foreign aid to different South American countries. Occasionally, we disagreed about particular occurrences – like how much and when and what had happened to the aid. But these differences were easily settled. What became increasingly clear was that, in all the relevant cases, our disagreements had nothing to do with these factual 'happenings'. Avi would interpret them in one way, and I would interpret them in the opposite way. The differences in our narratives were subtle, yet significant, so that the stories we were telling became like mirror images, like Gestalt reversals. They were closely related yet different, like left-handed and right-handed interpretations.

"In the early stages our exchanges were quite friendly and collegial, but before long, quite unexpectedly, they became rather nasty and, occasionally, personally offensive. The phenomenon was that the other's interpretation was literally experienced as outrageous – insulting. Indeed, we exchanged harsh words on a number of occasions, degenerating into name-calling. But we liked each other and neither one of us actually had an ideological agenda, so there were smiles as we parted. This was an exercise, an abstract meta-exploration of whether we could settle what seemed to be a factual issue based on clear and agreed upon citations of specific facts.

"Avi and I didn't reflect openly about what we were doing in terms of these philosophical issues until late in the dialogue. We were just having fun intellectually poking each other, a sort of philosophical entertainment. It was important that we engaged each other in terms of the political position that we had each adopted: competitive free-market capitalism for me and cooperative socialism for Avi. Clearly, in the background for each of us, was the recognition that significant numbers of people in the world held each of these opposite positions in what they took to be the scientifically correct, rational understanding of events. And on the flipside, significant numbers of people held each of our positions to be irrational or at least naive, confused and ill-informed. But neither of us was naïve, and there were no informational issues that we hadn't resolved, hadn't agreed about.

"I want to emphasize the immediate experiential – emotional – outrage that each of us felt when listening to the other's interpretation and representation of the 'facts.' Even though we were role-playing, perhaps, I honestly began to feel outraged and angry. And Avi clearly had the same experience. What was happening here? The 'facts' were not at issue and yet we, as philosophers of science, were presupposing that factuality was precisely what could resolve the issues.

"In support of our general theories – our political paradigms – each of us could cite a wide range of observations, supportive verifications – exemplars. Responding to Avi's charge that capitalists were universally selfish exploiters, I would give examples of where capital investment in poor South American countries had been valuable to the overall population, without much, if any, real return on the investment, without any selfish payback. "Very altruistic," I suggested. Avi would counter with clear examples of resource exploitation and CIA support of corrupt South American dictators. Admitted. Admitted. There were a few 'apparent' exceptions. 'But our intensions had been honorable,' I suggested.

"We didn't typically cite exactly the same collection of 'facts', because we were telling different stories, giving different narratives in advancing or defending our theories. I had my favorite 'positive examples', which Avi would summarily 'explain away', and Avi had his favorite 'positive examples', which I would creatively reinterpret. On the negative side, I would point out some obvious failing of socialism, for which Avi would have some 'reasonable' excuse, defending his theory against my critiques with equally creative auxiliary hypotheses.

"Our agreement about the 'facts' was real in a certain sense and yet illusory and misleading. We were each 'reading the facts' differently. The transfer of funds from one bank to another was a seemingly neutral 'happening', but with respect to our theories it was also a 'doing' – an action, an act with alternatively interpretable intensions. It was a fact that fit into each story, each account, differently – reflecting the theory-ladenness, the narrative-ladenness, of our representations. My reading of the act was that it was positive and well-intentioned. Avi's reading was that it was pay-offs to elites in exchange for the exploitation of the country's resources, resources that Avi took to belong naturally to the people of the country. It was our theory-laden 'readings of the facts' that were different, opposite, incommensurable, allowing us to talk past each other, never actually striking a decisive blow against the other's theory. In the sense relevant to the correctness of our theories, our different perceptions, observations, of the 'facts', the actions of the U.S. in South America, were conceptually different.

"How was it that our theories didn't conflict scientifically, on the evidence, and yet we almost totally disagreed as to the objective, rational structure of the realities?

"Although it remained unspoken, by the middle period dialogues the central focus had shifted from who was unquestionably right or wrong to a fascination with the ability of each of us to advance and defend our opposite theories in entirely reasonable, rationally plausible, ways, at least as reasoned from within our individual perspectives. We each began to search for 'the forcing argument' – some fact, some evidence, that would defeat the other's theory once and for all; something for which there was no comeback, no opposite reading. We both tacitly expected that there must be such an argument, based on scientific facts. But it remained frustratingly elusive. And as we each looked more aggressively for the knockout punch, the creative defenses offered up by the other seemed all the more intensely outrageous.

"I remember at one point, late in the dialogues, where I made a spirited, and what I took to be an entirely reasonable, defense of something that I really didn't believe. But believing it wasn't the point. Most of what each of us had put forth and defended was well beyond our personal beliefs. The question was academic. The question was whether we could, in principle, arrive at a rational, scientific settlement. Avi was visibly outraged by my defense. I mean, he was angry – really upset. He accused me of being disingenuous, which was, of course, true. But that wasn't the point. My defense was entirely rational and consistent with my way of reading the facts. Avi lost it and began a genuinely slanderous characterization of me and all those who could even possibly think like me – in terms of what I had just put forth.

"To Avi's bewilderment, I began to giggle. Then we both began to laugh. There was no resolution. It had become obvious at that point that no matter how many and how detailed the 'facts' it was always possible to offer a reading of the facts that produced a rational defense of either of our positions from within those positions. That we were each, actually, visibly outraged with the other's theory now brought forth a sort of humor – cosmic laughter. And why should we find this surprising? What we had engaged in was one of the core political dialogues in the history of civilization. How could one otherwise explain the perennial existence of a political right and a political left? By and large they had the same factual base, they just read the facts in different, rationally incommensurable ways.

"Moreover, and this was the kicker: we were talking politics – how we should live – our own position appeared to each of us to be morally uplifted, while the other's position appeared to be morally degenerate. What, as philosophers of science, we had presupposed to be a value-free debate about a matter of fact. Something we overtly tried to ignore was that the real conversation, more fully understood, had an irreducible moral component. What was offensive was not simply that the interpretations that each of us gave implied that the other's interpretation was confused, ill informed, irrational or systematically incoherent. The offensiveness, I believe, arose, quite unexpectedly, out of a sort of moral outrage. Each position tacitly embodied a value system – a version of what each supposed to be right and good. The unfriendliness wasn't about the factual analytics it was about something unspoken, and wholly unexpected within our scientific paradigm. It was about the moral antagonism.

"I could understand the words – what he was saying – it just didn't make any sense in terms of the presuppositions of my theory's framework. Think of a Marxist listening to a free-market capitalist's representation of reality, or the capitalist listening to the Marxist version of reality. Each might use the term 'justice', but they didn't mean the same thing. Their one-sided, ideological conceptions are logically discontinuous.

"What was most fascinating to us was the pertinence of all this to issues in the philosophy of science. The dominant presupposition in modern science was that all meaningful questions or issues could be decided factually, on the basis of the evidence, at least in principle. So either these issues simply weren't scientific, or real science was more like politics than it had been represented to be. What made it all the more intellectually disturbing was the clear sense that what we were debating was a factual evaluation of the world – how the world really worked and the corresponding 'one right way' to understand it.

"Reflecting back over nearly two decades, I began to see the Avinash Dialogues as identical in form to the First Impressions thought experiment I had discussed with Hawking and the crew in Oregon the year before. The 'official' first impression that I had adopted was that the United States was treating the South American countries as partners, cooperatively, on the same team. Avi's first impression was that the United States was competitively, selfishly exploiting and taking advantage of them.

"The provocative and discomforting suggestion of the Avinash Dialogues was that real inquiry about nature – about supposedly objective reality – was, in fact, just like our politico-economic debate. Each side could cite copious confirming evidence – for instance, in favor of the idea that the electron was a particle and, equally, in favor of the idea that the electron was a wave – each with plausible ways of explaining away apparent counter-evidence. The decision, the theory-choice, as to whether the 'real' economy was 'objectively' (i.e. universally) cooperative or competitive, socialistic or individualistic appeared to be – by the very nature of reality – inherently undecidable.

Lateral Conversions

"I had already become fairly adept at arguing either side of the competitive-cooperative positions, taking a position depending on the group I was with. Avi, who was equally adept, and I could have switched positions. Our debate wasn't about what we honestly believed but about what could be decided by evidence on the basis of seemingly logical, scientific reasoning.

"For many people, the switch from one side of a debate to the other is a major intellectual and spiritual event in their lives. Switching from one side of a perennial controversy to the other is referred to in both the technical and nontechnical literature as a *conversion*.

"Philosopher of science Thomas Kuhn, in his 1962 culture-changing book, *The Structure of Scientific Revolutions*, made the case for conversions in the history of science. Kuhn argued that advances in the history of science were often not based on logical (rational) arguments from *within* the earlier theory, rationally moving us to the later theory. Rather, they involved a sort of logically and conceptually discontinuous 'conversion.' What I am suggesting is that these conversions in the history of science were intellectually – logically and evidentially – quite _analogous_ to lateral conversions between first impressions, as in moving across the selfish-selfless divide, as in moving across the competitive-cooperative divide – moving between contrary positions.

"The claim that major advances in the history of science were discontinuous conversions didn't seem particularly plausible to people in the beginning. Science was supposed to be conceptually continuous, logical and systematic. But Kuhn's arguments were powerful and persuasive. And when critics responded by checking back to the historical cases trying to produce the logical links, the rational or empirical arguments leading from earlier to later better theories, it became clear that, in at least many of the most prominent and dramatic cases, in the so-called revolutions, Kuhn was right. Kuhn was suggesting that these advances in the history of science were _analogous_ to someone converting from a politically right-wing worldview to a politically left-wing worldview. Either way, right to left or left to right, these were shifts in what one considered the rational way to understand. The difference was that Kuhn's conversions were 'vertical', to a better understanding. Conversions between opposites were 'lateral' conversions. Both vertical and lateral conversions were 'conversions' in that they were conceptually, logically and evidentially *discontinuous*. Kuhn was primarily concerned with vertical advances in scientific understanding from an earlier to a later, better theory – not with flips from left to right. However, Kuhn's critique of the textbook interpretation of what had happened in the new physics suggested that instead of making a clear and distinct vertical conversion, the conservative supporters of the Mechanical Philosophy, faced with the enigma of complementary objectivities, had attempted a lateral conversion from the Newtonian-type of objectivity to the Maxwellian-type of objectivity.

"Let me give you an example of an analogous, 'revolutionary' lateral conversion between opposites. This one was talked about and quite popular in the 1960s and 1970s – part of the emerging feminist movement. Traditionally, women were given, and I think largely accepted, an official line that men were their best friends and protectors. Women were the homemakers and men were the breadwinners, and it was all this nice, happy model of harmony. We were all on the same cooperative team. But a lot of women were finding it difficult in the modern era to make sense of their lives within this model of reality. The most widely prescribed drug in the 1960 was valium, a tranquilizer – also known as 'mother's little helper' – for women who were finding it difficult to make sense of, to rationalize, their lives within this dominant model of happy harmony.

"The conversion was called the 'click experience' – like 'click', now I see the real truth. The name derives from an article by Jane O'Reilly entitled, 'The Housewife's Moment of Truth', in the 1972, premiere issue of *Ms Magazine*. O'Reilly described the 'click experience' as a conversion in how a large collection of experiences were understood – "a sudden moment of awareness when a women realizes she is being dismissed or oppressed by attitudes and actions she once accepted as a matter of course." The evidence hadn't changed, but it is now experienced, or re-experienced, even in memory, differently. This was a first-impression-like conversion, where the evidence that had appeared to support the happy harmony of the classical relation, all of a sudden, was seen as misleading, a misunderstanding, a misreading of the facts. The real relationship between men and women was actually sinister. Men were selfishly exploiting women. In the worst version, the men were viewed as engaged in an overall conspiracy quite analogous to the Marxist theme, where the upper classes were conspiring to exploit, oppress and enslave the lower classes.

Recalling another portion of the dialogue in the van: "Men always try to give you this grand-protector image," says Diana, "then you learn."

"First impressions," says Sue.

"Precisely," I say, "the feminist 'click' experience is a large-scale example of the conversion from a positive first impression to a negative impression."

"Not all women were converted to the feminist perspective – to that reinterpretation – where males were oppressing females. Indeed, over the years there was a significant division among women – as well as men, for that matter – on a number of feminist issues. Those who had made the conversion thought that their unconverted friends perhaps just needed to be educated. But the division wasn't about the supposedly 'neutral evidence' – to the extent that it can be described in terms of neutral 'happenings.' Notice that this is the same thing that happened when the Marxists concluded that they needed to educate – to radicalize, to convert – the working class, many of whom didn't feel at all oppressed. They had felt 'part of the team' in their work relationships.

"O'Reilly commented in her famous article: "It was heady stuff, recognizing ourselves as an oppressed class, but the level of discussion was poor. We explained systemic discrimination, and men looked pretty confused and said: 'But, I like women.'"

Back in the van: "These are like Gestalt switches. You know, where you look at a picture one way and you see a vase, but when you look at it another way and it's two faces looking at each other," said Pam.

"Is that what you mean? Are those conversions?" asks Pam.

"Yeah. I've seen those sorts of pictures," says Diana.

"One of the marks of a conversion is that the evidence, the actual observations, remain largely, in some sense, the same. In the Copernican Revolution from an Earth-Centered to a Sun-Centered model of the Solar System, observations of the sky before and after are 'the same' in some sense and yet 'not the same'. During the 19th century, as we investigated the relationship between light and the electron, various accomplished researchers switched allegiances, converting back and forth between the light-is-a-particle and the light-is-a-wave research programs.

The Underwood Story

"Another quite compelling example of conversions that I studied is the evangelical's 'born again' conversion. One extreme example occurred with the Moonies – associated with Reverend Moon's Unification Church, originating in South Korea. There were all sorts of charges that Moon's organization was a cult and that his followers were 'brain-washing' young college-aged people to join the cult. Moonies, of course, viewed the facts differently.

"There is a fantastic book written by a mother-daughter pair, describing their experience with the Moonies (cf. Underwood, Barbara & Betty, *Hostage to Heaven: Four years in the Unification Church by an Ex Moonie and the mother who fought to free her* (1979)). It contains both the daughter's and the mother's experiences. The daughter's perspective begins as that of a converted follower and, then, as a highly effective recruiter of new converts. Her mother's perspective was that of the committed rescuer, who had the help of a professional 'cult-deprogrammer.' The daughter's description of how she organized her recruiting, under Reverend Moon's guidance, is fascinating. She was a master converter. What she looked for were college-aged students who were disenchanted with the 'hard cruel world' – the selfishly competitive 'hard cruel world.' These targeted students would often be either without friends and a supportive community or, at least, would tend to look at the world as a competitive jungle – not caring for the individual. For them friendships seemed only superficial, really, illusory, associations of convenience in the competitive 'rat race' of life.

"Barbara would invite these students a few at a time to join her and some of her Moonie friends for a weekend retreat at the beach. It was all truly up-and-up, chaperoned by adults, separate sleeping quarters for males and females, nice and wholesome. Around the bonfire in the evening, Barbara and friends would talk about another way to see the world. More than that, she described another way to live and to think, to be and to act in the world – as part of a loving, cooperative, supportive community; a community that was real and didn't end – eternal. Barbara talks about engineering the emotional experience, how it builds and culminates with lots of hugs and tears as the newly converted recruit accepts the invitation to join the community.

"Now this doesn't sound so bad – so far. However, from the critic's perspective, one of the consequences dawns on the new recruit only later. When you join the Unification community, you become an integral part of the community. (By analogy think of joining a Newtonian clockwork universe. You become a valued working part in an eternal, rationally ordered community.) What you give up – in this extreme case at least – is your individuality. For instance, as a new convert, you might suggest to your new friends that everyone go to a movie together this weekend. But you learn that that is not the plan. That is not what Reverend Moon, our wise and loving spiritual leader, has in mind for us. Instead, we will be asking for donations at the airport and looking for more recruits. To the extent that you try to resist this agenda or to move away from the group, your new friends pull you back – with both love and the suggestion that the path to love is only through a complete identification with the group – and Reverend Moon's spiritual agenda.

"Barbara Underwood talks about how the strength of the emotional experience in that first evening at the beach represented the depth and effectiveness of the hook – the catch. If the conversion was powerful, the likelihood of any subsequent resistance to 'the group will' was negligible. Street beggars? Yes. But they were together in the loving service of God.

"When parents, appalled at their children becoming street beggars for Reverend Moon and leading their lives in an apparent cult, tried to convince their children to come home, to leave the group, the parents were vilified. The children turned against their parents and were encouraged to actively criticize their parents and to stop seeing them. 'This is your eternal loving community. It is just too bad that your parents can't see it, can't see that Reverend Moon is the savior, Jesus Christ reborn.'

"Not all parents took this response from their children passively. "They [the parents] were torn between their desire to respect their daughter, her integrity, her civil rights, her freedom to make her own decisions, on the one hand, and, on the other, their growing conviction that she had endured a form of brainwashing that effectively prevented her from making any free choices."

"Barbara's mother sought out a professional 'deprogrammer' to un-convert her daughter. To get even reasonable access for enough time to effect the deprogramming, it might be necessary to physically kidnap her daughter and to restrain her for several days. Obviously, from a legal standpoint, this was a questionable approach. Several of the parents instead sought and ob-tained an unprecedented, court order, restraining the Moon Organization and allowing the parents limited access to their children for three days.

"Over these days of talking with the deprogrammer – a sort of enlighten-ing psychotherapy – Barbara was led to an epiphany about herself, her God and the Reverend Moon. In her own words, she came to realize, through a careful rereading of the Bible with her 'deprogrammer', that God actu-ally wanted her to be 'an individual within a community.' Conceptually, of course, this can be viewed by ideologues from either side as oxymoronic, irreconcilably different ways of believing and behaving – opposite, contrary commitments. But Barbara accepted this conflict – an unresolved and un-resolving, critically dynamic middle-ground between the individual and the group perspectives.

"Kuhn uses the expression – 'essential tension' – in describing the healthy structure of the scientific community, the way it ought to be: natu-rally contentious, an argumentative yet cooperative dialogue between each competitive inquirer and the community. As in a courtroom, there is an on-going point and counterpoint between each individual and the democratic community. There are numerous specific local decisions, but no definitive, one-version-fits-all-issues way to believe or inquire. There is no 'rationally objective' universal theory that dictates what all practicing scientists should believe.

"In the midst of the deprogramming process, Barbara literally experi-ences the conversion to a new 'impression', to a new way of understanding: "My darkest enemies were metamorphosing into my liberating friends; what an ironic twist, I thought." (page 244)

"Reflecting on her feelings the first morning after her firm decision to leave the Moonies, Barbara comments on a disorienting feeling of loss: "Bad morning. Buried pangs of remorse for a lost quality, yet what was lost? Answers, pure and simple. A small point of pure love, detailed charts of right and wrong, power over people, over the world, over the historical future." (page 246)

"What Reverend Moon had offered to Barbara Underwood and the other Moonies was an escape from uncertainty, a single unambiguous co-herence, an ideology, like a Newtonian clockwork. All 'apparent' ambiguity and counter-evidence is reduced by 'ideological rationalization' to One uni-fied worldview, to one single rationality, to the 'one right answer'.

"The Underwoods are from Oregon and the book cover has the follow-ing review comment from then U.S. Senator Mark O. Hatfield (Oregon): "A moving and gripping account of a young woman's journey to define mean-ing and structure in her life, and in the universe, but at the expense of her individual and intellectual freedom."

"The happy culmination of the Underwood story is that Barbara ended up marrying her deprogrammer.

"Once, when I was in London, this sort of evangelical conversion really came home to me. I was introduced to a German woman, a college student – flaming red hair, very serious. She paced back and forth in the apartment as she told me and a few friends, with great ferment and intensity, the story of her conversion to Marxism. This was the first time I had met a real live Marxist, since it is way too politically incorrect to be an outspoken, identifi-able Marxist in the United States."

"There are plenty in Britain, Terry," Jonathan remarked.

"What struck me the most was that her story was almost identical – structurally at least – to that of Barbara Underwood's description of con-versions to the Moonies, from a view of the world as uncaring competitive individualism to a world of an extreme cooperative community. What made her story rather unappealing was that it implied that all the rest of us who didn't completely and unreservedly accept the Marxist perspective were somehow deluded – thinking and acting irrationally. At least she was moti-vated to 'educate' us, to convert us, to the true beliefs.

"There is a subtle difference between a balanced, reflexively self-critical religious community and a cult. Similarly, there's a subtle difference between the balanced 'essential tension' of a middle-ground society and the monoculture of an ideological society. Amongst those who have really thought about this, the dark side centers on the same issue – ideologies – where people imagine that they have, either in practice or in principle, the 'one right answer', the one and only right way of understanding.

In the conversion to the new physics there was – and still is – a struggle as to whether the transition should be understood as a continuous, inductive generalization to a new 'objectivity', thereby remaining within the classical 'one right answer' Scientific Research Program, or whether the new physics required a logico-mathematically discontinuous, revolutionary move, forcing us out of the Scientific Research Program to a More General Research Program, one that embraced the oxymoronic, post-scientific middle-ground of Bohr's complementarity.

Although there was a revolutionary discontinuity in the embrace of the new physics, it was hidden from sight for a long time. The momentum of the Mechanical Philosophy served to resist the advance to a new post-scientific physics.

What happened was that in an effort to retain the classical Mechanical Philosophy there was a 'lateral' conversion from Newtonian mechanics to the strange new Maxwellian mechanics. Newtonian particles were to be newly understood in terms of electromagnetic radiation – as waves. Bohr and his rebel followers, however, pushed for the 'essential tension' of the middle ground, embracing of the complementarity of these complementary (lateral) options. They sought a new *type* of vertical advance, a vertical conversion, to a new *type* of More General Theory that could subsume and supersede both the Newtonian particle program and the Maxwellian wave program as limited special cases.

The common strategy of those deeply committed to any ideology is either to convince the opposition or, perhaps more simply, to eliminate them – sometimes overtly, sometimes covertly. The whole objectivist enterprise is ideological in so far as it presupposes that there is always and everywhere, at least in principle, only 'one right answer.' Rejecting the complete validity of all ideologies, the middle-grounders are in the ambiguous middle, neither completely rejecting nor completely accepting. By embracing the limited validity of both, indeed all, Mechanics, the middle-grounders understand themselves as peace-makers with a progressive agenda. The middle way strategy seeks a More General Theory where the limited validity and limited value of each perspective is understandable. 'Oh I see what you mean.'

20

The Perennial Oppositions and
The Parliamentary Attitude

"I had a good friend in London, Howard Steers, who was studying for his Ph.D. in the famous London Peace Studies Program – formally referred to as the London War Studies Program. Howard was fresh from three full combat tours in Vietnam, having requested and volunteered for the second and third tours. He was a United States Marine, I think a Captain by then. He told me a few stories, such as when Viet Cong crawled inside his platoon's perimeter at midnight. Exciting! Howard was a combat marine, not a desk marine.

"Howard invited me to join him to see a movie, one that he had recommended earlier, the classic Russian war movie, *Alexander Nevsky*. Howard had already seen it several times. On the way, he confided to me that he was having trouble adjusting to the University of London graduate studies community.

"We both lived in London House, a special residency for graduate students created by the City of London exclusively for students from countries that 'had helped Britain win the Second World War,' – lots of Americans, Canadians, Australians, New Zealanders and white South Africans. Howard told me that in Vietnam he had come to size people up, to evaluate them, judge their character and worth as human beings in terms of whether he would want to share a fox hole with them during combat, a time when you put your life in their hands, and trust in their courage and fortitude. That was his rather severe 'First Impressions' criterion. His problem in London, living in an academic community of people studying for their Ph.D.s in all sorts of subjects – ranging from physics and biochemistry to psychology and philosophy to economics and medicine to music and art history – was that no one measured up. No one passed his first impressions criterion.

"My initial reaction was to feel rather insecure, since he hadn't explicitly suggested that I was, or might potentially be, an exception. But I figured that he had invited me to see *Nevsky* because he thought there might be hope for me. I think that he also thought I might appreciate the movie, and that it would help me to understand him and his perspective. Howard was Jesuit educated and from New York with the full accent. The Jesuit view, he explained, was that life was war. "Notice, that all the values of life – courage, honor, discipline, integrity… are values of war, of the warrior," he said, and elaborated.

"My only, largely anemic, response to his first impression of the London graduate community was to propose an auxiliary hypothesis – an extenuating circumstance – that most of these students, if transported to the combat zone in Vietnam, a rather different circumstance, would not only look different, but would actually behave quite differently.

"Over the years, I kept in contact with Howard. After finishing his Ph.D., he taught for a stretch at West Point, The United States Military Academy. "I had trouble there," he told me. "They didn't want to hear about the ambiguities, about alternative interpretations." They wanted to be clear – 'crystal clear' – when they were given an order. A few years later Howard joined the U.S. State Department. And it was around that time that he explained to me the fundamental difference between the mindset of the Defense Department and the mindset of the State Department.

"The Defense Department people take their job very seriously and so they tend to assume the worst. Even if someone appears to be your friend, you need to assume, as a matter of prudent caution, that they might be trying to deceive you in order to gain some advantageous position, after which they could attack and destroy you. Defense sees their job as dictating that they should always assume the worst. Andrew Grove, once CEO of computer chip maker Intel, offered a similar attitude, succinctly, in the title of his memoir, *Only the Paranoid Survive*. And so it is natural for them to see the world as fundamentally competitive. Countries that get out of hand or are potential threats need to be disciplined. Like the Jesuits, Defense tends to see everything in terms of the competitive war perspective – good versus evil.

"The State Department mindset is complementary. Its job is to trust and make friends and to take risks in order to build trust to bring everyone into one, mutually supportive community. One of the best ways to build trust is to engage diplomatically in cooperative, interdependent social and economic activities. The State Department presupposes that what the people of the world want – really deep down – and are actually trying to build, is a cooperative world community. When some country seems to be moving in the wrong direction, the appropriate State Department response is to try to understand them sympathetically, to talk and negotiate, perhaps to compromise, to find common ground and to build toward a mutually beneficial, win-win, future cooperation. President Richard Nixon's economic engagement with big bad communist China is an exemplar of the friendly, yet risky, diplomatic trust-building State Department approach.

"Both the descriptive perceptions and the prescriptive policies of Defense and State are theory-laden. Their contrary problem solving policy paradigms are driven by opposite, complementary belief systems. There is no one, universal, objective right answer or approach. There is no uniquely rational, right policy, with applicability to all possible circumstances. Is there a crucial experiment that would decide which of these worldviews is the right one? If you choose one or the other exclusively, are you naive? Irrational?

Soccer Thought Experiment

"All this talk in terms of ideologies tends to obscure the common sense of the middle-ground. In my attempts to make sense of complementarity in physics I had regularly tried to translate the issue into more day-to-day situations. That was the background to the analogy I proposed on the long-drive to the Oregon Coast in 1992: the selfish-selfless issue as analogous to the particle-wave issue. It is all too easy to lose track of common sense when issues are formulated in terms of quantum physics. The following thought experiment about complementarity and the middle ground is, I think, revealing, giving an image of what reality looks like, how we might make sense of reality, once complementarity is embraced.

"Consider sports. The game of soccer, for instance, is a popular sport played enthusiastically in almost every culture on Earth. Now consider whether when you are watching a soccer game are you observing competitive or cooperative behavior? Personally, when I think back on my high school sports career, it is occasionally about whether we won or lost a particular game, about the competition. More often it is about the cooperation, the teamwork, the camaraderie and the lasting friendships.

"When you play on a soccer team, at virtually any level, there is the danger that one of the players on your team is a 'ball hog' or a 'glory hound'. The former always wants to have the ball to himself and the latter always wants to be the one who shoots, making the highest point score, and so, hopefully, to be selected for the all-star team. Selfish ball hogs and glory hounds weaken the competitive power of the team. The famous and celebrated UCLA coach, John Wooden, was known to preach a philosophically sophisticated hybrid of competition and cooperation. Teamwork, the selfless orientation of the individual players maximizing the competitive effectiveness of the team, was considered inseparable from the personal self-development of the individual players, encouraged through 'mutual striving.' Wooden's message was that highly disciplined cooperation made for the most powerfully competitive team, as well as for the most personally accomplished individuals.

"You can't have a modern competitive team sport without an enormous amount of cooperation. If someone were to suggest that we have a game of soccer but suspend all the rules, including time and place, size of the field, specifics of the goal structure and on and on, there wouldn't be a soccer game anymore. There would be no sense in which the actions of the supposed 'players' could be understood as playing the same game. Without a cooperatively agreed upon set of rules there isn't a game. Good rules define good games. If the rules on fouls were removed, the first thing one might do, as coach, would be to have several of his players physically disable the star player of the other team. Eliminating the leader of an oppositional political movement is a well-known, often highly successful, win-lose strategy.

"By defining the game, even building the stadium and agreeing on the rules, cooperatively, we enable the players to excel individually as competitors. Rules, like physical laws, can be thought of as restrictive and constraining, decreasing the individual's 'freedom to act', and yet here we see that a balanced system of limited rules creates a platform that actually enables and develops the opportunity for individual creative action. If supposedly deterministic laws are all inherently limited in scope, organized and structured as in a game, they can be empowering, enabling actions that develop a better future. Taken to the extreme, laws and rules that completely define and determine every aspect of every player's actions, as in a Newtonian clockwork, wouldn't allow any individual free action. No game. On the other extreme, with no laws, no rules, everything is undetermined, random and chaotic. No game.

"A middle-ground interpretation recognizes that our individual choices are constrained but also understands the constraints in a new way, as structures and processes that enable the actions that can bring forth a more desirable future.

"The root of the Ancient Greek word for competition translates as 'mutual striving.' In this view competitive sport is not 'really' – by its very nature – about who wins and loses, it's about the benefits of challenging and inspiring each other to greater performance, benefiting everyone. Competition enables and encourages the achievement of individual excellence within the community serving the greater common good. The Ancient Greek philosopher, Hesiod, in his *Works and Days*, speaks of a sort of revelation that explicitly distinguishes this 'mutual striving' sense of competition from the destructive, winner takes all, zero-sum game, might-makes-right mentality of warfare.

"Thinking in terms of these analogies it seems that it would be naive for anyone to think that either ideological extreme – the 'no-rules free-market' or 'centralized command-and-control corporate, or socialist state' – could be the unique, universally correct, best way to live, or that either ideology could give a complete and consistent account of, make sense of, how the whole system works – of how the universe works.

Oakeshott's Wisdom and The Parliamentary Attitude

"Occasionally a stray idea enters one's mind, one's intellectual milieu, un-ceremoniously, casually, approved at the time of entry as genuine, interest-ing, possibly important, but without any clear sense of connections to other thoughts and beliefs. Later in new settings the idea re-awakens in the flow of thought, recalled unexpectedly. With each such recollection new connec-tions develop, and the sense of its importance may grow.

"In a conversation in London with a fellow graduate student in philoso-phy, whom I met only once and can't place exactly when or where, I picked up the following insight, attributed to Michael Oakeshott, a Professor of Political Science at London School of Economics, University of London. Oakeshott, I was told, maintained that the debate about whether society 'really' was or should be organized either as completely and consistently competitive or as completely and consistently cooperative – roughly, the de-bate between the extreme political right and the extreme political left – was fundamentally misguided. All working and workable societies, Oakeshott maintained, had essential – irreducible and unavoidable – components of both competition and cooperation, somehow along the lines of the 'sporting event thought experiment.' No ideologically pure, uniformly competitive society and no ideologically pure, uniformly cooperative society had ever existed or could exist. All human activity was a complex combination of both. Although competition and cooperation involved – at least taken by themselves – irreconcilably different types of organization, they were always balancing and rebalancing in a dynamic tension. I never took a class from Oakeshott, never met him. But I recognized something deep and profound in these remarks and carried them with me. Then, reflection-by-reflection, they helped to clarify my understanding of complementarity. Following the sporting metaphor, the games of politics and the games of economics always seemed to be unavoidably organized, irreducibly, to involve and promote both competition and cooperation.

"If you want to understand the literal operating structure of a society, its institutions and processes, its policies and values, you simply can't make sense of it in terms of just one or the other of these ideologies: selfish, competitive, free-market individualism or selfless, cooperative socialism. Oakeshott's contribution here, at least as I understood it, was, in the first instance, about the explanation of the purely descriptive; about matters of fact, observations of diverse types of structures and processes. You simply could not make sense, *completely and consistently,* of the structures and processes of any actual working society in the terms of One purely competitive or purely cooperative model. In using either approach exclusively, certain structures and processes, certain phenomena, simply wouldn't make sense and would appear to be irrational. All predictions, based on one or the other type of description, would be inherently incomplete and so the predicted outcomes necessarily involve inherent, irreducible uncertainty. The structures and processes identifiable by each ideology in their partial description, the partial conception, under-determined the overall action of the whole system.

"I imagined that the same sort of situation arose for a physicist unable to arrive at a comprehensive understanding of the processes and organizational structure of any actual physical system in terms of a purely particle or purely wave approach. From either approach alone certain structures and processes simply couldn't be made sense of, couldn't be described and would appear, in those terms, to be non-law-governed nonsense – irrational. Because physical systems embody these complementary types of structures and processes – wave and particle – each opposite paradigmatic (ideological) approach to inquiry would be inherently unable to arrive at a single universal, logico-mathematical – complete and consistent – description, explanation or understanding, of the whole system.

"Oakeshott was also, at least implicitly, pointing at a prescriptive lesson. His point wasn't just about the description of how we live, it wasn't just about how things are actually structured and operating at the moment. His further point is about how we should try to live and about how to make things better, about how to develop and possibly improve the structure and operation of society. But how are we to understand this suggestion of a prescriptive lesson?

"It may have been in that same earlier, half-remembered conversation that introduced me to Oakeshott that I was gifted by another one of these ideas that enters almost as if intellectually asleep but gradually rouses to a position of prominence, maturing through repeated partial recollections, stirred and re-stirred, making connections piecemeal."

"So what is the topic of your research?" I asked a fellow graduate student.

"I want to develop the philosophy behind the parliamentary concept of the loyal opposition," he said and continued, "No one has done that yet. Despite the history and importance of the concept and associated practices in modern parliaments, it doesn't have a philosophical grounding. Its origin is vague. The reasoning behind it hasn't been articulated or, at least, widely appreciated."

"I remember thinking that it was certainly odd, if indeed correct, that such a fundamentally important concept of modern parliamentary governance as 'the loyal opposition' lacked any rational justification.

"What was sleeping here was a reflection upon the very notion of political rationality. If there are real 'objective' distinctions between right and wrong, between good and bad, then there is no rational reason that right should tolerate wrong, or that good should tolerate bad or, for that matter, to the point, that rational should tolerate irrational. Historically, in both princely and religious reasoning – in Western Civilization at least – the law defining right and wrong, good and bad, is seen as being handed down from on high, either as the objective truths of religion or as the socially defining declarations of the ruling prince. No exceptions. There is a litany. What is right and wrong, good and bad, is not simply a matter of opinion. If disagreement arises on some question, if there is any uncertainty or ambiguity, then a hierarchical structure exists to settle the matter – from the priest to the bishop, bishop to the archbishop, archbishop to the cardinal, and, in exceptionally difficult or novel matters, the Pope is consulted to break any ties. Similarly, issues might arise from local state magistrates up the hierarchy to the prince. In nonreligious politics the foundation traces back to the rights of power, the rights of the sovereign. 'Might makes right.' The local power decides – and enforces.

"The reasoning behind the concept of the loyal opposition does not arise logically from within one of the perennial social-political-economic philosophies; does not arise from within the ideology of universal competition or from within the ideology of universal cooperation. Neither of these ideologies, taken by itself, has a rational place, rational room, for a contrary opposition that appears not to make sense, indeed, appears to be irrational. Religions and state monarchies are known for repressing opposition – not encouraging it. The concept of a loyal opposition cannot have been derived, logically reasoned, from any ideological concept of one universal complete and consistent rationality. Ideologies, by their very nature, reject deviations from their way of understanding reality. 'This is the way to think and that is that. Do not deviate!'

"The idea of institutionalizing a system of loyal oppositions is more of a meta-concept – a concept that could only arise from a self-reflexively critical appreciation of the limits of the different ideological concepts of political rationality. The idea of a loyal opposition involves the hypothesis that there might be some (More General) 'higher rationality' of a different *type*, one that naturally encompasses and encourages qualitatively diverse rationalities, embracing and embodying an 'essential tension.'

"The concept of a loyal opposition is not based on any claim to 'knowledge' in any objectivist sense. Rather, it arises as some sort of self-critical, social-political 'wisdom.' The reason I mention all this here is that I suspect that the concept of a loyal opposition is closely allied with the notion of complementarity.

"The concept of a loyal opposition only makes sense if there is more than one right and good and successful way to understand and deal with reality.

Both the competitive and cooperative policy approaches to understanding and improving society might be viewed, from a more general meta-perspective, as 'compatible and competitive' alternatives in each historically local specific situation. Faced with a new problem or opportunity, the wise society looks at the particulars and decides critically and reflectively – perhaps only after further experimentation – on the best approach here and now in this situation. In policy matters, competition and cooperation, the right and the left, would be seen prescriptively as local, complementary strategies – 'meta-competitive locally and yet compatible globally'.

"Advancing one approach as universally applicable, as always the best choice – perhaps even trying to eliminate the ideological opposition – abandons the loyal opposition policy.

"In a *Scientific American* article many years ago the authors commented that we were fortunate in the scientific community that the question of whether the electron was a particle or a wave was never settled, since great and wonderful technologies have been developed based on each of these contrary hypotheses, based on each of these contrary research programs. If there had been an ideological consensus among the established physicists – those who controlled research policy and funding – one or the other line of research and development might have been terminated.

"The reasoning leading to the 'balance of powers' structure of modern constitutional democracies is closely related to the loyal opposition way of understanding. Roman reflections on the 'balance of powers' were influential on the founders of the American Constitution. The inability of the Roman Senate to administer (viz. rule by committee) pointed to the need for a separate, but limited, administrator. Similarly, an independent judiciary, where counter-positions and policies are formally presented and democratically evaluated is now considered an essential but limited, power-balancing component of the modern state. Instead of 'one right answer' there is a dynamic, mutually respectful tension between competing perspectives and approaches.

"In the dialogue leading up to the 10th Amendment to the U.S. Constitution (about decentralizing power), John Adams wrote to Thomas

Jefferson: "My dear sir: I doubt me that any body of men, including this Congress of yours, would be so wise as to correctly resolve every issue with which it is presented. I therefore think that you would give too much power to these gentlemen." Jefferson replied: "My dear friend: I believe you are right. I shall therefore propose that substantial powers be reserved to the state and local governments, that they may serve as civic laboratories to address such problems as the future will confront us with." (Federalist Papers No. 54) The natural extension is to systems that embrace an essential tension balancing the state, the team and the individual.

"The philosophy of the Parliamentary Attitude is to honor, respect and encourage the opposition, even though the opposition may seem to be advancing beliefs and policies that, from the current governing party's point of view, are not only wrong (viz. or 'not even wrong') but also 'don't make sense.' The opposition's views are not so much experienced as wrong; they are experienced as irrational.

A difficulty with tolerance and encouragement of diverse opinions, as Lakatos emphasized in applying similar reasoning to scientific research policy – but really applicable to all policy forums – is that while listening to some new policy idea that doesn't make sense in your terms, you are faced with the problem of distinguishing 'the potential genius from the crank.'

"The Parliamentary Attitude embodies a partially constrained, partially rule-governed milieu of non-ideological reasoning. The Parliamentary Attitude embodies a 'higher rationality' that makes sense in a post-ideological, post-objectivist, post-scientific More General Understanding of a developing reality. The Parliamentary Attitude is middle-ground, per hypothesis, presupposing that both the perennial competitive and cooperative approaches have value in serving a more general ('higher order') common agenda, wherein competition and cooperation are limited, irreducible components.

"If major political ideologies are complementary, and if, as Oakeshott suggested, in all real, surviving societies these contraries are both contributory, then the actual history, the actual evolution of society can't be made sense of in terms any one of the ideologies alone. History doesn't proceed according to one consistent, ideologically pure rationality – in terms of one logico-rational order. From the point of view of each complement's rationality, the course of history is under-determined. There may be periods of the relative dominance of one or another ideology, but eventually, for societies that survive and develop, there will be a plurality of independently successful, compatible, yet competing, problem solving approaches. Oakeshott's analysis would expect that the political push towards ideological purity – eliminating the complementary opposite and the diversity of the middle-ground ideas and approaches – would gradually self-destruct. Successfully developing societies must emerge, qualitatively, creatively constructing progressive win-win middle-ground relationships. The process of 'creative compromise' is logically and mechanically under-determined such that the progressive outcomes are unpredictable from within either ideological position. The search for a constructive middle-ground, win-win creative compromise is the same as the search for a More General Theory – a new understanding that subsumes and supersedes the value, the limited value, of each of the complementary policy agendas.

"Should we, following this reasoning and in accord with Thomas Kuhn's revelations, postulate that the history of science develops similarly? Should we, following this reasoning and in accord with a sort of Parallel Hypothesis, postulate that the history of the universe develops similarly – through the progressive 'interplay of opposites.'

The Parliamentary Attitude in Science

"If major scientific research programs are complementary – like Newton's particle program and Maxwell's wave program – and if, in any 'healthy and productive' research community contraries are in a dynamic interplay – 'an essential tension' – then the actual history of science, the actual evolution of the research community, actual theories and practices, including actual scientific method itself, can't be made sense of in terms of one research program alone. The shocking revelation of Thomas Kuhn's *The Structure of Scientific Revolutions* was that major scientific advances could not be made sense of in terms of one conceptually consistent rationality, in terms of a conception of One timeless order governing the universe. 'Better' theories must be qualitatively better – emergent, innovative and conceptually creative. New knowledge, genuinely new discovery, is not simply a systematic, consistent logico-mathematical extension of what was known before. New knowledge, genuinely new discovery, is not simply a conceptually continuous, uniform extension of the previous conceptual understanding.

"Popper had also addressed the research policy question as part of his rejection of the Logical Positivist's representation of inquiry. Popper argued that peer review of research policy proposals would retard advances in science, since the peers – the established scientific elite – would tend to favor research proposals that were 'the most probable', the most likely to succeed, based on what is currently understood, based on what 'makes sense' in terms of their established successes.

"Popper rather dramatically suggested a balancing with opposition policy. He first noted that the most momentous scientific discoveries, the most significant, had always been the most unexpected, the ones that made the least sense in terms of current understanding. The best research policy, he proposed, if your goal is to maximize the conceptual advance of scientific understanding, would be to fund the research proposals that seemed to the established peers to be the least likely to succeed. When one of these 'bold hypotheses,' as Popper called them – challenging prevailing scientific beliefs – succeeds, we realize the greatest advances, and the most is learned. Popper seemed to be suggesting that we should encourage criticism; indeed, that we should adopt a Parliamentary Attitude and actually encourage heretical opposition. Such an attitude would be in the interests of promoting real, open scientific inquiry, preventing ideological behavior, preventing ideological dominance of the point of view of recent past successes. Popper was suggesting that we should not only tolerate the opposition, viewing them as 'a loyal opposition', but that we should actually encourage and fund innovative opposition ideas that don't make sense in terms of current demonstrated theory.

"The Copernican Sun-Centered proposal was not only opposed by the Church. Equally, and perhaps to a greater extent, it was opposed by the then current scientific establishment, thoroughly convinced by the great, centuries-long successes of Earth-Centered Ptolemaic astronomy. The deeper you question, challenging ever closer to the central tenets that define a research program at the core of the web of beliefs and theories, the greater the potential you have to learn something genuinely new. In the spirit of Popper's 'bold hypothesis' proposal, there are researchers who take a deliberately contrarian approach. I have met them. They aren't typically well funded, but nonetheless engage enthusiastically in a sort of antithetical research, looking for evidence – verifications of contrary phenomena – in support of beliefs that quite clearly don't make sense in terms of the current dominant beliefs of the establishment scientific community. They consider themselves scientists, part of a loyal, even if sometimes 'radical' opposition, as Kuhnian revolutionaries. Often they initiate their research programs by reversing core assumptions of the currently received, official views of the established scientific community. They 'stir the coals,' challenging us and making us think – at least those of us who are willing to listen. Some of them, of course, are cranks. Some of them – like Copernicus and Einstein – are eventually anointed as geniuses.

"In a seminar in London in 1970, co-taught by Lakatos and Feyerabend, Paul advocated an 'anarchist' approach to scientific research policy, a sort of 'anything goes' policy. Lakatos countered, defending a single-minded authoritarian peer review system – 'scientific fascism', as he enjoyed calling it. And yet each of these close friends insisted that their two policies, although competitive, were compatible. Feyerabend told me that his 'anything goes' anarchism was broad enough to include brief, limited periods of fascism. Lakatos argued that his fascism was 'enlightened' enough to be quite tolerant of numerous oppositional ideas, with the caveat that the authorities reserved the right to decide just 'who was a genius and who was a crank.' Feyerabend might well be thought of as favoring creative and innovative scientific individualism and Lakatos as favoring the organizational coherence found in either corporate research and/or in a peer reviewed socialist system.

"What they were each pointing to was the un-resolving and unresolvable 'essential tension', the mix of order and disorder, of a middle-ground reality and logically under-determined, middle-way research policy. Each of them expressing this in terms of their opposite emphases, reflecting their personal assessments of the current, historically local, cultural balance (or imbalance) in the scientific community.

"Recall my perplexity that questions about the nature of science and scientific rationality were not themselves scientific questions. Such questions do not arise naturally from within any *one* of the individual, ideologically represented, scientific research programs. The practitioner within one research program, one Kuhnian paradigm, seeking new confirmations has no occasion to raise critically reflective questions of the core defining presuppositions of the paradigm. Similarly, the questions of the nature of socio-political rationality don't arise from within *one* of the ideological approaches. Only when it is realized that there is more than one, essentially different, contrary type of rational approach that actually works, that is successful, do these reflexive questions seem to arise.

"Beyond the sciences, could it be that there is an even broader context wherein there might be a mutually respectful Parliamentary Attitude, a loyal opposition relationship between the sciences and the humanities, between fact-oriented and value-oriented perspectives, between the descriptive and the prescriptive, between the Spectator and the Participant?

21

Quine, Popper and the Other
Path to Complementarity

There were two paths to complementarity in the 20th century academic community. One was the path of the new physics with its embrace, in quantum theory, of the essential complementarity of particle and wave phenomena. The other path, less well known, or at least less well-appreciated as such, was in the philosophy of science. This latter path arose as a critique of Logical Positivism and the associated Mechanical Philosophy. Just as the new physics arose through a critique of the limits of classical physics, most of 20th century philosophy of science can be understood as arising from a rebellious critique of the Logical Positivist representation of classical science.

Over many years of reading Popper and the other rebels, I concluded that they hadn't started with any clear sense of the correct representation of science, but, instead, are best understood by their readers as having backed into their best insights through a relentless critique of the Logical Positivist representation. They arrived at their insights and alternative representations through a process of critical hindsight. The more they criticized the inadequacies of the Logical Positivist representation – that 'stood to reason' in the Mechanical Philosophy – the clearer and more mature their emerging new philosophy of science became.

The parallel between the development of complementarity in the new physics, championed by Bohr, and the development of complementarity by the rebels in the philosophy of science has, I think, so far, been underappreciated.

One of the seminal and representative characters in the 'hindsight' development of complementarity in the philosophy of science was Willard Quine (1908–2000). Quine spent his entire career at Harvard, from student to Professor, and is 'officially' recognized as one of the five most influential philosophers of the 20th century. Quine is an excellent example of someone whose work – in particular his link to complementarity – grew from his encounter with, and ultimate rejection of, the Logical Positivist representation of science. Quine spent time in Vienna with the early leaders of the Positivist movement. As the implications of the Logical Positivist model were drawn out, its inadequacies became increasingly apparent, and Quine emerged as one of 'the dissatisfied.' Quine rebelled.

Willard Quine

In his ongoing critique, Quine's most significant contribution arose from questioning the Positivist account of the relationship between theory and evidence. In the Positivist model, evidence pointed to theory and theory pointed to evidence – no ambiguity. Moving from theory to predictions was simply deductive. Specific observational consequences deduced from the theory should clearly confirm the theory. If any of these deduced consequences of the theory didn't check out, that would constitute counter-evidence and grounds for rejecting the theory. The simple description of the procedure was: from your hypothesis (or theory) you deduce consequences, and if they don't check out experimentally, you reject the hypothesis. That litany was what came to be called, by the rebels, naive falsificationism.

Quine focused even more on the reasoning in the other direction – from evidence to theory. The Positivists naturally endorsed the method of inductive generalization, where the investigators discern or discover regularities in their observations that, per hypothesis, reveal that the phenomena under consideration are governed by a regular causal relationship. The scientist then makes the inference, an inductive generalization, proposing that a general, time-space invariant law governs the causal relationships. The admittedly limited number of observations of the regularity is taken to be 'a representative sampling' of the general time-space invariant regularity – the causal relationship.

In principle at least, the relation between theory and evidence in the Logical Positivist model was a clean, two-way, inductive-deductive, logico-mathematical relation. The accumulated observational evidence must always, at least in the long run, 'in principle', determine a unique theory, the unique theory corresponding to the One, unique, objective reality.

Quine's contrary insight was that 'the accumulated evidence' never fully determines one, unique theory. Evidence always 'under-determines' the choice between many possible alternative theories. 'Theory choice', what the rational individual should choose to believe, was never, as the Positivist model had imagined, completely determined by the evidence. According to Quine, the evidence never pointed unambiguously to one unique theory, never pointed to one unique reading of the facts. On the other hand, he was not taking the extreme opposite position. Rational choice was not entirely undetermined, not entirely arbitrary, not entirely irrational. Complete undetermination would mean that any belief or theory one might choose in light of some body of evidence would be 'rationally justified.' Quine's under-determinism thesis argued that rational choice was certainly constrained by the evidence, just not uniquely constrained. Rational choice was 'somehow' middle-ground.

Quine's reasoning had, at least to some extent, the same character as what, in the new quantum physics, had led to the embrace of complementarity. The problem was that there were too many 'objectivities.' Quine noted and emphasized that there wasn't just one right, good and successful way to understand and successfully operate in reality. There were many. There is always a plurality of conceptually different, yet quite demonstrably successful, alternatives. Under-determinism was a 'middle-ground' position between extreme objectivity – where the evidence fully and uniquely determines the one and only one rational theory choice and extreme relativism – where the accumulated evidence and experimental demonstrations put no constraints at all on rational theory choice.

The thesis of under-determinism was that there is now and will always be *uncertainty* associated with any supposed, unique theory of the nature of reality. Quine was arguing that we could never arrive at one single, complete and consistent, objective Theory of Everything.

What is crucial to understand is that this irreducible uncertainty isn't because we couldn't, in practice, collect all the evidence or because of some lack of observational technology. This was an 'in principle' claim. Quine's under-determinism is making a positive, post-scientific claim about the actual nature of reality.

Quine's argument is that you can't find a single, unique understanding of objective reality because there isn't a single, unique objective reality. Quine might be understood as saying that reality has more characteristics, is more conceptually ample, more diverse, than any one theory can capture.

Quine's under-determinism thesis entails that for any successful theory supposedly explaining a body of evidence or the behavior of some system there are always, come what may, alternative, incommensurable, qualitatively discontinuous, successful theories. There are always alternative rational ways to understand and successfully work in the system under study.

By implication, reasoning along the lines of the Parallel Hypothesis, Quine's critique of the Positivist philosophy of science is saying something both about inquiry and about the nature of reality. Inquiry under-determines the possibility of One, unique type of scientific theory, because reality is not governed by One unique type of mechanism. For Quine the nature of reality is mechanically under-determined. The nature of reality is middle-ground between being fully determined, time and space invariant, and completely un-determined.

Quine's initial argument is that my experience of reality – the evidence – under-determines my choice of what to believe – theory-choice – and how I should act. The Parallel Hypothesis implication is that the future state of any system is causally, mechanically, under-determined by its present state.

More specifically, the Parallel Hypothesis is that 'the future state of any system' is under-determined by the causal factors identifiable by any one conceptually consistent description of the current state of the system. This means that the current state of a system is under-determined by any possible One (complete and consistent) mechanics. The evolution of any specified system is not uniquely determined by any finite set of specifiable factors, by any possible mechanics. The evolution of any specified system is therefore mechanically under-determined, unpredictable to some irreducible extent, in terms of any specifiable set of causal laws, in terms of any One mechanical characterization.

Wolfgang Pauli (1900-1958), one of the founders of quantum theory and 1945 Nobel Laureate in Physics, offered an equivalent statement of under-determinism in terms of the new physics: "As this indeterminacy is an unavoidable element of every initial state of a system that is at all possible according to the new laws of nature, the development of the system can never be determined as was the case in classical mechanics... Like an ultimate fact without the cause, the individual outcome of a measurement is, however, in general not comprehended by laws. This must necessarily be the case... The probabilities occurring in the new laws have then to be considered to be primary, which means not deducible from deterministic laws." (Pauli, Wolfgang, "Matter", in *Writings on Physics and Philosophy* (1994), edited by Charles P. Enz and Karl von Meyenn, page 32)

Additional support for under-determinism came unexpectedly from the mathematicians. There is always, they pointed out, more than one mathematical description of any collection of evidence. Even if mathematics were the language of nature, as Positivism insisted, there were still always 'mathematically rational' alternative theories – descriptions – of any finite collection of evidence, of any specified system.

But Quine was making an even more radical point than is captured in the quantitative, mathematician's formulation of under-determinism. Quine maintained that there were always conceptually distinct, logico-mathematically (and mechanically) discontinuous alternative ways to understand any system; for instance, wave and particle approaches to understanding the electron, or the Newtonian and Maxwellian approaches to understanding the whole universe. These are incommensurable, qualitatively distinct, conceptual alternatives. Formally, the wave and particle concepts can be said to be qualitatively distinct, discontinuous, if you can't generate one from the other, if you can't logico-mathematically express one in terms of the other.

From within any one ideology the qualitatively opposite, complementary type of evidence can perhaps be 'experienced', paradoxically, as an incoherent phenomenon, as a logico-mathematically 'irrational' phenomenon – not of primary interest to one's current research efforts, to be dealt with later under a Promissory Note.

Quine's point, although not limited to this argument, is the same as saying that the observations of the soccer game or Oakeshott's society can't be completely and consistently understood in either competitive or cooperative terms alone – simply because they aren't completely and consistently ordered either way.

The Logical Positivist's hypothesis of a unique relation between evidence and theory implied that the totality of evidence would uniquely determine one complete and consistent Theory of Everything describing objective reality. And plausibly, the Positivist should naturally expect that accumulating evidence, being mechanically uniform and consistent, should at least be pointing in the direction of the unique theory that would be only perfectly expressed in the eventual objective Theory of Everything. Under-determinism on the other hand expected that even with the totality of all evidence you wouldn't be able to find one complete and consistent description, one unified Theory of Everything. Even with the totality of all possible phenomena you wouldn't be able to find a way to understand them completely and consistently in terms of one *type* of phenomenon and one *type* of law – precisely because there is more than one *type* of phenomenon and more than one *type* of law.

Quine extended the consequences of under-determination to question the relations between the traditional scientific disciplines. What reason do we have to believe that biology, chemistry and physics deal with the same *types* of phenomena? The Positivist approach had offered a Promissory Note – reductionism – the hypothesis that all these different scientific disciplines would eventually be unified by the One 'truly objective' description of reality at the lowest possible, microscopic level, where, per hypothesis, all phenomena could be understood in terms of only one *type* of phenomenon. The 'atomistic' microphysical description was to be the 'real' description of reality and all the other sciences would be reduced to, and eventually turn out to be, mathematically calculable from that One, lowest-level, conceptual understanding. For the Positivists, per their hypothesis, all the sciences would be deducible from the final Theory of Everything.

Under-determinism, like complementarity in the new physics, entailed the inherent incompleteness of any proposed 'one type,' 'one right answer' reductionism. Under-determinism rejected the reductionist's conceptual unification and rejected the complete and consistent, fully deterministic, logico-mathematical uniformity called for in the Mechanical Philosophy.

Quantum theory challenged, indeed demolished, the traditional notion that there could be knowledge of a unique, objective, material substratum – a stuff, perhaps atomistic, of which the scientific universe was 'supposedly' composed. As Bohr emphasized, there simply is no observer-independent 'objective' quantum reality for everything else to be reduced to. There simply is no unambiguous lowest level 'there' there. Like Bohr's complementarity, Quine's under-determinism broadens the challenge to all knowledge, forcing us to question what it means to have 'knowledge of reality.'

What to believe – theory-choice – then is always under-determined. If there is a way to make a choice between demonstrably successful alternatives, it can't be based on the fact that your proposed rational choice is the One universally true, complete and consistent alternative. There are always rationally defensible, demonstrably successful alternatives. If one chooses to believe, ideologically, that there is just 'one right theory' one is necessarily making a choice that can never be fully justified by the evidence.

There is, in Quine's under-determinism thesis, a message of tolerance for alternative beliefs and belief systems. Quine was quite explicit in reopening questions the Logical Positivists had hoped to close: questions about values, ethics, the intelligence of the emotions and religion. Quine argued that one's choice of theories, even between animate and inanimate views of reality, is under-determined by the evidence, by all possible evidence, come what may.

Popper's Question and Complementarity

Popper's Question provides an even more explicit and recognizable path to complementarity in 20th century philosophy of science. Whereas Quine's under-determinism argues for real and different choices of how one might observe and understand a system, there is the remaining 'hope' – the reductionist Promissory Note – that these might all eventually be reduced to the 'one right way.'

Popper's Question and the developments leading to quantum theory are more decisive because they ask for a *formal* specification of the relation between the different types of phenomena, between the different choices of how one might observe and understand a system.

Popper's Question and quantum theory address the question of how, in what way, and on the basis of what evidence, these alternatives actually represent *formally* 'irreconcilably, irreducibly, different types of theories and phenomena'.

Popper's initial attempt at demarcation, resulting in what came to be called naive falsificationism, suggested that as soon as there was any counter-evidence, as soon as a theory failed to predict anything correctly, just once, the theory should be discarded. The defense of a core belief or theory in the face of even 'apparent' counter-evidence was initially thought to be the mark of uncritical pseudo-science. However, such defensive practices failed to distinguish science from pseudo-science, since both science and the supposed pseudo-science used auxiliary hypotheses to defend the core of their Research Programs. Indeed, the use of auxiliary hypotheses could potentially serve to advance and develop a theory's overall understanding of the world. The discovery of an eighth planet disturbing the orbit of the seventh outer planet, in Lakatos's thought experiment, would have been a positive addition to that Research Program's understanding of the solar system while retaining some sort of partial continuity with the original core theory of planetary motions.

To deal with Popper's actual formative problem – the uncritical behavior of the Marxists – what Popper's Question asked for was a 'prior specification' of a type of counter-evidence that could not be 'explained away' after-the-fact of the observations and experiments. He was asking for a type of counter-evidence that could not even be 'made sense of' from within the research program, come what may. In the mature understanding of Popper's Question this is equivalent to asking the proponents to specify how to verify *a contrary type of phenomenon*. In the mature understanding of Popper's Question this is equivalent to asking the proponents to specify how to verify *a complementary phenomenon*. The Surprising Answer to Popper's Question is that they can't answer the question from *within* – in terms of the One (unified and uniform) conceptual system of the research program being questioned. The contrary type of evidence or behavior can't be made sense of, is conceptually incoherent – 'irrational' – in terms of the queried research program. Selfless behaviors are incoherent, incomprehensible – irrational – in terms of the rational world posited by Universal Selfishness.

Popper's mature challenge was to say that if you couldn't answer this question then you weren't making a distinction. And if you weren't making a distinction you weren't saying anything, you weren't making a meaningful statement. If you couldn't answer the question, if you couldn't specify how to verify a complementary type of phenomenon, then your theory isn't falsifiable. If your theory isn't falsifiable, then you are not making a distinction and your 'proposed' statement about the world doesn't say anything, lacks content, isn't meaningful and should not be taken seriously.

The mature Popper's Question formulation presents a different, more powerful formulation of what it means for a theory to be falsifiable, more powerful than was suggested in the naive falsificationism version of Demarcation.

Popper's Question entails that all meaningful theories must have meaningful alternatives. Every meaningful theory – any theory that has content, making a meaningful statement about the world – must involve, by its very nature, some sort of limiting idealization. There can be no meaningful theory that is universally true, 'objectively true', able to provide a complete and consistent description of reality, of all possible types of phenomena. For every meaningful theory that says something meaningful about reality there must be a logico-mathematically incommensurable, conceptually discontinuous, yet equally meaningful, theory. The inescapable conclusion is that all scientific theories must be false in objectivist terms. They must be false in the sense of involving, by their very nature, idealizations – incomplete descriptions, qualitatively, conceptually incomplete conceptions of reality.

Just as the wave-particle question couldn't be settled objectively, so too, according to under-determinism, it would be impossible to arrive at scientifically objective answers to questions such as whether human nature is selfish or selfless, whether socio-political reality is completely and consistently competitive or cooperative. This was not because we needed greater observational perspicuity, but 'in principle,' because reality isn't objective, isn't universally, uniformly one way or the other, because people aren't universally, uniformly one way or the other. Everything is middle-ground.

Whether it is 'better' to understand a particular problematic circumstance in terms of one perspective or the other, or whether it is 'better' to understand a particular experimental result in terms of the particle or wave conceptual frameworks can be compellingly obvious in specific situations. These distinct alternatives can be locally competitive and still be globally compatible. Local successes don't force us to extrapolate to the conclusion that just this one of the alternative perspectives provides us with a universal, complete and consistent account of all possible phenomena in all possible circumstances. The post-scientific position must be that each successful alternative must be a limited, special case within some More General understanding of reality.

One crucial point about under-determinism and complementarity, and the source of much confusion and frustration, is that these claims can't be made sense of in traditional, classical, scientific terms. For instance, it is incorrect to say that the quantum 'uncertainty' is 'objective' because, if the uncertainty is irreducibly 'real', it undermines the possibility of making sense of it from within a Research Program that is defined by the presupposition that reality is One, uniformly 'objective' mechanism. Under-determinism and complementarity can't be understood as claims about 'objectivist' reality, as commensurable claims, from *within*, in terms of, the Scientific Research Program and the Mechanical Philosophy. To understand under-determinism and complementarity coherently we need a post-scientific, post-mechanical More General Theory.

Although it is difficult for a thoroughly committed ideologue, most of us can experience the limits of our favored, or traditional, way of understanding reality and our place in it. Most of us live in the uncomfortable middle-ground of aporia, puzzled by the diversity and changeableness of both reality and our beliefs. It seems, at least on good days, that there must be a More General Theory, some more general understanding of reality and some more general understanding of our place in it.

22

Complements, Conjugates and Contraries – Yin-Yang

To say that all scientific theories, indeed, that all fixed ideological belief systems, are limited seems naturally to suggest that they are limited understandings of something more general. It is worthwhile forming an initial hypothesis as to how it might be possible to understand, in a More General Theory, how and why all scientific theories must be incomplete (viz. false in the objectivist sense); limited, special cases, always involving some idealization.

To observe, to focus, to pay attention, to think in one way may well, per hypothesis, preclude observing in another, different, incommensurable way; may well preclude observing in another complementary way. It seems common sense that when we observe, we observe less than the whole of the universe, in both spatial and temporal senses, making all actual observations piecemeal and selective. All observation methods separate out what we observe from the whole and in so doing isolate that portion of the universe from what is outside of that portion. This selective isolation constitutes, by its very nature, an idealization. Furthermore, to observe in some specific way, by some specific method, is to understand, to conceptualize, the isolated system in a particular way. To observe is to conceptualize, is to idealize – is to falsify, in the sense that the nature of what is observed is not completely captured by that one conceptualization, by that one way of observing. Per hypothesis, if observation is conceptually and qualitatively selective, then the *type* of each observation, by its very nature, is not representative of the nature of the whole system, is not representative of the nature of the whole universe.

No supposedly observable 'isolated system' is ever completely isolated spatially and temporally, except, perhaps one might imagine observing the entire universe all at once as an 'isolated system'; as in the Spectator's idealized perspective as a completely detached observer. Isolation, per hypothesis, is always an idealization. And the way we choose to observe, the experimental setup we use, the way we seek to understand, to observe a system, is never the only way. How we choose to observe and understand a situation, a system – possibly imagined, per hypothesis, to be 'out there,' detached, 'in itself' – is under-determined by the nature of the system, by the nature of its 'in itself' reality.

The notion that there are different types of observation, different types of evidence and that no one of them is representative of the whole, isn't a new idea. In a parable originating in ancient India, several blind men are observing an elephant. Being blind, none of them can see the whole elephant. One feels the side of the elephant and concludes that the elephant is like a wall. Another feels a leg and concludes that the elephant is like a large tree. Another feels the trunk and concludes that the elephant is like a snake. Another feels the tail and concludes that the elephant is like a rope. The reports of each of the blind men are true and repeatable, but being limited, their observations are not representative of the whole. Each blind man's experience is, by its very nature, characterized by how and where he observes. Each of the blind men's observations is of a different *type*, leading each of them (reasonably) to a different conception of the whole. The parable also tacitly suggests that each blind man, at least initially, presupposes that there is One, correct understanding of the whole, One, correct conception of the whole. Each tacitly presumes that there is 'one right way' to observe, that there is one *type* of observation, that there is one uniform and consistent *type* of phenomena, that there is 'one right answer.'

In the Modern Science Tradition, prior to quantum theory, it was imagined that scientific theories, as evidenced by the success of their predictions, were advancing by consistently converging, closer and closer, as approximations, to One, objective, uniformly coherent reality. In the tradition that supposed that the language of nature is logico-mathematical, the approximations, the betterness of the theories, was imagined to be expressible quantitatively, evidenced by the increasing quantitative accuracy of various predictions. Then complementarity showed that there wasn't just one path to an understanding of reality, and, consequently, that there wasn't just One uniform, quantitative measure of the betterness of our theoretical understanding. This undermined the Spectator representation of inquiry as one single Research Program, uniquely, quantitatively, and continuously approximating, converging to One, observer-independent, 'objective' reality.

Once complementarity is embraced, 'objectivity', in the sense of 'the only right way' to observe or to conceptualize reality, is lost, so that what you observe, what you experience, depends, to some irreducible extent, on when, where and how you observe. All observation is now, somehow, to some irreducible extent, to be understood as inherently observer-dependent, although not completely observer-determined.

Complements as Conjugate

In the development of quantum theory, Heisenberg was, eventually, clear and explicit that the Uncertainty Principle could be derived directly from the fact that theoretical variables – observables, measurables – were conjugates. The most well known conjugates in quantum theory are position and momentum. To observe one member of a conjugate pair precludes, or at least inherently restricts, observation of the other member of the pair. There is a trade-off between observing position and observing momentum. You can't observe both with complete accuracy at the same time. How strange!

Having made a concerted effort to study all the sciences, encountering all sorts of observables, and having heard Heisenberg's remark about conjugate variables, it struck me as curious that anyone would ever have suggested that variables – measureables, types of observation – are conjugated, are oppositely paired. I wondered, 'where did this notion that observables are conjugated come from, when did it start? In the parable of the blind men and the elephant, the observations are different but not oppositely paired. My initial presumption had been that it had started with quantum theory. But Heisenberg's remark seemed to be appealing to some more general context.

So wearing my hat as a historian of physics, over a number of years, I kept looking into the origin of the idea that observables were conjugates. I asked my physicist colleagues. No one was clear. 'Maybe it started with Hamilton,' several responded. I was genuinely surprised that even among physicists no one seemed to be clear as to the origin of the notion of conjugates.

Kuhn had warned that the real history of ideas and learning might be contrary to the official pronouncements of the Scientific Litany. Kuhn went so far as to suggest that being educated into the tradition of Modern Science was a process of indoctrination. This process applied equally to scientists and professionals in the history and philosophy of modern science.

The breakthrough, for me, came as part of my obsessive questioning of my physics colleagues. In a conversation with Portland State University physicist, Jack Semura, he casually responded, as if it should be obvious, that conjugates weren't new to the modern era, didn't start with the quantum theory or with anything in the new physics. Semura pointed out, "All the measureables in Newtonian physics are conjugates." The fact that many of the most fundamental observables, characteristics of reality, are conjugated, and that this didn't start with the new physics was an important clue. Conjugates turned out to have a remarkably long history. Ancient Greek scientists called them 'contraries.' For instance, the modern contrariness of rest and motion, as well as of constancy and change, were clearly recognized by the ancients. The Ancient Greek scientists had lists of contraries. Although not always easily correlated with the concepts of the Ancients, the new physics likewise has lists of conjugate variables. The recognition that the notion of contraries in nature was ancient also supported Capra's thesis in his *Tao of Physics:* that the recognition of complementarity in quantum theory was a re-discovery of ancient understandings. The Eastern characterizations of the yin-yang as 'interpenetrating opposites' strongly suggests that they are co-defined conjugates in the sense adopted by de Broglie, Heisenberg and Bohr in the formulation of complementarity in quantum theory. In the ancient Western tradition the same insights are prominent in Pythagoras and Heraclitus, and, are expressed quite generally in the Western aphorisms as to the wisdom of 'balance' (viz. nothing in excess).

What was perhaps completely new and striking with Heisenberg was his realization that the *methods of observing* a system are themselves conjugated. To observe in one way – making one type of measurement, using one *method of observation* – somehow inherently limited the possibility of simultaneously observing the system by the opposite conjugate *method of observation*. That the products of observing – the observations, the measurements – are conjugate was surprising to Modern Science. But what is particularly remarkable, the much more important point to be grasped, is that the actual methods of observation, the specific experimental setups and procedures, are themselves conjugated. In other words, *the organizational structure* of the experimental setups are conjugated, are complementary. The very actions involved in setting up these opposite experimental arrangements are conjugated, are complementary.

A plausible guess as to why all this was surprising, why it was there (as Semura pointed out) and yet seriously underappreciated, is that conjugate variables are not reducible one to the other, not translatable into one conception. Consequently, conjugate variables can't be made sense of in terms of the presuppositions of the single uniform, logico-mathematically continuous reality expected in the Mechanical Philosophy. The further expectation is, then, that if there is 'one right way to understand', then there must be 'one right way to observe' (viz. in principle, in the final analysis). There should be one universally right *type* of scientific method, one right (logico-mathematically consistent) path to the truth, represented by the Scientific Research Program. The reason, I suspect, that most of my physics colleagues were largely clueless about the centrality of conjugates, was that, as Kuhn expressed it, they had been indoctrinated into the 'enforced' litany of the Scientific Research Program.

It is crucial to grasp that the claim that measurables are conjugated, that the methods of observation are conjugated, cannot be made sense of as a claim about 'objective' reality, 'out there'. It cannot be made sense of within a framework defined by the presupposition of a uniform, 'objective' reality. It cannot be made sense of in the Mechanical Philosophy. Rather, it is somehow a novel realization and self-realization about the nature of measurement and observation, about how we observe and act in reality, as Participant observers and actors embodied and embedded in reality.

For traditional 'objectivist' Modern Science, Heisenberg's revelation was that all observation necessarily requires the observer to make a selective choice of how to observe among real, conceptually distinct, options. Quantum theory cannot be made sense of from a Spectator perspective. Accepting the embrace of complementarity in quantum theory forces us toward a Participant-observer framework. To make sense of the nature of reality, what it is and how it develops, as observer-dependent to some irreducible extent, requires a More General Theory.

If both the results of observation and methods of observation are inherently conjugate then any actual observation always requires the observer to choose, and the choice constitutes a bias. Since the conjugates, the complementary alternatives, are conceptually discontinuous, qualitatively different, the choice is a qualitative choice. To observe is to manifest a value preference, actualizing one type of observation, bringing about, actualizing, one *type* of future rather than another.

Heisenberg's teacher and colleague, Niels Bohr, certainly didn't miss the tacit implication of the broader and grander context of complementarity. In being honored for his monumental contributions to quantum theory, Bohr was inducted into the Order of the Elephant by Danish king, Frederick IX. As part of this process Bohr was to design his own family coat of arms. As the central image Bohr chose the ancient Taoist symbol of the yin and yang and the motto – in Latin *contraria sunt complementa* – contraries are complementary. Fritjof Capra was not the first to develop the theme that the modern embrace of complementarity in the new physics was a sort of rediscovery of ancient Eastern wisdom that contraries – yin and yang – are fundamental aspects of the overall structure and function of reality. My eventual realization that contraries – conjugates, complements – were also a core-defining theme in the Western tradition in Ancient Greek Science (and Philosophy) simply broadens the sense that complementarity is somehow a fundamental characteristic of the structure and functioning of reality.

Bohr's Coat of Arms

Popper's Question, in its mature version, beyond naïve falsification-ism, is simply asking for an acknowledgement of your bias, asking you to reveal the bias of your choice by specifying how one might observe, how one might verify, a different, formally opposite, type of bias. If you are telling me that your understanding of reality does not involve a bias you are telling me nothing, not making a meaningful claim. Popper's Question is asking, perhaps, for the yin to your yang orientation or the yang to your yin orientation; animate to inanimate; male to female. Trying to convince me that you are observing in the 'one right way', revealing of the complete representative conception of the nature of objective reality, isn't credible. That proposed type of claim isn't acceptable by Popper's criterion as a meaningful claim about reality. Complementarity undermines the 'one right answer' premise of the Scientific Hypothesis and the universal applicability of the Law of Excluded Middle. Complementarity is what forces us to accept that the observer has a choice, must choose, is always, naturally, necessarily choosing.

All Theories are False

I shared with Lakatos my initial shock and dismay with the rebel's notion that all scientific theories are false. How could we know that without testing them? Lakatos took me the next step, the step beyond the naive falsificationist sense of refutation and falsification. "Every theory is false at conception," he said. "If you waited for a perfect theory, one without any counter-evidence, you would never get out of the laboratory to propose anything. Every theory, at the time it is proposed, has counter-evidence against it." This makes naive falsificationism sound rather ridiculous, with its expectation that we should take a theory to be true only until proven false by some one test that it fails. They are all false – that is, they are all incomplete as representatives of 'objectivist' universal truth – from the beginning, by their very nature. All observation and all conceptualization involves idealization, how to observe and how to understand a system (or person), involves a choice, a conjugate bias, that entails incompleteness and embodies uncertainty in the theory's limited conception of reality.

If you think of the universe as analogous to a single electron, the point is that you can make sense of the particle aspect of the universe in a particle research program, but you can't make sense of the wave aspect of the universe in a particle program. And likewise, you can make sense of the wave aspect in a wave research program, but you can't make sense of the particle aspect in a wave research program. Similarly, thinking of the soccer game or Oakeshott's society, you can make sense of the competitive aspect within a competitive research program, but you can't make sense of the cooperative aspect within a competitive research program. And likewise, you can make sense of the cooperative aspect within a cooperative research program, but you can't make sense of the competitive aspect within a cooperative research program.

The Actuality of Complementarity and The Parallel Hypothesis

Just as there were two paths to complementarity in the 20th Century, one in philosophy of science and one in the new physics, so too there were two mutually supportive paths to what I have referred to as the Parallel Hypothesis: that there is a parallel between one's philosophy of science and one's theory of the nature of reality.

Popper's Question is not merely an abstract consideration, as if we were standing outside the universe, like a Spectator. The Surprising Answer to Popper's Question makes, or at least entails, a 'factual' claim about reality, about the universe, about any universe where meaningful theories are possible; about any universe where falsifiable theories are possible. For a theory to be meaningful, the possibility of the experimental demonstration of the limits of the theory must be *actual*. The possibility of providing an answer, specifying how to confirm a complementary phenomenon, a complementary type of measureable must be actual, an *actual* characteristic of the nature of reality. Popper's Question asks you to specify how one could *actually* verify a type of phenomenon that cannot be made sense of in terms of one's initial, possibly ideological, type of observation and theory. Popper's Question asks you to pre-specify a contrary, conjugate, complementary type of phenomenon.

To say that reality must actually have this complementary characteristic is equivalent to, or entails that all observation, all conception, all theories involve a choice. And according to the reasoning of Popper's Question, in any universe where meaningful, falsifiable theories are possible, the nature of reality must actually have this complementary characteristic and observational choice must be possible. If complementarity is a ubiquitous aspect of reality, then choice – however it is to be understood – is also ubiquitous.

This reasoning from Popper's Question to the nature of reality constitutes one path to what I have referred to as the Parallel Hypothesis. The question of the nature of reality had previously been thought of as an 'objective' scientific question. In clarifying the requirements of meaningful inquiry – of falsifiability – Popper's Question entails that reality must actually have an irreducibly complementary nature.

The Other Path to the Parallel Hypothesis

Quantum theory isn't just about the nature of reality. The Parallel Hypothesis can be reasoned from within the new physics by pointing out the implications of the embrace of complementarity for the philosophy of science. Complementarity in quantum theory requires a new understanding of the nature and representation of inquiry and a new understanding of the nature and representation of the inquirer. Complementarity forces us away from the Spectator representation of inquiry and away from the scientific understanding of the nature of the universe and our place in it – pointing us unavoidably toward a Participant Theory. Inquiry into the nature of reality must now, somehow, self-referentially include inquiry about the inquirer, inquiry about the place of both inquiry and the inquirer in the new post-objectivist understanding of the nature of reality.

The reason you can't find an 'objective' (observer-independent) universally true scientific Theory of Everything is because reality isn't like that. The limitation (the uncertainty) isn't because of any lack of evidence or due to the inadequacies of observational techniques.

The most striking implication of quantum theory for the philosophy of science has to do with choice. The new enigmatic image is of a universe composed of particle phenomena, wave phenomena and Participant observers necessarily choosing among available alternative ways of observing. Embracing complementarity, the quantum 'situation', makes choice a necessity. The choice is, by its very nature, under-determined in terms of any one of the alternative, classical scientific ways of understanding reality. Complementarity, as it addresses the Promissory Note Argument, requires that the sum of all possibly successful, yet incomplete, scientific ways describing reality can't produce 'One' scientific coherence – can't result in 'One' conceptually continuous Scientific Theory of Everything; 'One' logico-mathematical consistency that could erase the essential difference, the essential opposition of the complements.

The 'choice', however it is to be understood and explained, cannot, by its very nature, be understood or explained scientifically, as being 'caused' by some one or the sum of several mechanical 'causes'. The choice must be, to some irreducible extent, mechanically arbitrary, possibly chance-governed, at least from the point of view of all possible scientific theories. The choice must be – always to some extent, by its very nature – mechanically under-determined, causally free.

Paradoxical Nature of Advances in History of Science

Both Bohr and Heisenberg emphasize the conjugate – the complementary – nature of observation. Since observing in one way precludes, or at least restricts, observing in other ways, and since characteristic observations and techniques literally define paradigms, as Kuhn emphasized, the actual history of successful inquiry, per hypothesis, the actual history of science, must be more complex and nuanced than the logico-mathematically consistent and conceptually continuous expectation of the Logical Positivists and the Mechanical Philosophy.

What has been demonstrated in the new quantum physics – that there are at least two complementary orders governing 'reality' – is precisely the evidence that shows us *NOW, come what may,* that the Scientific Research Program is inherently incomplete, inherently limited. This entails that the Logical Positivist's attempt to 'rationally reconstruct' the history of science ('real' inquiry) as one logico-mathematically consistent and conceptually continuous enterprise must fail.

The embrace of complementarity suggests that the demonstration of the limit of a research program, of a successful way of understanding and inquiring, must arise as a confusion, as a paradox – as a situation of systematic, irreducible incoherence. In other words, the counter-evidence to a coherent, rational order must appear initially as an incoherent, irrational order. The moment of recognition of the limit comes with the acceptance that there are other, conceptually discontinuous, yet successful ways of understanding and inquiring. In the history of the new physics it is the moment of recognizing that we have two highly successful complementary rational orders – the Newtonian and the Maxwellian – that cannot make sense of each other. Complementarity is, then, in a sense, more of a negative statement – denying and rejecting the notion of One objective reality. Oxymoronically, paradoxically, we have 'two objectivities'; we have two 'one right answers.' There is no common denominator, no possibility of a rational 'reduction' to some new One 'objective' order. Lacking the possibility of a reduction to a new type of 'objective' mechanical order, the only plausible line of theoretical inquiry is to seek a More General Theory that is able to understand these 'putative objectivities' as limited special cases.

But where is the new positive understanding, the new More General Theory? Struggling with the challenge to come up with the new understanding – along with many, many others – I tried to give at least an interim description of the 'situation.' My tentative formulation was that: Apparently you can *experience* and even demonstrate the limit of a way of understanding reality in terms of that way of understanding (viz. paradoxically) – without being able *to understand* the limit, without being able to make sense of the limit (viz. resolving the paradox) in terms of the original way of understanding. You can experience the limit paradoxically, but you need a More General Theory to resolve the paradox and understand the limit. But this doesn't seem to automatically generate a new theory of reality, a new understanding. From the scientific perspective, we are stuck.

In trying to move forward it is important to recognize that complementarity can't be made sense of as an 'objective' feature of an 'objective reality' – 'out there'. From within the scientific conceptual system there is no longer any way to completely and consistently – unambiguously – refer to 'reality', to refer to 'the universe'. The specific scientific, mechanical nature of reality – particle, wave or whatever – is now, as in the perennial disputes, 'an essentially contestable concept.' The 'objectivist' question is formally undecidable.

With the embrace of complementarity my initial Spectator representation of Hawking's activity as a scientist must be transformed, reconsidered and understood in a new way. Hawking can't be discovering the objective laws governing all phenomena in the universe, determining the inevitable course of events for the simple reason that the new physics has demonstrated there are no such laws. Even if Hawking thinks that is what he is doing, it can't be what he is 'really' doing. With the embrace of complementarity, 'real inquiry', Hawking's inquiry, his personal history of science, cannot be understood, cannot be rationally reconstructed, either as an attempt to converge or as an actually successful convergence to a complete and consistent understanding of an observer-independent 'objective' reality.

The resolution of the question of how to make sense of the two, apparently incompatible, ways of understanding Stephen Hawking (and ourselves as inquirers) seems increasingly to be inseparable from the discovery of the More General Theory – one that can provide a new coherent understanding of both the Spectator and Participant representations.

23

Talking Across
Complementary Paradigms

In my original formulation of the question of the limits of science, applying Popper's Question to the Scientific Research Program, I was asking what type of evidence, if it were to occur, could not possibly be explained by any future scientific theory, by any theory that was consistent with the research program's core belief, with the Scientific Hypothesis: that all phenomena are governed by One time-space invariant order.

This was the approach I had proposed to Jonathan on the ride to Eugene. I want to know *NOW*: what evidence (the nature/type of the evidence), if it were to occur, would force any reasonable person to abandon the claim that his core hypothesis represented a complete understanding of the situation or system in question. What I challenged Jonathan (and Hawking) to come up with was the nature of the evidence that would force them to conclude that the Scientific Research Program was inherently incomplete.

What Popper's Question is asking for here then is a phenomenon that, by its very nature, couldn't be explained, predicted or even made sense of within the conceptual framework of the Scientific Research Program; couldn't be explained or made sense of by any possible future scientific theory – come what may.

Popper's Question is asking for a 'prior specification' of how to verify a contrary *type* of phenomenon, a complementary phenomenon. The demonstration of a complementary phenomenon would address the original

Promissory Note of the Scientific Hypothesis. The Promissory Note asserted a 'faith' that there would always be a *scientific* theory that could supersede and correct the failings of any prior scientific theory. In principle, this sequence culminates with the scientific Theory of Everything that would be able to explain all phenomena as being of One logico-mathematical type, all phenomena as being governed by One universal, mechanical order.

But in asking for specification of how to verify a phenomenon that is contrary to *all possible scientifically explainable phenomena*, it sounds like Popper's Question is asking for a type of evidence that is non-scientific evidence – whatever that might mean.

The reasoning, from the Surprising Answer to Popper's Question, helps us understand that contrary phenomena are conceptually discontinuous so that one cannot be understood in terms of the other; one cannot be reduced to, generated from, or expressed in terms of the other. A convenient way to say this is that the irrational cannot be expressed in terms of the rational. What is irrational cannot be 'made sense of' in terms of what is rational.

Since the Scientific Hypothesis postulates that all phenomena are governed by One universal mechanical order – the scientific rational order – phenomena that are scientifically irrational would be phenomena that, by their very nature, are non-law-governed. Evidence of non-scientific phenomena would be evidence for non-law-governed and, at least plausibly, chance-governed phenomena.

For a phenomenon to be 'law-governed' means that it is 'governed by a regular, causal relation.' 'Regular' means repeatable, means space-time invariant, as in the clockwork image. 'Chance-governed' is one natural way to express the conceptual contrary, the irrational opposite, of 'law-governed'. The demonstration of the limit of science, of the inherent incompleteness of the Scientific Hypothesis, would be verification of phenomena that are – by their very nature – causally independent, chance-governed, as perceived, as experienced, as 'understood' as the opposite of law-governed; that is, from *within* a clockwork causality point of view. This is one way to formulate the answer to Popper's Question as applied to the Scientific Hypothesis.

Since 'chance-governed' is the conceptual opposite of 'law-governed', a chance-governed relation is – in clockwork causal terms – an incoherent, 'irrational' relation. Chance-governed phenomena cannot be made sense of in a completely and consistently law-governed universe. And law-governed phenomena could not be made sense of in a completely and consistently chance-governed universe.

Then, in the research leading to quantum theory, chance-governed phenomena were demonstrated to be 'real' – undeniably, irreducibly. What happened in quantum theory was indeed 'surprising' in terms of classical, law-governed, clockwork physics. De Broglie, commenting on the research leading up to quantum theory, said: 'In our search for the classical causal laws governing material reality, we were forced to the conclusion that there were phenomena that – by their very nature – were governed by chance.'

The proponents of the classical 'causal' Scientific Hypothesis offered a 'faith', a Promissory Note, that all these 'apparently' chance-governed phenomena could eventually be understood, could eventually be explained and could eventually be predicted, at least in principle, in terms of some future classically causal, law-governed clockwork-like Theory of Everything. There would then be no 'real' chance-governed phenomena. There couldn't be; not in any universe completely governed by the universal causal maxim; not in any universe consistent with the Scientific Hypothesis.

What actually happened in physics, leading to quantum theory, was the result of the failure of the interface of two, largely independent, highly successful research programs: Newtonian particle mechanics and Maxwellian wave mechanics. In order to be able to integrate these two programs one would need to find a type of phenomenon that could serve as a sort of 'common denominator' of both particle phenomena and wave phenomena. However, it soon became clear that these two types of phenomena were contraries, inter-defined opposites – a wave is a non-particle and a particle is a non-wave. The wave is non-local (viz. locally indefinite; distributed) in both space and time, whereas a particle is local (viz. locally definite) in both space and time. Particles do things that waves cannot and waves do things that particles cannot. A wave is not a type of particle and a particle is not a type of wave.

The reasoning from Popper's Question helps us understand the implications of these phenomena being conceptual complements. The Newtonian and Maxwellian phenomena are 'irrational opposites', so that one cannot be expressed in terms of the other. Since, from the point of view of Newtonian clockwork causality, the Maxwellian wave phenomena were formally incoherent and causally 'irrational', Max Born recognized that they could be represented as non-law-governed (viz. in the Newtonian sense), and so chance-governed. In the maturing years of quantum theory, Born proposed to reinterpret the Maxwellian electromagnetic wave phenomena in terms of probabilistically distributed charged particles (viz. electrons). Accordingly, the electron, could, in practice, be represented as a non-localized particle. When using Schrodinger's wave equation, this distributed particle, this 'irrational opposite' had to be represented, at least prior to being observed, as being probabilistically distributed in both space *and time*. According to this representation there is a certain probability of an observer finding (actualizing) the particle as a Newtonian-like particle – localized – in one or another 'location' in the space-time field. Trying to give a clear understanding of reality in these terms remains a challenge. If you find this approach to representing reality confusing – welcome to the club.

Of Clouds and Clocks

Another way to 'understand' the quantum 'situation' along these lines is to consider how one might represent the opposite of the classical scientific clockwork image. In a clockwork, everything is connected, correlated and coordinated and synchronized. The scientific maxim 'same cause, same effect' holds everywhere and always in space and time.

The opposite of the clock image is cloud-like where the parts, the particles, are causally independent; the motions of the individual particles are uncorrelated, uncoordinated and non-synchronized. In such an idealized cloud, the motions of the particles in relation to each other is random –

non-causal in the clockwork sense. Such idealized, cloud-like motion cannot be made sense of in clockwork terms.

Using the images of clouds and clocks, Popper rejected the Law of Excluded Middle approach to the order-disorder discontinuity, suggesting a possible middle-ground: "My clouds are intended to represent physical systems which, like gases, are highly irregular, disorderly, and more or less unpredictable. I shall assume that we have before us a schema or arrangement in which a very disturbed or disorderly cloud is placed on the left. On the other extreme of our arrangement, on its right, we may place a very reliable pendulum clock, a precision clock, intended to represent physical systems which are regular, orderly, and highly predictable in their behavior. ... [The] ontological thesis that there cannot exist anything intermediate between chance and determinism seems to me not only highly dogmatic (not to say doctrinaire) but clearly absurd; and it is understandable only on the assumption that they believed in a complete determinism in which chance has no status except as a symptom of our ignorance." (Popper, Karl, "Of Clouds and Clocks: an approach to the problem of rationality and the freedom of man", The second Arthur Holly Compton Memorial Lecture, presented at Washington University on 21 April, 1965. Reprinted in *Objective Knowledge* (1973) pages 207, 228-229).

The crucial point is that the 'particles' in an idealized cloud must 'really' be as causally independent as the parts of an idealized clockwork must 'really' be causally dependent. A thought experiment helps. Consider a room full of coin flippers. Each outcome – heads or tails – for each flipper is independent of each flipper's last outcome (historically in time). Moreover, and this is the key point, the pattern of the sequences of the different individuals are uncorrelated. If the person next to me flips a heads, this is completely independent of whether I just flipped a heads or tails. The individual outcomes of the flippers are causally independent – unconnected, uncorrelated, uncoordinated and non-synchronized.

If we try to think of the cloud as the complete and consistent objective representation of reality, it isn't clear that we are talking about the behavior of 'particles' anymore. In the Newtonian representation particles, by their very nature, have definite locations and definite trajectories. A Newtonian particle, by its very nature, can't be either non-local, probabilistically distributed or moving about randomly with an indefinite trajectory. In the idealized cloud, the action of one particle cannot have any definite predictable causal effect on any other particle. This is the same as saying that the behavior of one coin flipper doesn't causally influence, can't be used to predict, the outcomes of the others. One is inclined to conclude that in an objective, complete and consistent cloud the 'particles' don't *causally* interact at all – at least in any way that can be understood in classical Newtonian terms. And, lacking definite predictable locations, moment-to-moment and place-to-place the particles seem to dissolve into a probabilistically distributed field-like reality.

The cloud is very similar to the paradigmatic image of 'the ideal gas'. There is an ongoing ambiguity in the ideal gas model as to whether the particles are 'really' behaving randomly or whether we just treat them 'as if' they were behaving this way because of the difficulty of observing each of the enormous number of 'supposedly' Newtonian clockwork trajectories. If they are truly random then 'in principle' even a Laplacean Supermind could not follow the trajectories and predict where each particle was to be in the future or retrodict where it had been in the past.

Popper's Question suggests that the evidence for the limit of the Newtonian clockwork should be the demonstration of a type of phenomenon that can't be made sense of in definite particle terms, in clockwork-like causal terms. If the Scientific Hypothesis presupposes causality – the causal maxim that every phenomenon has a specific cause – then the opposite type of phenomenon is one that, by its very nature, doesn't have a Newtonian-like cause, and the opposite type of universe can reasonably be represented as one where everything is causally unrelated, causally independent. This line of reasoning leads us from a Newtonian clockwork to the irrational opposite, expressed incoherently in Newtonian terms, as a cloud, a sort of irrational, incoherent clock, one that doesn't give us One universal, objective time; one that doesn't have One definite, spatial structure.

The Promissory Note from the Scientific Hypothesis, expressed in the Newtonian Research Program, was that all 'apparent' chance-governed phenomena could eventually be understood in clockwork terms. However, the new physics has demonstrated that there can be no common denominator to reduce the highly successful Newtonian and Maxwellian Research Programs to either one, unified, causally dependent, or one, unified, causally independent whole. Since the Maxwellian program is formally 'irrational' in Newtonian law-governed particle terms, it was reasonable for Born to introduce an idealization, to interpret, in practice, Maxwell's field of electromagnetic waves as probabilistically, space and time, chance-distributed, charged particles.

Chance became real for the Scientific Research Program, if one begins with the classical Newtonian formulation, with its encounter with the demonstrably successful, per hypothesis complementary, Maxwellian Research Program. This is one way of explaining the consensus conclusion that, with the new physics, the Scientific Research Program discovered 'an irreducible probabilistic' component of reality. And this 'irreducible chance' component could not be explained or understood 'rationally' in terms of the commitment of the Scientific Hypothesis to deterministic law-governed, clockwork causality.

What is perhaps most important to emphasize here is that these claims are not claims about 'objective' reality. De Broglie's claim that we discovered 'phenomena that are, by their very nature, chance-governed' is not claiming that 'chance-governed phenomena' are 'objectively' real. And quantum theory does not force us to conclude that the 'irreducible probabilistic component' is an 'objective' feature of reality. The quantum 'situation' embraces the strange middle-ground of irreducible order and irreducible disorder. To make sense of the middle-ground we need a post-scientific, post-objectivist More General Theory.

Bucciarelli's Example: Real World Consequences of Complementarity

All this talk of the complementarity of the Newtonian and Maxwellian Research Programs will appear rather abstract – 'philosophical' – to many readers. So I want to draw out and illustrate one very real and practical implication.

Just as political left and right, proponents of cooperation and competition have difficulty communicating, tending to talk past each other, as if speaking different languages, the same sort of difficulty arises between people who speak Newtonian and people who speak Maxwellian.

Take this one step further into the context of real world research and engineering. MIT Professor of Engineering Lawrence Bucciarelli, in his book, *Engineering Philosophy* (DUP Satellite, 2003, pages 15-16), notes that there is a problem of communication between mechanical engineers and electrical engineers. He points out that, as a result of their education and training each sees the world in terms of his corresponding knowledge base. "Each 'object world' language of an engineer is rooted in a particular scientific paradigm, which serves as a basis for conjecture, analysis, testing and designing within that world... The languages of different object worlds are different; their proper languages are different.... The structural engineer speaks of stress and strain; of displacement, stiffness, and load path." Mechanical engineers in general have a traditional Newtonian technological worldview. "In another world... the electronics engineer speaks of power, voltages and currents, analogue and digital, resistance and capacitance." The electrical engineer has a Maxwellian, electromagnetic, technological worldview.

These technological worldviews don't translate into one another. These different worldviews have conceptually different languages, and correspondingly, different approaches to problem solving. In Kuhn's terminology each of these worldviews, embodying different techniques and technologies, operates in a different paradigm. Bucciarelli goes on: "The mathematics may appear similar – there are strict analogies that apply in some instances – but the world of electronics is different, populated by different variables, time scales, units, scientific law and principles of operation. So too, different kinds of heuristics, metaphors, norms and knowledge are codified as tacit know-how."

The problem is not simply subjective in the minds of the engineers. The problem of communication is embodied, in that the languages are embodied in the instrumentation, in the design of the experimental apparatuses and in the conceptual understanding of the observables. Bucciarelli notes: "The elements of an object world language are more than words, more than the symbols and tokens of a proper language displayed by a particular scientific paradigm. I have already pointed to specialized instruments, prototypical bits of hardware, tools, ways of graphically representing states and processes as ingredients of object worlds. These all can be considered linguistic elements for that is how they function."

The practical consequence is that mechanical and electrical engineers have a fundamental difficulty talking to each other. This inability to communicate presents profound problems when they are trying to work together to design electro-mechanical devises. Bucciarelli argues that there is no one elite "that knows the full meaning, has a god's eye view, of the object of design and another group with a less sophisticated, common understanding of the design task. Rather, there are multiple elites, each with its own proper languages. It is in this sense that different participants within different object worlds with different competencies, responsibilities and interests speak different languages. Crudely put, one speaks structures, another electronics, another manufacturing, still another marketing, etc."

A few years ago, I asked Ron Adams, then Dean of Engineering at Oregon State University, about Bucciarelli's characterization. He immediately recognized the theme and told me stories of how this had been a serious problem at Tektronix (viz. Oregon's famous maker of oscilloscopes) where he had worked for a number of years. "The few people who had degrees in both mechanical and electrical engineering were invaluable. They were the only ones who could begin, if not to fully translate, at least to build bridges," he said. The analogy here is that the classical Newtonian mechanical world doesn't make sense, conceptually, in terms of the classical Maxwellian electromagnetic world. You can't logically generate or derive electromagnetic technology from classical mechanical technology – either one from the other. There is no 'God's eye view' procedural manual that would allow you to reason, or calculate, your way from one to the other. They don't translate. They are conceptually and logico-mathematically discontinuous.

This is a real world macroscopic consequence of complementarity. The observations, as well as the practical developments of each research program, don't evidentially conflict because they don't make sense in terms of each other, because, being complementary, they are conceptually discontinuous. Similarly, if all meaningful scientific theories are incomplete, all meaningful scientific theories – by their very nature – must be special cases, having limited, non-universal applicability. They can only be understood in some new way as unified, possibly, per hypothesis, from the perspective of a More General, post-scientific understanding of the nature of reality and our place, the place of the Participant inquirer, in that reality.

If one adopts this understanding of Popper's Question as entailing actual complementarity, the various sorts of scientific knowledge would each be technologically, experimentally 'repeatable' only in a limited range of applicability, not true in all possible observational setups, not true in terms of all ways of observing. This injects an irreducible element of relativity – observer-dependence – such that your experience of the universe will vary depending on where and when and how you observe. Abandoning the notion of 'One universal objective order' in favor of multiple, conceptually discontinuous 'objectivities' seems to push toward a non-universal time aspect and a non-universal space aspect to all observation. In other words, what you can observe depends to some irreducible extent on where and when you observe.

Born's reinterpretation of the laws and phenomena of Maxwell's Research Program in probabilistic terms is certainly 'understandable' in light of the philosophical understanding of complementarity. But it sounds, by analogy, rather like saying that there is a chance-governed cloud of cooperation surrounding every definitely competitive behavior, or a chance-governed cloud of competition associated with every definitely cooperative relationship. Curiously, these actually might be helpful ways to express the relation between cooperation and competition in the soccer game or in Oakeshott's middle-ground society, or in the economic theories of cooperation and competition.

The More General Theory, called for by complementarity, needs to be able to 'make sense' of the successes of both the Newtonian and Maxwellian research programs, but can't itself be a 'unified' mechanical theory. Reality for the More General Theory can't be made sense of as any sort of simple sum of opposite types of mechanics. Einstein pointed out that the Newtonian and Maxwellian Research Programs were presupposing opposite types of reality: "Physics is an attempt conceptually to grasp reality as it is thought independently of its being observed. In this sense one speaks of 'physical reality'. In pre-quantum physics there was no doubt as to how this was to be understood. In Newton's theory reality was determined by a material point in space and time; in Maxwell's theory, by the field in space and time. In quantum mechanics it is not so easily seen." (Einstein, Albert, "Autobiographical Notes", in *Albert Einstein: Philosopher-Scientist*, edited by Paul Arthur Schilpp, Volume One, page 81-83 (Harper Torchbooks edition (1959))

24

Spectators Turned
Participants at Microsoft

The Participant side of Stephen Hawking's life, invisible to the public, became apparent in the development of the arrangements for the Seattle and Vancouver lectures. Hawking's strong connections to Microsoft, as with his relation to Intel, has served as a focus for advancing computing power for people with disabilities. Because of his personal relations, Stephen served as a crucial point of reference and inspiration for Microsoft's early and considerable commitment to developing accessibility features (of its otherwise mass-market software) for persons with disabilities.

I don't recall exactly when I first heard about, or at least took any notice of, Nathan Myhrvold. I think it was Sue Masey who mentioned his name. Myhrvold is one of those names someone speaks, and you hear and could roughly repeat phonetically, but couldn't spell if asked. Stephen had just agreed to the second half of my original invitation for a series of four public lectures in the Pacific Northwest. Portland and Eugene were in early 1992. Now Seattle and Vancouver, British Columbia were scheduled for the summer of 1993.

"Seattle! That's wonderful," said Sue Masey. "Oh, be sure to let Nathan know about it."

"You'll want to be sure to give him tickets," she added.

"Who's Nathan?" I asked.

"You know Nathan Myhrvold," she said.

"Sue, I don't know…" I said.

"Nathan is a friend of Stephen's, and Nathan and his brother Cameron are something or other – I think important – at Microsoft," she said.

"OK, yeah, of course, I'll make sure they get complimentary tickets," I said, and left it at that.

A couple of weeks later, I was talking to Martin Middlewood, my account executive at the high-tech public relations firm, Waggener Edstrom, in Portland. I had been assigned to Martin, or more properly, the Institute had been assigned. Martin was handling Waggener Edstrom's support of the Science, Technology and Society Program in Portland. A couple of years earlier – Waggener Edstrom was still very young at the time – when searching for corporate cosponsors for the STS program I had managed to arrange a lunch with Jody Peake, one of the Vice Presidents. I pitched the STS program in brief. Jody listened and said, "We have been looking for ways to enhance public awareness of our clients' value to the community-at-large." We were definitely on the same page, understanding that new technologies were the driving force transforming modern civilization. I truly valued her attitude and let her know it. It was business but with an uncommonly broad perspective. Waggener Edstrom wasn't just about pushing product and promoting companies; they had the larger vision. Bill Gates, co-founder of Microsoft, is often quoted as saying that the real drive and ambition of the high-tech community – per hypothesis, the Participant-engineering community – is in helping to shape and bring about a fantastic new era in the history of human civilization; a better, more desirable, future. It was that same vision that I was hearing from Jody Peake, and I loved it.

Waggener Edstrom is perhaps best known as the longstanding public relations firm for Microsoft. Melissa Waggener and Bill Gates go way back – to promotional tours for Intel. Melissa Waggener founded the Waggener Group in 1983, working at the time with Microsoft and other technology companies. Melissa had been the one, so the story goes, to get Bill Gates' picture on the cover of Time Magazine in April, 1984. Pam Edstrom joined the Waggener Group that same year, as an agency partner to Melissa Waggener, and the firm was renamed Waggener Edstrom in 1990. Since then Waggener Edstrom has become one of the largest privately owned, full-service public relations agencies in the world. 'WE provides communications services to global organizations focused on innovation.' The agency broadened its client portfolio and range of services with specialized practices for bioscience, corporate, consumer and public affairs issues. In 2005, the agency was renamed Waggener Edstrom Worldwide to better reflect its global operations across the U.S., Europe and Asia.

With Jody Peake's endorsement, Waggener Edstrom became a major, long-term co-sponsor of the Science, Technology and Society series in Portland, Eugene, Seattle and Vancouver, BC. Jody had assigned Martin to be my handler.

Martin was just about the only male I encountered at Waggener Edstrom. My impression was that it was nearly an all female organization. Occasionally, I suspected that Martin was kept on as their token male, and I used to tease him about that. An exceptionally self-effacing sort of person, he joked that he suspected that occasionally as well. There was always an erotic undercurrent in a visit to the Waggener Edstrom offices. Surely, for a Public Relations firm that dealt with a lot of often socially awkward techies, this was no accident. To begin with, the receptionist was always a goddess of the first order.

As I waited in the reception area for my appointment, my mind would wander. I imagined a soft and beautiful voice: 'Welcome. You have been cleared for entry into paradise. You're home now. Relax. Please feel free to enjoy the slight wisp of perfume and the appearance of ridiculously gorgeous women, dressed like models with perfect make-up, walking here and there through the reception area.'

When I was there to talk to Martin, his appearance at reception to lead me back to his office was rather like a splash of cold water in the face. "Wake up, Bristol, this isn't reality," Martin would say. I want to emphasize here that the atmosphere was not sexual, but erotic in a classically Greek way. There is an enormous difference. This wasn't like walking into the Playboy Mansion. The dress and the manner were quite different. It was business – professional.

My agenda in meeting with Martin was to see if Waggener Edstrom, Microsoft's primary public relations firm, could assist me in any way to secure Microsoft as a co-sponsor of Stephen's public lecture in Seattle. Martin made it clear that Wagg Ed could not serve as my conduit to Microsoft. "It's just not politically viable. The relationship doesn't permit us to make that sort of request. They have their policy, and it's their policy," said Martin.

Martin explained to me that there was a delicate relation between Microsoft's internal public relations people and Microsoft's external public relations people – the latter being the folks at Waggener Edstrom. They were all friends and colleagues of course, on the same team, but when the budget and formal responsibilities were assigned there was, unavoidably, a little competitive tension. So Martin explained very politely that there really wasn't anything that he or Waggener Edstrom, in any capacity, could do to help make a connection.

"So who is Nathan Myhrvold?" I asked.

Martin, who had been glancing at something on his computer screen, jerked around in his chair, visibly stunned by the question.

"How do you know Nathan?" he asks in a voice that reflected a genuine and focused curiosity.

"I don't. Stephen does. They're friends I guess. I don't know the whole story," I said.

"Well, if Stephen knows Nathan then that might be a different story," said Martin.

"I can't help you approach Microsoft for a co-sponsorship directly through Waggener Edstrom, but I might be able to help you connect with them if Nathan is favorable," he offered.

"Well, I have been instructed to be sure to let Nathan know…" I said.

"So who is Nathan anyway?" I asked.

"Suffice it to say that he is in Bill Gates' inner circle – say, one among the closest ten. So if you have access to Nathan, you have a chance with Microsoft," said Martin.

Martin promised to check with Jody, Pam and Melissa about the contact and any possible official involvement on their part. I was never privy to the exchanges, but there were conversations. About a week later, Martin got back to me with the name of someone in Public Affairs at Microsoft headquarters in Redmond, Washington and let me know that they had been advised of Nathan's interest in having Microsoft support Stephen's lecture in Seattle. I was to call and make an appointment to visit to discuss details. – Score!

Running the gauntlet at Microsoft.

Microsoft, like most large technology start-up corporations, had adopted a policy of avoiding local sponsorships; indeed, to politely bypass any involvement with the several thousand cultural and non-profit organizations that annually seek corporate contributions. I didn't understand this immediately. I figured that as a substantial corporate presence in the Seattle area one would find Microsoft sponsoring a whole host of events and activities: symphony, opera, theatre and a diversity of functions of charitable non-profits.

But the reasoning became clear. I had encountered it at Intel two years earlier. You can bet that a large portion of the tens of thousands of cultural and charitable organizations in Oregon and Washington approach Intel and Microsoft every year. To adequately respond to these thousands of requests you need to create an entire department and staff it. But that's not the main reason. What is to be avoided is the draining away of time and energy of front-line employees. Everyone has a favorite charity or cultural enterprise. Once a company starts down this road, employees are diverted as they try to influence decisions on who gets funded. Conflicts and competition having nothing to do with the corporate mission begin to cloud the corporate culture. Furthermore, and this is particularly true of fast growing, entrepreneurial companies like Intel and Microsoft, they want their employees' full attention – hearts and minds. So, when I asked whether Microsoft had sponsored any other public lectures in the past, I wasn't surprised to find that the answer was 'no'. – Well, there had been one, an exception to the rule. A couple of years earlier Microsoft had cosponsored a public presentation by

Cokie Roberts – at the time, just about the hottest media commentator on what was happening and what was about to happen. But Roberts had been the sole exception over the last, roughly ten years. Tight rule.

Even then, in the early 1990s, the Microsoft campus was beautiful. By comparison, the Intel campus in Hillsboro was primitive and barracks-like. Of course, Intel in Oregon was a frontier outpost. Intel's main campus was in California's Silicon Valley. As I drove onto the Microsoft campus on a beautiful, sunny, fall morning I felt like thousands of others must have felt – like, 'this would be a nice place to work.'

Every large corporate campus has an atmosphere. The experience, of course, (and your observations) depends on who you are and why you have come there. On this occasion, I felt as though I was suspect. I represented a violation of the 'no cultural distractions' policy – as in 'who is this guy and how did he penetrate the outer fence.' Microsoft doesn't co-sponsor public science lectures.'

I had been given clear directions to find my way to where I was to park and to the right building, where I encounter the first level of security. I am greeted by a young woman with a very serious, no nonsense countenance. 'Who am I? Picture ID please.' Scanned and recorded. "Sign-in here, please. Whom have you come to see?" she asks. "Public Affairs," I say. "Yes, but do you have a name?" she asks. I can't recall now, but I did and provided it. "OK. Thank you. It won't be a minute," she says. She is doing something with a computer interface. Apparently I am not on any serious watch list because no alarms sound.

I was provided with a personalized name badge that identified me prominently as a 'Visitor'. Within three or four minutes, another young, slightly more cheerful woman arrived in this outer security area to greet me – I mean, I am barely out of the parking lot so far. She led me through the next level of security, where her ID was checked as carefully as mine. Following her down a couple of long corridors, we arrived at a medium sized conference room, large enough to hold maybe sixteen to twenty people comfortably. "Wait here. The others are on their way," she said. "Could I get you coffee or a soft drink?" she asked.

Over the next five minutes, the room filled with about a dozen people. Small talk was brief and we were down to business. Of the group I assessed that there were just one or two principals and the rest were their functionaries. I was introduced around. It was all quite pleasant. They were smiling and welcoming. Now that I was inside and expected, I was in the presence of people who enjoyed being part of Microsoft. They made me feel right at home.

In the meeting, my exceptional status was explicitly acknowledged quite openly serving to reduce my paranoia. The feeling was – 'we know who you are and how you got in here – and it's OK.' We all had a bit of a laugh. I thought, 'these are people I can work with.' Despite all the pressure to perform, these people were having fun. It made me think, again, that it would be rather enjoyable to work at Microsoft.

"So..." I start, "I am hoping that Microsoft is interested in cosponsoring Stephen Hawking's presentation at the Seattle Opera House."

"I think that is settled. Yeah, we are in," said one of the principals.

Microsoft's involvement was due to a high-level executive override of the basic policy. Love those 'high-level executive overrides'. That is exactly what Martin had anticipated that Nathan could accomplish.

"Are you in for Vancouver too?" I said. I thought this was an opportune moment to bring that up.

Blank looks. –– "What is happening in Vancouver?" the same principal asked.

I explain that there is a second lecture in Vancouver, British Columbia. They were unaware.

"I think that would need to be Microsoft Canada's decision. All we know about is the support for the Seattle event," said the principal.

There is a little discussion about this. Finally, I am told that I should contact Microsoft Canada directly about an event in Vancouver.

I had sent a formal invitation a few days earlier, covering the Seattle event only, specifying the dollar amount I was seeking and what I had to offer in exchange. The 'benefits to Microsoft' list included a full color, full page, space-ad, featuring Microsoft in the sixteen-page program-magazine to be handed out to the expected 3000 attendees on the evening, a ticket discount for Microsoft employees and the opportunity to introduce Hawking at the event. Finally I mentioned that Stephen would potentially attend a Microsoft VIP dinner. Alternatively, Microsoft could cosponsor a meeting with a group of students with disabilities.

The full page in the program-magazine was acknowledged and instructions were given there and then to one of the functionaries whose job it would be to see that it happened. There was confusion about how to handle a ticket discount for Microsoft employees, but I was assured they would contact someone who knew something about whether that was workable and how to implement it.

"And would someone from Microsoft like to introduce Stephen at the event?" I asked.

"Yes, well, it is our understanding that Nathan will do that," he responded.

"Great. Now there is the possibility of arranging a VIP dinner that Stephen would attend or the meeting with students with disabilities. We have done both these sorts of things in the past," I said.

"Actually, that is already arranged, I believe. Nathan has been talking to Stephen about a dinner and they have something already in the works," said the principal.

I suddenly had this sinking feeling. I thought I was the go-between – the negotiator. But it was now clear that Nathan had already contacted Stephen and arranged everything. I wasn't there to sell Microsoft or to negotiate, but merely to confirm and coordinate.

Nathan, being exceptional and a prominent deviation from the 'organization man' concept, had decided over the past couple of years to become a gourmet chef – in a serious way. He had been taking classes (more like personal training) from Thierry Rautureau, the world famous Chef at Rover's Restaurant. For the uninitiated, Rover's Restaurant was featured in the movie *Sleepless in Seattle*, where Tom Hanks takes his date – the one with the funny laugh – before he later connects magically with Meg Ryan.

Hailing from the Muscadet region of France, Thierry Rautureau, known as 'the Chef in the Hat' because of his ever-present fedora, moved to Seattle and opened Rover's in August, 1987. According to the restaurant's promotional material, "Chef Rautureau's vision is a warm, comfortable dining environment, similar to dining at a friend's house. Rover's is dedicated to professional service, exquisite wine and food, and an intimate environment." Expensive.

After I left Microsoft, it struck me that not only wasn't I organizing the dinner, it was not at all clear that I was invited. I vented this concern to Sue Masey a few weeks later. "I mean, I am Stephen's official host while he is in Seattle. Don't you think that I should be able to attend the dinner?" I said. "Sounds appropriate to me. I will bring it up with Stephen with the proper subtlety," said Sue. A few days later I received an official invitation to Nathan's dinner for Stephen at Rover's.

This was a pattern over the years. Even though I was an outsider, a plebeian in many gatherings, Stephen always made sure I was included. He also made sure that his entire crew, the nurses and assistants were included, not just the shift on duty at the time, but everyone. This was also true whenever we did anything fun, any adventures. It was Stephen's way. Whenever possible almost everyone was treated as family. All that was part of the more desirable future that Stephen Hawking – the Participant – was working to bring about.

Microsoft Canada and the Science World Meeting

When I first contacted the folks at the Public Broadcasting Station in Seattle, KCTS/9, in 1991, they had indicated that they would be even more interested in being a media co-sponsor if we could have a second event in Vancouver, British Columbia. Why? Because forty percent of the KCTS/9 audience was actually in British Columbia, centered around Vancouver. Canadian provincial television rebroadcasts KCTS/9's U.S. programming both directly and through the cable system. That is how the original, beginning vision for three lectures in the Pacific Northwest had become the plan for four. Stephen agreed to both lecture venues, Seattle and Vancouver.

Although only 300 miles north of Seattle, Vancouver was another world. The folks at Microsoft in Redmond had also made it clear that, 'Microsoft Canada is really a separate entity.' "We don't have any direct ties from public affairs in Redmond. So you are basically on your own," they had said.

So I just did the old 'cold call' to Microsoft Canada. I explained what I was up to and Jackie Slemko became my key contact. She listened. I sent her a proposal, and we arranged a meeting.

The Microsoft Canada offices in Vancouver were right downtown, in the BC Gas building. Modest. I had expected something grandiose. The main office, I learned, was in Montreal.

The proposed benefits were essentially the same as proposed to Microsoft in Redmond: cosponsor recognition, full-page in the program-magazine, the introduction and discounted tickets. And finally there was the possibility of either a VIP dinner or a visit with a group of students with disabilities.

Jackie's eyes lit up.

After only a moment's reflection, she responded, "No question that we would be very excited about facilitating a meeting with students with disabilities." After I had outlined the types of events we had arranged in the past, Jackie asked if she could have a couple of days to talk with people she knew in the Vancouver School District and the Provincial Educational Authority to see how they might develop the idea.

Slemko was most definitely tuned into one local controversial issue in the British Columbia educational system: the recent decision to 'mainstream' students with disabilities. Historically, students with disabilities were sequestered in 'special-needs schools', where they could receive – as the thinking had been – special attention, special education. Makes sense – 'stands to reason'. However, these students and many of their parents had a different attitude. These 'special education' students lost all contact with their generational peers. The small special education schools were isolated – social-cultural set-asides, almost like concentration camps. Also when out of sight, they were out of mind. But it wasn't obvious to everyone that a student with a disability – this in-itself being quite a diverse category – would benefit considerably by being included in the larger regular schools.

'Mainstreaming', the countermovement to isolated special education schools, had been gaining ground over the years, and had just recently been implemented by the British Columbia Ministry of Education. Under this new policy, students with disabilities were to be included in the regular schools and frequently in the regular classrooms. The non-disabled students didn't mind and, in fact, enjoyed and preferred having the students with disabilities with them. It had heart and soul – 'he ain't heavy, he's my brother.'

On the other hand, aside from heart and soul there was a concern that the students with disabilities took a disproportionate amount of the regular classroom teacher's time, and, as a result, deprived the regular students. Whereas some parents viewed the schools and education as a cooperative community – as in, 'we're all family here, on the same team' – others tended to the view that schools were intellectual playing fields, competitive, where the Darwinian principle of 'the survival of the fittest' ruled.

You could still feel the heat of the political battle that had taken place in the Vancouver School District. When the topic would come up explicitly in my meetings with Vancouver School District educators, they would either look away or maybe roll their eyes. They didn't want to argue about it any more: "We're going 'mainstream' and that's that."

Trumpets blaring – enter Dr. Stephen Hawking. When Jackie Slemko had contacted the Ministry of Education it hadn't taken two seconds for them to see the opportunity. They had taken substantial flack and vehement criticism from parents who didn't want to sully the intellectual playing field with students with disabilities. Here was an opportunity for a counterplay. Here was Stephen Hawking arguably more severely disabled than most of the students with disabilities who had been sequestered in isolated special education schools. Here was Stephen Hawking arguably one of the most accomplished physicists of his era, continuing to produce leading-edge research, despite all. Here was Stephen Hawking who had a bit of an attitude about how people with disabilities were treated in schools and in society more generally. Stephen's hometown, the City of Cambridge, was an early adopter in England of wheelchair ramped curbing on all the street corners. Need I say more?

So the Ministry of Education realized in a flash that this was the perfect opportunity. 'So you thought students with disabilities couldn't accomplish much huh! Consider this: STEPHEN HAWKING!!!!!' There was to be no mercy.

Jackie put me in contact with someone from the Ministry of Education, someone in the Minister's Office – a big deal that I didn't recognize at first. I had gone right to the top – instantly. So we had a meeting. "Do you think Dr. Hawking would be open to the possibility of …" I was asked. I told them what I knew of his victorious battles in Cambridge and described the powerful meeting with students in Portland the year before. They smiled and I think they were mentally rubbing their hands together, thinking, 'Oh, boy!'

By the end of the week I was presented with the proposed scenario: Stephen would meet with a group of 200 students with disabilities, brought in from The Greater Vancouver Area. The gathering was to be held at Science World, Vancouver's premier high-tech science museum. The British Columbia Minister of Education herself would moderate. Stephen would be introduced by an accomplished celebrity with disability, Rick Hansen, well-known in Canada as the 'Man in Motion'. And just to make sure that no one missed the message, the Minister had arranged to have the event televised live to all nine major school districts throughout the Province of British Columbia.

"Do you think that Dr. Hawking will agree to all this?" she asked.

I had no doubts at all, but being polite and cordial and prudent, I said, "Well, it sounds fine to me, but I will need to check with him directly, of course."

"Oh, of course," she said.

Hawking, the Participant, is perceived by the organizers of the Vancouver meeting with the students with disabilities as a motive force helping to bring about a more desirable future. They are concerned with the history of the universe and they clearly don't see it as having a pre-determined outcome.

Mainstreaming is a creative middle-ground policy, working on a sort of Parliamentary Attitude of mutual respect for opposing views, seeking to implement a novel policy, an inspired marriage of the competitive and co-operative extremes – to bring about a more desirable future.

25

The Tiger and the Shark

Vancouver has the feel of an international city even though the formal population is only about 600,000. The closely surrounding metropolitan area has nearly two and a half million. The geographical region had been populated for at least 9,000 years. Then it was 'discovered' and described to European civilization by Captain George Vancouver, a British explorer of the 1790s. The population is ethnically diverse: over half the residents have a first language other than English. The downtown is built on a promontory of land pushing into the Strait of Georgia, the last mile of which is dedicated to an enormous recreation area – Stanley Park. The overall setting of the city is bordered on one side by the 12,000-plus foot mountains, a portion of the Pacific Coast Range that reaches from the Alaska Range in the north, through Canada and Washington to include the Cascades in Oregon, continuing south into California as the Sierra Nevadas. Vancouver is certainly one of the most beautiful cities in the world and regularly judged to be so by international surveys.

Hawking and his crew flew into Vancouver while I drove up from Portland, and with my assistants from Seattle to meet them at the airport. Hawking's crew had become rather sophisticated about dealing with the airlines. They had tried boarding Stephen first, before other passengers. This is the normal sequence used by most airlines: families with children and anyone who might need more time boards first. However, since Stephen's personal wheelchair needed to be dismantled after the transfers to an airline seat, they had found that it was best for Stephen to board last. People were much more tolerant of waiting a few extra minutes in their seats on the plane as opposed to waiting a few extra minutes to board.

Stephen and one nurse sit in first class where there is adequate room for Stephen's medical equipment, always to be ready at hand. The rest of the crew sits in regular economy. During the flight, depending on the length of the flight, the nurses would switch seats as one would go off-duty and another come on-duty. The off-duty nurse in the back is, nonetheless, always 'on-call' in case of an emergency. With similar reasoning, it was common for Stephen to be the last to disembark the plane. Let everyone else off first. Stephen last.

Apparently, in the early years of Stephen's airline travel there had been several occasions when unexpected complications had resulted in delaying the flight for half an hour or more. No one, of course, was about to jeopardize health and safety to get into the air a few minutes sooner – on time, that is. As Stephen traveled more and more, the nurses and graduate assistant became increasingly adept, optimizing the loading and unloading procedures. Efficiency, it had been discovered, was enhanced by prior dialogue with the airline and, in effect, training airline staff on what Hawking's team had found to be the best approach.

Stephen saw himself as a trailblazer, forging a new path for travelers with disabilities such as his own. Stephen understood what he was doing. It wasn't just about him wanting to travel. He was deliberately expanding capabilities and access for everyone with a disability. "Many people with disabilities are discouraged to fly because they, or the airline, see it as too inconvenient and disruptive." By working through procedures that allowed Stephen to travel easily and efficiency, everyone learned, and everyone's confidence grew.

Again, this is Stephen Hawking the Participant – something of a rebel innovator.

Hawking does not travel light. The nurses and graduate assistant, of course, had their own normal clothes and sundries luggage. Then there was Stephen's stuff. There were the day-to-day medical supplies and equipment; then the 'just in case', level-one medical equipment – mostly having to do with clearing his lungs, intubating occasionally, since he doesn't cough strongly. And finally, there was the 'just in case', level-two for anything more serious that might, but never did, happen. All this equipment was under the careful oversight and management of the nurses.

The crew on this trip had two components. The head nurse was Joan Godwin, Hawking's longest serving nurse, and apparently the overseer of all of Stephen's nursing staff. Joan was by now one of Stephen's closest day-to-day friends and confidants – trusted. Joan, a few years older than Hawking, always greeted you with a cheery smile and a willing cooperative attitude. Stephen's needs always came first but Joan was magically able to make sure that he was ready and on time for all his outside commitments. Joan exuded an air of professional confidence and integrity. She was a nurse's nurse – respected; the cheerful matriarchal commander of Stephen's Nursing Corps. None of the other regular nurses, such as Pam Benson, was available for the current trip, so Joan asked a personal friend, one of her close nursing colleagues, Joan Grant to join her for this trip. The Joans' friendship went way back to their student days. "Joan and I trained together at St. Bartholomew's Hospital, London in the fifties," she told me. Subsequently, I learned that Joan Grant later went on to become 'a high-powered' Nursing Advisor to the Health Authority for the United Kingdom.

Hawking's new graduate assistant, replacing Jonathan Brenchley, was one Timothy Hunt. Tim was a tall, slim, good-looking guy – not the image of your typical mathematical physics nerd. Tim had a delightful time flirting with my

The Joans, Tim and Stephen

two assistants, Christy Richardson and Jennifer Lund, two staffers from the Paramount Theatre in Seattle with whom I had worked in the past. I had

273

asked them to help in both Vancouver and Seattle and they had instantly agreed. Christy was the Manager of the Paramount Theater in Seattle, gorgeous, straight-talking, with a cheerful professional presence. Dark-haired Jennifer was ridiculously rock star cute and we teased her about starting a personal fan club.

The second component of Hawking's crew for this trip was the duo of Sue Masey, the DAMTP Administrator, joined by Andrew Dunn, one of Stephen's earlier graduate assistants. Andrew was just completing his Ph.D. thesis – something about galaxy formation. Gradually, it becomes clear and explicit that Sue and Andrew are a couple. She was quite a bit older than Andrew. She was divorced, raising two teenage girls. So it was necessary to tease her, in private, about 'robbing the cradle' and to raise silly questions about university ethics policy on staff sleeping with students. It was a little side entertainment in a group that was continually teasing and joking and laughing. It was Stephen's culture. Sue and Andrew were with us only for the Vancouver leg of the journey. Their real mission and the principle justification for their joining us was that they were on their way to Japan to check up on arrangements for Stephen's planned visit there a few months hence.

After the plane landed, early morning local time, and the other passengers had disembarked, I wandered down the gangway. I watched as Stephen's chair was brought up from the plane's luggage compartment. Tim is reassembling the dozen components. The nurses, one on each side, locking hands beneath to form a sort of sling, carry Stephen from his seat on the plane a few dozen feet to his waiting wheelchair, just outside the airplane door. I could see that the level-one medical equipment accompanied Stephen, packaged in a small, clear plastic suitcase, about 12-by-12-by-6 inches, hanging from a strap slung over the back of the chair. Once Stephen was in the chair, Tim finished the electronic portions of the assembly process attaching and activating Stephen's computer system. I had nodded to Stephen and said hello once he was in the chair. With the activation of the computer system, more formal greetings were in order.

"Hello, Stephen. Welcome to Vancouver," I say, adding, "Good to see you again. You are looking quite fit." I pause. A normal response to such a comment would have been immediate. But anticipating that it may take Stephen a moment or two to respond, I keep chatting.

"I trust your flight was reasonable," I say, looking first at Stephen and then turning to Joan Godwin indicating that the question was general.

"Everything went just fine. It was a beautiful flight," says Joan. "Stephen really enjoyed it," she adds, glancing at Stephen for approval of that assessment. Stephen might have given her an 'eyes-up' affirmation, but I didn't catch it.

Stephen's eyes had been diverted to the computer screen for a few seconds as it was booting up, but now he looks up, straight into my eyes as the voice synthesizer booms: "Hello. How are you?"

Once the system was fully activated, Stephen's responses were rapid. As I learned later, he didn't have to construct it word for word. Ready-to-use phrases were pre-constructed and stored in a special 'common phrases' section of his computer system's dictionary.

"I'm great. I have a taxi-van plus my SUV to transport everything to the hotel. Nice place. You'll like it. The manager is a Brit and a fan of yours," I say.

"Do you think you have enough room for all the luggage and everyone?" asks Joan.

"Yeah, I think so. But we can hire another taxi if we need to," I add.

"How far is the hotel?" Joan asks.

"I think it will take about 20 minutes from the airport to the hotel, depending on traffic."

I didn't realize it at the time but Joan is thinking about scheduling. The nurses have a series of daily, every other day and weekly medical and quasi-medical routines for Stephen. Most are time-flexible: minor medical procedures, meals with vitamins, toileting, cleanup, bathing and so forth. I am not privy to much of this, just aware that whenever I want Stephen to be somewhere for some event, I need to be coordinated with the nurses' routines.

We all move slowly to the baggage claim area. On the way I brief Stephen on my understanding of the public events.

"Tomorrow you give a presentation to 200 students with disabilities at Science World, Vancouver's science museum; very cool, huge geodesic dome," I say.

"I'm afraid the initial plan has gotten a little out of hand," I say parenthetically.

I relate the story of Microsoft Canada and the Ministry of Education and the fact that there has been controversy brewing over the new policy to mainstream students with disabilities into the regular classrooms.

"But I figured you would be up for helping the Ministry of Education kick a little ass on this issue. I told them the story you had told me about how you dealt with the Cambridge City Council on modifying street corner curbs to allow for wheelchair access," I say.

"What I need to ask you about… They just asked me if it is OK to televise this and make it available to other locations," I say.

"Is that all right, Stephen?" Joan asks directly, looking into Stephen's face as we walk through the airport lobby. I sense from Joan that Stephen has some concern.

"National?" Stephen asks.

"All over Canada?" Joan reiterates.

"No, just around British Columbia – you know, just this western most Province. They've gotten a little carried away. We started planning for you to meet with 20 students at the hotel. Now we are at Science World, 200 students, three-camera production, introductions by the Minister of Education herself…" I say.

We all agree that we can talk about the details when we are settled in the hotel.

Once in the airport elevator, I continue, "Then that evening we have your public lecture at the Vancouver Paramount Theatre. Apparently, it is already sold out. And there is a VIP reception to follow."

Recalling the story, which Pam had apparently shared, about the chocolate covered strawberries from the Portland reception, Joan remarks, "Well, Stephen should be looking forward to that. Any goodies?" Of course, it isn't just about the 'goodies'. Stephen genuinely enjoys these crowd interactions – a time to be social.

We arrive at baggage claim and continue to chat. Tim gradually assembles the 22 bags, double-checking from a list to make sure everything is accounted for.

Individual, personal bags are a minor portion. To support Stephen's chair there are tools and backup repair modules that fill two full-sized, hard-shell suitcases – think of two particularly large toolboxes. One of these hard-shell cases contains a backup computer and all the electronic equipment to diagnose and repair the increasingly complex electro-mechanical Earth-ship that Stephen drives. Another contains replacement parts and tools for the non-electronic parts of the chair: wheels, seats, handles and so forth.

Power. That is a separate issue and there is a separate world of batteries, chargers and transformers. For this trip, fortunately both the Canadian and U.S. electric systems operate on 110 voltage.

There are three large, heavy batteries. Each one is roughly double to triple the size of a large automobile battery, fitted with a helpful carrying strap. Stephen had given the different batteries names. The names are either hand-written in white lettering on the otherwise totally black individual batteries, or have attached name labels.: Beethoven, Brahms, Puccini.

Each battery weighed in the vicinity of thirty-forty pounds, maybe more. The composers were normally charged in Cambridge on the United Kingdom's 220-volt system. So in Canada and the United States the charging system needed transformers. Management of all this equipment was the one of Tim's primary assignments. This was the first opportunity I had to meet Tim in the flesh. I had spoken with him numerous times on the phone, coordinating, sharing and anticipating.

"What happens the next day, after the evening lecture?" Joan asks.

"Ah! Adventure Time," I muse.

"What does that mean?" Joan asks with a coy, slightly suspicious, still cheery smile.

"Instead of a boring plane flight to Seattle, we are taking the ferry to Vancouver Island, through the Strait of Juan De Fuca to Victoria, where we will tour the famous Butchart Gardens, possibly have a meal there and then take the fast hydro-foil south to Seattle," I say.

Vancouver is by far the largest city in British Columbia and is located on the mainland. Then, just to be confusing, the Canadians placed the capital of British Columbia, the City of Victoria, on a wholly separate island – called Vancouver Island.

"Terry, that sounds grand," says Joan. "You are so nice to work with."

"I neglected to mention the 'hidden agenda' for the trip over to Vancouver Island," I say.

"Which is?" Joan asks.

"Stephen will have an opportunity to commune with the Orcas – concerning the nature of space and time… You know, like a Spockian mind-meld," I say. I had just made all this up and I receive bewildered looks. I figure that when reality is a little sparse, one should always look for a good fantasy for accompaniment.

The Hotel Vancouver is one of the majestic railway hotels built across the country by the Canadian Pacific and the Canadian National Railways. Built in the first half of the 20th century, these authoritative structures were marvels of construction and engineering in their time. Materials and decor, everything was the biggest and best. Located in the very heart of downtown Vancouver, the hotel is a living part of British Columbian heritage and identity. Conveniently, the hotel is only a few blocks from the Paramount Theatre, where the lecture and VIP reception were to be held. The General Manager at the Hotel Vancouver, very English (right and proper) from the best tradition of British public administration, had accepted our invitation to be a co-sponsor of Hawking's public lecture in Vancouver, in trade for an almost free stay for a dozen people for two nights. The General Manager was enthusiastic, recognizing Hawking both as the famous cosmologist and as a fellow countryman.

Although 'officially' wheelchair accessible the hotel wasn't completely retrofitted. After unloading in the drive-through passage with double-sliding glass doors, it was necessary to veer off from the normal path of the grand entrance, down a side-passage, past a few shops selling newspapers, magazines and chocolates, on one side, frightfully expensive ladies' hand bags, jewelry and perfumes, on the other – only then, finally, emerging into the majestic lobby.

As we move across the huge lobby with its floor of brown, cream and black inlaid marble, towards the registration desk, scatterings of rose petals are oddly strewn here and there. I can't help but look up, two stories, at the vaulted ceiling, cathedral-like. Glancing around the extensive lobby complex, there are a dozen magnificent, towering flower arrangements animating the space, complementing the inanimate stone setting. Full-sized, museum-quality, human sculptures, suggestive of Ancient Greek mythology, highlight a half-dozen smaller, individualized subsections of the football-field sized main lobby. Each subsection is semi-secluded for semi-privacy while still open to the main lobby. One, slightly elevated, plaza-like area is where guests might enjoy afternoon tea – a decidedly British accouterment.

And then there is Stephen Hawking. Into this grandiose setting, where one might naturally expect a well-dressed, rather formal clientele, rolls a slightly rumpled Professor Hawking and his ragtag band of adventurers – slowly traversing the lobby, feeling very small, surveying the surroundings, awestruck by the extravagance.

I asked one of the hotel staff about the rose petals and was told that the Dalai Lama had checked out late the night before. Devotees had garnished the path of his departure. 'Too bad,' I thought. At an earlier meeting, the Hotel General Manager had mentioned the close encounter in their schedules. I had made some inquiries in hopes of arranging a meeting: Stephen Hawking and the Dalai Lama. That would have been fun.

Check-in was easy and everyone migrated to his or her rooms.

University of British Columbia Museum of Anthropology

From an outsider's perspective, there seems to be a fairly common, natural intuition that a person with Stephen's level of disability must spend a lot of time doing nothing. I mean he is so disabled and all. Ha, ha, ha. One of my favorite Hawking quotables is: "Why should I worry about what I can't do when I don't have enough time to do all the things that I can do?" This just barely begins to point toward Hawking's activist attitude and lifestyle. When I mentioned that there were no plans for the afternoon, I was immediately asked about tourist options. Personally, I had imagined a rest or a quiet time after the long flight. – Forget that!

When he isn't working on cosmology questions, meeting with or corresponding with his graduate students and colleagues, Stephen is nearly always doing something socially or culturally imaginative. Whenever possible, we were looking for the best, most attractive restaurants. He travels extensively – to the U.S. regularly now, to Germany, France and Italy, South Africa, Australia, Asia and ... Have I forgotten anywhere? In each of these locations he becomes the consummate tourist, seeking out the best and most intriguing. Museums and local music, including nightclubs, are favorites. If he is stuck inside somewhere for an afternoon, you can expect him to be listening to an opera – at high volume.

Stephen decided that we should visit the renowned Museum of Anthropology, located on the extensive University Endowment Lands of the University of British Columbia. From the museum property there is also a fabulous view of Vancouver Harbor. Everybody went, nearly a dozen of us. Once through the entrance, we split up into smaller subgroups. The Joans stayed with Stephen, who had his own ideas about what he wanted to see first.

I had mentioned to Stephen my favorite piece in the museum that I thought he would enjoy. So after about an hour, I sought him out. Finding him among a collection of giant totem poles, I showed him the way. The immense, three-dimensional, several ton stone carving is framed in its own side room and features an enormous raven, standing atop a partially open clamshell, from which six naked human figures are struggling to emerge. The sculpture illustrates the legend of Raven and the First Humans: "One day after the great flood, Raven was walking along the beach at Rose Spit, in the Queen Charlotte Islands, when he heard a sound emanating from a clamshell at his feet. He looked more closely and saw that the shell was full of small humans. He coaxed, cajoled and coerced them to come out and play in the wonderful, new world. Some immediately scurried back into the shell, but eventually curiosity overcame caution and they all clambered out. From these little dwellers came the original Haidas, the first humans. The sculpture rests on a bed of sand brought to the Museum in 1980, by children from the village on Haida Gwaii, where this event is said to have taken place."

Throughout the museum were images of orcas and wolves. As travel guide, I offered another image – an image of complementarity. In the tales of the early Pacific Northwest coastal tribes – the First Nations, as the Canadians called them – wolf and orca are actually the same: an orca can become a wolf and a wolf can become an orca. Like the human hunters, both orcas and wolves are intelligent and cooperative, hunting in packs. Although different species, in this core behavior they are all the same. In summer, these orca-wolf hunters appear in the form of wolf; in winter, in the form of orca. Orcas reportedly help the human hunters on the sea, cooperatively driving walrus. First Nation boats were often embellished with images of the orca, and wooden carvings of orcas hung from the native hunters' belts. Small sacrifices could be offered to orcas: tobacco, for instance, was thrown into the sea. The orca-wolf also reportedly helps the human hunters in its guise of wolf. The wolves drive the reindeer into positions where the human hunters could more easily kill them. Orcas, wolves and humans work together. There is, per hypothesis, per tradition, a common brotherhood of pack hunters.

I wasn't the first to play on such metaphorical analogies. Bruce Wheaton, author of an earlier study of the experimental origins of the wave-particle duality, chose as the title of his book, *The Tiger and the Shark*, based on a quote from legendary British experimentalist Sir Joseph John Thompson, who received the Nobel Prize in 1906 for the discovery of the electron: "The position is thus that all the optical effects point to the undulatory [wave] theory, all the electrical ones to something like the corpuscular [particle] theory; the contest is something like one between a tiger and a shark, each supreme in its own element but helpless in that of the other." (Thompson, Sir Joseph John, *The Structure of Light: The Fison Memorial Lecture, 1925*, Cambridge University Press (1925) page 15).

Orcas and dolphins are known to be exceptionally smart and to communicate amongst themselves in sophisticated ways, not understood, not translating into our human forms of communications. So throwing in a little science fiction fantasy, I hypothesized to Hawking that orcas might have unique insights into the nature of the universe – from their distinctive perspective. Spockian mind-meld, anyone?

All this was preamble to my discussion with Hawking, carefully planned for the ferry trip to Victoria in two days.

Dinner at the Tower Restaurant first night

Arriving back at the Hotel Vancouver and in tune with Stephen's active lifestyle, I proposed that we go out for dinner that evening, preferably at a restaurant with a view of the City. Vancouver is framed to the north by the spectacular North Shore Mountains. Stanley Park, one of the largest urban parks in North America, projects from the base of the city center into Vancouver's international seaport.

The natural choice was the Tower Restaurant, only a dozen blocks away, but too far to walk. The plan had all seemed simple when we left the hotel. However it turned out to be difficult to find a place for the van to unload Stephen and his chair. Then the route from the street to the restaurant's elevator was strange and convoluted. These are the common day-to-day annoyances of people with disabilities. Next the elevator was terribly slow in arriving to take us up. From the street to the restaurant alone must have taken fifteen minutes.

These small but accumulating inconveniences were threatening to dampened enthusiasm for the evening. But everyone worked to keep spirits up with a constant teasing and attempts at the humor of 'British understatement'. Someone speculated that the slow elevator was part of an overall plot to reduce food consumption in British Columbia – suggesting parenthetically that Vancouver, being so far north with a shorter growing season, was probably short on food. Contributions ranged from usually silly to occasionally clever. The whole scene would have fit well into a Monte Python (the British comedy troupe) sketch – 'Stephen Hawking Goes to Dinner'.

Finally we were seated. The promised view was certainly worth the wait, although I imagine that it would have been better during the daylight. The waiter went around the table taking individual orders. Joan Grant was on duty with Stephen and asked about how certain things were prepared – concerned about whether gluten had been used to thicken the soup and about which oils were used here and there. I don't know if I assumed that Stephen would eat special meals, since he isn't able to feed himself directly. I hadn't actually paid much attention. So it was unexpected when he ordered rack of lamb. I confess that I entered a mild state of shock when I heard this. "Rack of lamb?" I slowly questioned. "How would you like that prepared?" the waiter asks. "Rare," says Joan, glancing at Stephen for confirmation. Somehow my mind wasn't putting together the pieces of a picture of Stephen Hawking and a rack of lamb. For those of you who don't know, a rack of lamb is the lamb version of a cross between lamb chops and spare ribs. Less conservative people would finish such a meal by gnawing the meat right off the rib bones. That is the way I eat rack of lamb. But Stephen can't even lift his arms. Furthermore, he is generally fork or spoon fed by one of the nurses. Even if he were able to control his head movement enough to accomplish gnawing – which I seriously doubted – I couldn't conjure an image of Joan holding up a lamb rib while Stephen attacked it.

Stephen clearly thinks my perplexity is funny. "I am a carnivore," he says smiling widely.

"You can say that again," adds Joan Godwin. "Stephen is a major meat eater. And he likes it rare," she says.

"Blood rare," Stephen adds, giving me a stern, no nonsense look now. The feeling is that I am talking to Attila the Hun or Conan the Barbarian while I am looking at the disheveled body of Stephen Hawking.

Before actually dining with him, I vaguely imagined that he might eat a lot of mush and pureed vegetables; items easy to eat – a fully liquid diet, perhaps. So it is remarkable to find that he doesn't eat like that at all. Moreover, it is rather shocking to find that he eats pretty much whatever he wants, and what he often wants is lots of rare red meat.

I hadn't thought much about what Stephen was eating in Oregon in 1992. I was too busy and frantic. Thinking back to the run from the coast to the Eugene Hilton where we had ordered the dinner over the cell phone, I remember that he had ordered the steak – rare. The image is more than a little incongruous.

I had talked to Sue Masey much earlier about whether Stephen had any special dining needs. "Not really. Actually he prefers to eat out a lot at fine restaurants," she had said.

The fact is that Stephen does have some minor difficulties eating. First of all, since he has negligible strength in his arms he has to be fed by one of his nurses. They cut up any food as necessary to have reasonable sized bites. Once the food reaches Stephen's mouth, there is a second problem. He doesn't have great muscular control of his jaws. He lacks full control of both chewing and swallowing. The inevitable result is that a portion of the food that enters his mouth is eventually destined for his lap. The first time I saw him lose a mouthful, I wasn't sure whether to laugh or to be slightly put-off. Not being many years away from feeding my young children, I was inclined to think of it as rather funny. But you didn't want to laugh or joke because you knew he was trying. The best policy seemed to be to just ignore it, which became easier and easier to the point where, after a while, you didn't even notice anymore. The nurses, Sue Masey and Stephen's grad assistant, Tim, don't seem to notice anything out of the norm. Of course, in their experience, this is the norm.

Stephen does unexpectedly well with liquids – soups and tea. He drinks a lot of tea. Success depends on the skills of the nurse holding the cup of tea and Stephen's angle of attack. A small portion of the tea, nonetheless, almost always runs down into the bib, with its neat built-in pocket at bottom, conveniently catching whatever escapes consumption.

There are those who, if they had such difficulties keeping their meal fully in their mouth while dining, would be inclined to be embarrassed and consequently avoid eating in large groups or in public places – like in fine restaurants. Stephen Hawking is not one of those people.

There are those who, if they had such difficulties keeping their meal fully in their mouth while dining, would be worried that this consequence of their disability might upset or perhaps even offend other patrons dining in fine restaurants. Stephen Hawking is not one of those people. I surmise that one of the reasons Stephen goes out to restaurants to eat is the same as with flying. Many restaurants have never served anyone with a major disability. Stephen helps them understand that they can serve the disabled, and they learn how better to manage it. The same applies to hotels and museums and opera houses.

The idea that he would let some aspect of his disability lead him into isolation is simply not Stephen's way. Stephen confirms to me later that he sees himself as a trailblazer for everyone with a disability. He sees himself as helping to define a new cultural norm for people with disabilities. His outgoing lifestyle is self-consciously an encouragement to others with disabilities to reject the notion that they should hide away.

"Dessert anyone?" – Of course.

26

Making Sense of a Post-Scientific
More General Theory

As Kuhn's careful historical studies had served to warn us, demonstrations of the inadequacies of a research program do not naturally or automatically bring forth a new, better, superseding theory and research program. The new theory, the new approach to understanding, has to be discovered. The path to a new paradigm isn't logical, cannot be reasoned from within the current inadequate theory and research program. The path forward requires exploration and, as Einstein later reflected, 'an intuitive [conceptual] leap.'

One clue in beginning the search for a post-scientific understanding of reality is that any acceptable superseding More General understanding of reality must naturally subsume the successes of all meaningful scientific theories as limited special cases.

A second clue is that the embrace of complementarity entails the abandonment of the Spectator's universal objectivity, forcing us to adopt a Participant representation of inquiry, forcing us to recognize choice (viz. however that is to be understood eventually) as an irreducible aspect of reality.

Recently, Anton Zeilinger, one of the world's leading quantum experimentalists, emphasized: "This notion of complementarity was introduced by Niels Bohr as one of the great lessons we learn from quantum physics. Simply we are unable to know the world with complete precision... We always have to make our choice." (Zeilinger, Anton, "How We Became Certain About

Uncertainty", *Dance of the Photons: From Einstein to Quantum Teleportation*, (2010) Farrar, Straus and Giroux, page 32).

Zeilinger goes on: "We have learned … that the observer has a significant influence through his choice of the measurement instruments, through his decision of what to measure. The point is that his measurement instruments don't just influence or change the observed systems. That would still be acceptable [to classical thinking] in some way. But we have learned that the choice of measurement instrument actually defines the property of a quantum system that becomes realized as an experimental result." (Zeilinger, Anton, "What Does It All Mean?" *Dance of the Photons: From Einstein to Quantum Teleportation* (2010) page 32).

A third clue is that, in the history of 20th century physics, the complementarities seemed to arise, enigmatically, *within* the Scientific Research Program, presenting us with irreducibly opposite types of mechanical theories. Any superseding More General, post-scientific, post-mechanical theory that can make sense of these irreducible oppositions cannot be simply another mechanical theory. Furthermore, the More General Theory can't be any sort of simple (mechanical) sum of the opposite Newtonian and Maxwellian mechanics. The More General Theory that can make sense of contrary, formally complementary mechanical theories cannot itself be mechanically uniform with One universal type of order. By its very nature, complementarity cannot be made sense of in terms of One universal mechanics.

What then might a candidate More General Theory look like? I began to look for a theory of the nature of the universe that understands reality as mechanically incomplete – by its very nature. One approach is to embrace the notion that the nature of reality involves both order and disorder.

In the Scientific Worldview the original defining rational coherence was represented in Newtonian physics in the image of a clockwork universe. There was supposed to be one unified causal nexus – one clockwork coherence – evidenced by the repeatability and the regularity of causal relationships over time and space. The clockwork became the most popular enduring image of an objective mechanical universe.

From within the clockwork perspective disorder, the irrational opposite, might be expressed – in clockwork terms – as an irrational, incoherent, cloud-like, chance-governed universe.

With the embrace of the complementarity of competition and cooperation, we reasoned that just because cooperation appears to be irrational – an incoherent view of reality – from *within* the competitive perspective, that does not mean that cooperation is 'really' incoherent. There might be, – should be – there is – an opposite complementary type of coherence to cooperation.

By analogy, from the perspective of the Newtonian clockwork universe, the encounter with Maxwell's non-local electromagnetic fields presented an irrational, incoherent realm of phenomena. It should not be surprising then that Max Born proposed that Maxwell's electromagnetism could be represented as if it were a charged particle, like the electron, probabilistically (non-locally) distributed, like a cloud, in space and time. Quantum reality, under this representation, is composed of order and disorder; a mechanical rationality and a mechanical irrationality. Welcome to quantum theory's enigmatic – seemingly paradoxical – embrace of the complementary aspects of all phenomena. Many scientists objected that to allow a probabilistic, chance-governed aspect into the clockwork universe undermined the Scientific Hypothesis. Einstein, in particular, was stalwart, insisting that an irreducible probabilistic element in quantum theory meant that the theory was no longer a scientific theory. For Einstein, the proposed quantum theory was 'scientifically incomplete.' Einstein mused: "Quantum mechanics is certainly imposing. But an inner voice tells me that it is not yet the real thing. The theory says a lot, but does not really bring us any closer to the secret of the 'old one'. I, at any rate, am convinced that *He* does not throw dice. (Einstein, Albert, Letter to Max Born (4 December 1926); *The Born-Einstein Letters* (translated by Irene Born) (Walker and Company, New York, 1971).

Chance entered modern physics in at least two ways. First, as mentioned, it entered in the relation between the complementary Newtonian and Maxwellian Research Programs. Per hypothesis, this is analogous to the rational-irrational relationship between competitive and cooperative ideologies.

The second related entry of chance arose with the proposal for a superseding inductive generalization that would unify the order and the disorder and somehow 'resolve' the enigma of complementarity. Clearly, the Newtonian framework was inadequate to make sense of relativistic, electromagnetic phenomena. By analogy, this is like saying that the competitive framework is inadequate to make sense of cooperative phenomena. The temptation in quantum theory was to make a conversion to the electromagnetic framework as the 'new' objective framework. By analogy, this is like making the conversion from viewing reality as completely competitive to an opposite, completely cooperative, evangelical religious worldview or, similarly, to a completely cooperative, ideological, Marxist worldview. In the new physics, Heisenberg had shown that what had been understood previously as Newtonian matter (particulate) could be successfully understood – in many novel experimental settings – as radiation, as an electromagnetic phenomenon (wave). Since electromagnetic phenomena cannot be made sense of in the Newtonian framework, in the search for a new 'objective' reality, the temptation was to make a shift – a lateral conversion – to the complementary Maxwellian electromagnetic framework – with a tacit Promissory Note.

Another advantage of a conversion to the Maxwellian framework was that it returned us to One mechanics – one of the two competing, complementary mechanics. This suggested that we were still within the Scientific Research Program.

But there was an uncomfortable entailment. Since Max Born had attempted to retain conceptual continuity within the Scientific Hypothesis by proposing that Maxwell's electromagnetic field could be represented as if it were a charged particle, like the electron, probabilistically distributed in space and time, the conversion to the Maxwellian framework, as the new type of universal 'objective' reality, now seemed to imply that reality was chance-governed – a probabilistic particle-field distributed in both space and time. Ironically, Born's attempt to retain scientific 'objectivity' had led to the conclusion that the universe was not a Newtonian clockwork – but rather the cloud-like chance-governed conceptual opposite. The century of attempts to fulfill the Promissory Note and subsume and supersede the traditional (localized) particle order within the new (non-local, distributed) wave disorder is the history of what is referred to as Quantum Field Theory. It hasn't worked – even with the heroic attempts at 'renormalization'. Feyerabend suggested to me that 'renormalizing' Quantum Field Theory was a last gasp attempt to save the traditional Scientific Research Program.

So much for what happened and what has led to a seemingly endless dialogue about the nature of reality in light of quantum physics. (In subsequent chapters I will outline why I think Einstein's shift to Relativity was a similar lateral conversion – equally enigmatic – to the Maxwellian space-time framework.)

A Superseding Mechanics?

To find the More General Theory that can properly subsume and supersede both the Newtonian and Maxwellian Research Programs is to answer Popper's Question applied to all possible 'one right way', ideological, mechanical theories. What is potentially misleading is the suggestion that what I am looking for is only the evidence that can't be made sense of in the clockwork framework – traditional mechanics. This is potentially misleading because Maxwell's Research Program is just as mechanical as Newton's, albeit in its own very different, complementary, way.

Einstein captured the enigma: "In pre-quantum physics there was no doubt as to how this [reality] was to be understood. In Newton's theory reality was determined by a material point in space and time; in Maxwell's theory, by the field in space and time. In quantum mechanics it is not so easily seen." (Einstein, Albert, *Albert Einstein: Philosopher-Scientist*, edited by Paul Arthur Schilpp, Volume One, (Harper Torchbooks edition (1959) page 81-83).

Einstein's argument is that both the Newtonian and Maxwellian frameworks define reality in terms of a symmetry principle. For the Newtonian program, reality is completely and consistently *local* – the same (symmetric) every where/when. The Newtonian presupposition of 'absolute simultaneity' entails that everything happens at the same time (viz. according to the One universal clock time). This can only make sense (to Einstein) if everything happens in the same place – a dimensionless material point in both space and time. Newtonian reality is absolute locality in (of) both space and time – a point-reality.

For the Maxwellian program, the complementary reality is completely and consistently non-local – a field-reality, the same (symmetric) every where/when. The Maxwellian field-reality is universally distributed in both space and time – the same, in Maxwellian terms, every where/when. In Newtonian terms, the non-local Maxwellian field-reality is completely and consistently non-same (non-symmetric), lacking all 'real' locality. There are no non-local Maxwellian (distributed) aspects in the perfectly local Newtonian point-reality. There are no local Newtonian (non-distributed) aspects in the perfectly non-local Maxwellian universe (viz. no particle in the Maxwellian field). So, defined and well expressed by Einstein, the symmetry principles and the corresponding space-time frameworks of the Newtonian point-reality and the Maxwellian field-reality are, in Bohr's terms, complementary.

'Show Me the Evidence'

The demonstration and embrace of complementarity is one, albeit paradoxical, answer to Popper's Question as to what evidence would force you to abandon the Scientific Hypothesis that the universe is governed by One complete and consistent order. The limit of the scientific approach to understanding reality is 'experienced' in the enigma of complementary mechanics. The evidence for the limit of the Newtonian particle program is found in wave phenomena, and the evidence for the limit of the Maxwellian wave program is found in particle phenomena. Per hypothesis, there should be some evidence, some *type* of evidence that, by its very nature, can't be made sense of in *either* the Newtonian or the Maxwellian frameworks. Per hypothesis there should be some evidence, some *type* of evidence that, by its very nature, *can't be made sense of in terms of any possible mechanics*. And if we could find, come to understand, what this type of evidence is, it might provide a clue as to the defining nature of the More General Theory.

What is needed, then, to answer Popper's Question applied to the Scientific Hypothesis is to specify a *type* of evidence that could not be understood completely and consistently in either the Newtonian or the Maxwellian frameworks alone. Since both frameworks argue for a defining symmetry (viz. per hypothesis, the hallmark of every mechanics), what is needed, very simply, is evidence that demonstrates an irreducible difference – an irreducible asymmetry. Arguably, the type of evidence – the asymmetry – needed is, perhaps just as simply, the evidence for an irreducible complementarity; the evidence that wave and particle phenomena are complementary. This means, at least, following de Broglie's insight, that all wave phenomena have a particle aspect and all particle phenomena have a wave aspect, that all 'real' observations, all 'real' evidence must have complementary aspects. Great! However, this simple embrace of the complementary character of 'real' evidence does not automatically produce a new 'positive' – coherent – understanding of the nature of that evidence. This is evidence that can't be made sense of in all possible mechanics, in all possible worldviews defined by symmetry.

Per hypothesis, there should be more direct evidence that would appear to be incoherent in terms of *both* the Newtonian, the Maxwellian and the Scientific Research Program understood quite generally. Indeed, such evidence – by its very nature – must be incoherent for all mechanical research programs; must be incoherent for all objectivist Spectator research programs.

I began thinking about what this type of evidence might look like as an undergraduate, obsessed about it in graduate school and have lived with it ever since. In a sense, I knew the answer early on – as did most of the scientific community. It had to do with the historical character of the universe – with time's arrow. And it had to do with our Participant position in the universe. It had to do with choice – real, meaningful value choices about seeking to bring about a more desirable future. We just couldn't figure how to make sense of it while, at the same time, retaining what was clearly correct, albeit limited, about mechanics. We lacked the superseding More General framework. We lacked the new way of understanding.

What was needed was a way to find such evidence expressed in terms of the reasoning about the limit to the scientific hypothesis. Since the hypothesized evidence could not be made sense of in any mechanics, in general, come what may, it seemed reasonable to posit that the hypothesized evidence could not be space-time invariant. More colloquially, the evidence could not be – by its very nature – repeatable over changes in time and location. If there is a new type of non-mechanical order, it must have *an irreducible non-space-time invariant* aspect. It must be *historically unique* and non-repeatable everywhere and always in space and time.

It also seemed reasonable to suspect that since such evidence cannot, by its very nature, be made sense of in any mechanics, it should 'appear' through 'mechanical spectacles' (viz. from all mechanical perspectives) as a non-scientific, incoherent, irreducibly non-regular aspect to all phenomena. This would mean, of course, that it is not at all clear that such evidence, such phenomena, could even be 'experienced' from within a rigorously policed, thoroughly committed Scientific (Mechanical) Research Program.

The other hope in the background is that by finding and understanding this evidence that must be, per hypothesis, an aspect of all 'real' possible observations, it might provide a sense of the nature of the defining character, the defining coherence of what we are really after – the More General Theory.

27

Uniformity, Symmetry and
the Limits of Science

Since all mechanics are defined by some sort of symmetry principle and the proposed new type of evidence would appear in all possible mechanical perspectives as mechanically incoherent, I began by looking for clues for the nature of the evidence that would verify an essentially irreducible space-time asymmetry. I began looking at the characteristics of the non-regular, non-repeatable evidence for chance-governed events.

First of all, in a chance-governed process – think of a series of coin flips (non-repeatable) – you can't predict individual outcomes. And it has always seemed to me that 'what you can't predict you can't explain.' Consequently, you can't explain, logico-mathematically – scientifically – why any individual chance-governed event actually occurs. Individual outcomes in a 'chance-ordered sequence' can't be calculated, predicted or explained logico-mathematically – mechanically. The sequence is in this sense mathematically, logically – and mechanically – discontinuous, at least in the sense of classical causal clockwork mechanics. You can't generate a chance-governed sequence by any law-governed, rule-governed causal clockwork mechanical process. And you *can't make sense* of a mechanically discontinuous, chance-governed sequence in mechanical terms – as mechanically continuous.

Just to be clear, in the modern era, there is a seriously misleading common reference to mechanical 'random number generators'. Actually these are all 'pseudo-random' number generators. When you look for the formal, analytical definition of 'random' you find that, in this technical logico-mathematical sense, there isn't one. The definition of 'random' is at best a non-definition. The concept of random is a non-concept. Randomness is *inconceivable* – by its very nature.

To describe an actual chance-governed sequence requires an event-by-event recording, in contrast to a law-governed sequence that can be specified in 'short-hand' by a rule or law. For instance, like counting – Rule: add one to the last number. That rule can generate a law-governed sequence. A fully random, open-ended sequence is in this sense a uniquely historical sequence – requiring for its description a real-time recording of the individual event outcomes.

Virtually by definition there is no way to mechanically generate an 'objectively' random sequence. That is what a random sequence is – that which cannot be mechanically generated. From knowledge of one segment of a random sequence you cannot reliably predict, calculate or generate a later segment of that random sequence. In an idealized coin-flipping experiment, knowledge of the sequence of outcomes of the first 1000 tosses doesn't enable one in any way to predict, to calculate or to generate the event-by-event sequence of individual outcomes of the next 1000 tosses. There is an inherent lack of uniformity – a lack of order. Each sequence of unlimited length is historically unique. If there are 'real' chance-governed sequences, they cannot be made sense of in classical causal mechanical terms, as having been generated by a specifiable, objective logico-mathematical rule or law.

Consider, then, what a middle-ground reality might look like. In addition to the demonstrable timeless regularities, every place and time would have an irreducible timeless irregularity, an irreducible element of historical uniqueness. Although yesterday and today have many of the same regularities (samenesses), yesterday and today are also historically unique. Isn't this just common sense? Every place in both space and time appears to have some non-regular, non-repeatable unique characteristics. All humans have enormous similarities, and yet each of us is unique in time and place. Like all of reality, in some sense I am the same person over changes in time and space, and yet each 'point' in my, and the universe's, space-time history has an irreducible uniqueness.

If reality were universally, objectively symmetric in some sense, the same or 'uniform' every where and every when then, in that sense all 'real' differences, all discontinuity, all non-uniformity, all asymmetry and all uniqueness would be impossible. Claims of real 'objective' differences would be universally false. 'Universal' here means space-time invariant – 'over changes in time and location'. On the other hand, if there were a complete 'objective' continuity of differences, a complete continuity of non-uniformity, a complete 'objective' asymmetry, if every when and where were unique – objectively, universally – then all sameness and uniformity, all symmetry would be impossible. Claims of real 'objective' samenesses would be universally false. The universal 'objective' truth of one complement precludes even the 'appearance' of the other complementary type of order. Complete symmetry precludes any asymmetry. Complete asymmetry precludes any symmetry.

The middle-ground reality begins to make familiar sense when one reflects on a crucial and often overlooked feature of what we normally think of as chance-governed systems like a coin-flip. They are not 'completely' chance-governed, because there is always regularity, a symmetry, a constancy in the system. This constancy seems to be necessary in order to be able to meaningfully define an observable, chance-like sequence. The structural property of the coin must remain constant (viz. uniform and regular) in order to be able to generate the chance-like sequence (viz. a non-uniform and irregular sequence). In a dice-tossing apparatus, the structural characteristics of each die must remain constant. The chance-ness is not so much 'in relation to' the constancy, but more like a natural complementary property of the system. In both the coin and die systems there is a regularity and an irregularity; order over time and space and disorder over time and space. By this reasoning, the coin-flipping or die-tossing events are perhaps better understood as 'middle-ground', involving a sameness and a difference, a regularity and an irregularity, a symmetry and an asymmetry. Just as you can't describe a working society completely and consistently in either competitive or cooperative terms, similarly, you can't describe a series of coin-flips or dice throws completely and consistently in terms of either mechanical order or chance-governed mechanical disorder.

It seems reasonable to explore the generalization that all real object-actions have both aspects. What is special about the coin-flip and dice games is the rigorously engineered symmetries of the dice and the coins. If I throw my plastic coffee mug across the room, there is both order and disorder in both the sequence and the outcome. Per hypothesis it's all middle-ground. Observable relationships and actions over space and time have irreducible complementary aspects of both order and disorder, of both constancy and change, of both symmetry and asymmetry.

Scientific Idealization

If all real events have both a regular and an irregular aspect, then the standard, logico-mathematical representation of scientific reasoning involves an ideological idealization. Think of a series of repetitions of an experiment. Each repetition of the experiment is presumed, in scientific reasoning, to be the same, to be another instance of the experimental trial. And yet everyone in the scientific community, every experimenter knows, and in the background tacitly presupposes, that they are never completely the same, never identical. There is, consequently, in all scientific reasoning about repeatability and confirmation, always some degree of idealization – ignoring known differences. In a successful series of trials, the differences are considered to be causally irrelevant – not part of the 'relatively (relevantly) isolated' clockwork. And yet clearly, each actual experimental trial is part of a historically unique sequence. Each trial is uniquely recorded, time and place, in the laboratory notebook.

As graduate students attending Popper's Seminar in the Philosophy Department at the London School of Economics, we were told that Popper had made an edict that when discussing probability we were forbidden to use gambling examples. At the time, most of the graduate students suspected that Popper was just an ultraconservative, prudish type and forbade these examples because he thought gambling was morally degenerate. Now I understand that there was a deep insight behind his policy. Gambling examples and the reasoning involved in all mechanically idealized models of probability require that the possible *types* of outcomes of the system are space-time invariant – always the same from the beginning of time, for forever. The key point here is that in these idealized, isolated, mechanical (viz. gambling) systems there is *no historical development* of new types of outcomes, no revolutionary, qualitative emergence allowed.

In all the mechanical models of probability, 'the possible (types of) actions of the system' are assumed to be the same over time – uniform, conceptually continuous. No conceptually novel possibilities develop as the system goes through its mechanical combinations and permutations. There is the irreducible chance-governed aspect but it just repeats. The possibility space remains constant. No surprises. No qualitative discontinuities. No revolutions.

Real Historical Sequences

These considerations of the historical uniqueness of random-like sequences led me to reflect on the uniqueness of historical events and sequences more generally. When Stephen Jay Gould was visiting Oregon as a presenter in the Science, Technology and Society Program, I had asked him what he thought of the relations between the historical sciences, like his fields of geology and evolutionary biology and the hard, mechanical sciences (viz. ultimately, physics). He was rather emphatic; he didn't think that the historical phenomena studied by the historical sciences could ever be made sense of in terms of the hard sciences; indeed, in terms of any possible mechanics.

The question was whether the phenomena of observational cosmology, geology, evolutionary biology and human socio-technological development could ever be reduced to, understood, fully predicted or explained in terms of the hard – mechanical – sciences. Gould doubted the Promissory Note of the Scientific Hypothesis. The challenge for both of us was how to prove it – one-way or the other. What evidence, if it were to occur…?

The question wasn't new to me or to the scientific community in general as to whether there might be unique historical events or an irreducible historical uniqueness to all events. Indeed, in the history of science there had been one particularly illuminating debate of the core issue. It arose initially in geology as uniformitarianism versus catastrophism.

The uniformitarian position was most famously articulated by Charles Lyell in his *Principles of Geology*, subtitled *An attempt to explain the former changes of the Earth's surface by reference to causes now in operation*. In other words, Lyell wanted to explain *inductively* what had happened in the geological past in terms of currently observable geological processes. Lyell's approach required the presupposition that geological processes were uniform throughout time and location – space-time invariant. In brief, mountains raise volcanically, rain and wind wear them down, rivers cut canyons and valleys and so forth. Lyell's uniformitarian bias was motivated primarily by his concern for making geology into a science – a mechanical science. Secondarily, there was pressure to oppose speculations based on Biblical accounts of unique historical catastrophes – great floods and cataclysms. Real science for Lyell was mechanical, and the reasoning must be logico-mathematically inductive (viz. 'it stood to reason'). The phenomena to be explained (and predicted), were governed – always and everywhere – by the same, demonstrably repeatable, space-time invariant laws and processes.

It was William Whewell who actually coined the term 'uniformitarianism' in a review of Lyell's *Principles*. Interestingly, it was also Whewell who, in 1833, invented the English word 'scientist'. Before then the most common term used to characterize those who studied nature was 'natural philosopher'.

Gould pointed out that Lyell realized that to make geology a science uniformitarianism needed to be

Stephen Jay Gould

embraced as a general, methodological principle – indeed, a principle re-quired in all the sciences – in all 'real' (viz. mechanical) sciences. According to Gould, Lyell 'postulated the invariability of natural laws in space and time as a necessary condition to his contention that reference need only be made to [currently] observable processes in explaining past changes.' (Gould, Stephen Jay, "Is Uniformitarianism Necessary?", *American Journal of Science*, Vol. 263, March 1965, p224.)

The defining principle of Lyell's 'scientific' research program is captured by the expression: 'the present is the key to the past' (cf. Lyell's *Principles*). Gould went further: "Substantive [ideological] uniformitarianism (a test-able theory of geologic change postulating uniformity of rates or material conditions) is false and stifling to hypothesis formation."

The modern, generalized version of uniformitarianism is one of the defining presuppositions of the Scientific Hypothesis and the Scientific Research Program – that the laws governing all phenomena in the universe are space-time invariant. The nature (*types*) of phenomena and the laws governing all phenomena are presupposed to be completely and consis-tently (viz. uniformly) the same everywhere and always. All phenomena and their relationships are to be understood, predicted and explained in terms of regular, uniform, repeatable causes and processes. All this was captured in the eternal clockwork model: specifiable causes always and everywhere producing specifiable effects. In order for universal laws to be applicable to all phenomena, there must be a continuity of *type* in the phe-nomena, a conceptual continuity, entailing that the 'ultimate type of stuff' composing all the phenomena in the universe must be the same everywhere and always. Typically, and traditionally, in the Promissory Note reasoning, the ultimate stuff out of which all phenomena are composed, the ultimate stuff that the final Spectator's scientific Theory of Everything will refer to and, be applicable to, is some sort of eternal and indestructible 'atoms' or 'energy'.

In his later influential, yet underappreciated book, *Time's Arrow, Time's Cycle: Myth and Metaphor in the Discovery of Geological Time* (1987), Gould points out that Lyell and Whewell had realized that the presumption of the 'uniformity of nature' was needed to justify basic inductive generalizations. It was induction that allowed science to explain one period of time in terms of another or the patterns of phenomena in one spatial region in terms of those in another. "The assumption of spatial and temporal invariance of natural laws is by no means unique to geology since it amounts to a warrant for inductive inference which, as Bacon showed nearly four hundred years ago, is the basic mode of reasoning in empirical science. Without assuming this spatial and temporal invariance, we have no basis for extrapolating from the known to the unknown and, therefore, no way of reaching general conclusions from a finite number of observations." (Gould, Stephen Jay, "Is Uniformitarianism Necessary?", *American Journal of Science*, Vol. 263, March 1965, p223-228.)

Uniformity allows one to reason logico-mathematically, to calculate and predict, from one segment of space-time to another. The uniformity presumption was at the foundation of Laplace's claim that, at least in principle, complete knowledge of the state of the universe at any one instant would allow a Supermind (i.e. Laplace's Demon) to predict the future and to retrodict the past in complete detail with complete certainty.

The 'hidden' major premise that would make the inductive inference deductively valid (i.e. syllogistically) is 'The Uniformity of Nature', where the past, present and future are uniformly the same – logico-mathematically, conceptually continuous. 'The Uniformity of Nature' is what makes the universe *continuous* – in both space and time. 'The Uniformity of Nature' is what makes the universe *symmetric* – in both space and time. The rejection of induction as the logic of science entails a rejection of 'The Uniformity of Nature' and the rejection of the universal symmetry defining mechanics as the only path to making sense, the only path to the intelligibility of nature. This is an instance of what I have referred to as the Parallel Hypothesis, that there is a parallel between (1) what one takes to be the successful logic of learning and reasoning about reality and (2) what one takes to be the nature and structure of reality. In other words induction 'would be' the proper logic of science, of learning and reasoning, if the universe were uniformly, continuously mechanical; if all phenomena were governed by One universally uniform, space-time invariant order. Of course, if everything were comprehensively uniform, the same everywhere and always, it is not clear that there would be anything to learn. There wouldn't be any meaningful differences – at least by the criterion of Popper's Question.

I must confess that I and, I believe, most of the graduate students in the Philosophy Department at London School of Economics, were initially quite baffled by Popper's rejection of induction as the logic of science. Only later, when I realized that the evidence was overwhelming, that the biosphere, the Earth and, indeed, the cosmos must have an irreducible historical aspect – developing non-uniformly, non-mechanically – that I appreciated Popper's rejection of induction as the basis of the logic of learning.

What is at issue here is the intelligibility of reality and our place in it! Is reality understandable and explainable completely and consistently, universally, in terms of One uniform mechanical order?

The uniformitarian presupposition has often been represented in terms of a steady-state model of the universe. The early geologists reasoned that the Earth must be in a balanced steady-state and that it had always looked more or less the same: same processes, same results. Somewhat ironically, this steady-state balance is actually a dynamic balance, the result of contrary, possibly complementary processes.

In response to the growing evidence favoring the non-uniform, historical nature of the Big Bang model in cosmology, there was an effort to save Scientific Cosmology and the mechanical intelligibility of the cosmos. Cosmologists Fred Hoyle, Thomas Gold and Hermann Bondi famously formulated a Steady-State Cosmology – quite analogous to the steady-state model in early geology. New material particles emerged between galaxies as old particles were lost over the horizon of the expanding universe. As a result the universe would look, and always had looked, the same. Uniformity is another word for sameness and sameness is, in more modern parlance, about symmetry. Einstein had characterized the competing realities of the two mechanics – the uniformly local Newtonian point-reality and the uniformly non-local Maxwellian field-reality. Each of these realities is supposed to be 'objectively' uniform – symmetric – no differences of (the other) type anywhere in either type of reality. Each type of mechanical reality is symmetric – uniform – in its own opposite, complementary way. The Steady-State symmetry is different, a dynamic symmetry resulting from the action-reaction of 'mechanically' opposite processes. If the 'opposite processes' are not mechanically opposite, but rather complementary such that their interaction does not result in a net zero, then the 'interplay of these opposite types' must result in a net history.

Well into my second period of graduate study at University of London, my wife, Suzanne, and I rented a small vacation cottage over Christmas outside London, on the road to Cambridge. During an intense afternoon of study and exploration, trying to fit everything together, it finally just hit me. I remember sitting up in a sort of jump and saying to myself – out loud – "reality is just ambiguous!" The question I had been pressing was as to whether reality could be understood ultimately, exclusively, in either wave or particle terms, in either selfish or selfless terms, in terms of either of the several apparently timeless, mechanically understood, opposing orders. My epiphany was that the question was *undecidable* precisely because reality simply is not unambiguously – objectively, uniformly – one way or the other.

What the embrace of complementarity requires is a More General Theory that can make sense of a middle-ground reality with an irreducible aspect of both continuity (uniformity, symmetry) and discontinuity (non-uniformity, asymmetry). What is needed is a More General Theory that can make sense of an irreducible aspect of both constancy and change, that can make sense of an irreducible aspect of both order and disorder.

It is crucial for any acceptable More General Theory that neither the order nor the disorder aspects of reality can provide a comprehensive account of all aspects of all phenomena. Neither the order nor the disorder can be 'objective' – universal in the classical mechanical sense. Per hypothesis, with complementarity, every phenomenon, every measurement, as well as every observable system must have both a regular, uniform, non-historical aspect as well as a uniquely historical aspect.

28

Making Sense in the Middle-Ground – My First Hypothesis

In searching for the More General Theory, for the More General Theory's new way of understanding reality and our place in it, I surmised that I needed a new post-mechanical middle-ground way of understanding. My First Hypothesis as to the coherence of the More General Theory focused on the idea that there must be a non-mechanical historical aspect to the nature of reality.

Any such historical aspect, by its very nature, would appear to all possible time-invariant mechanics as mechanically incoherent, as a mechanically discontinuous aspect. Any such mechanically incoherent aspect would naturally appear, in mechanical, law-governed terms, to be an irreducible, non-law-governed, chance-like aspect.

However, in any acceptable More General Theory, such a non-mechanical historical aspect must be understood in some other way, in term of some other type of coherence.

Gould's reasoning about uniformity had been particularly revealing – so I pursued. Gould, trained initially as a geologist, carried his concerns and insights about uniformitarianism in geology further, into the arena of another of the 'historical sciences' – evolutionary biology. And clearly, Gould did not expect that the historical phenomena of evolutionary biology could be made sense of in terms of the classical mechanical framework of the hard sciences.

What had seemed significant and revolutionary to the early investigators of biological evolution was that there had been 'category changes', meaning changes of *type;* the categories of species in the world were not fixed once and for all and everywhere. The Evolutionary Hypothesis was about *developments* within a species, about *developments* from one species to another and about *developments* of the overall system. New types of phenomena (species) and new types of relationships between phenomena appeared to have evolved over time. Evolution had been a major focus of 18th and 19th century learned-dialogue much before Darwin. Evolution was about a progressive change involving some sort of qualitative improvement; some sort of betterment. There was no steady-state theory of 'evolution'.

'Mechanical change' in an isolated system is defined as a 'causally symmetric, law-governed rearrangement of the same basic type of stuff.' 'Mechanical change' is naturally non-progressive – symmetric. Mechanical change results in a steady-state rearrangement, preserving symmetry, preserving logico-mathematical uniformity and conceptual continuity. Recall Lyell's steady-state geology. There are no mechanically discontinuous qualitative changes in the *type* of laws or in the *type* of material, processes or phenomena. The classical conservation principles – Conservation of Energy and Conservation of Matter – are not discoveries, but, rather, are (logical) entailments of the Mechanical Philosophy's defining presupposition of symmetry. On the other hand, 'category change' suggested a qualitative change of *type* in reality. In terms of the laws governing phenomena, a category change would be a move from one type of regularity to another mechanically discontinuous type of regularity; where the latter could not be causally-mechanically generated from the former; where the latter could not be logico-mathematically reasoned or understood in terms of the former. A 'category change' would be a mechanically discontinuous 'jump' or 'leap' from one type of mechanical order to another.

The sense, the intuition, of the early evolutionary biology researchers, judging from the fossil record, was that the evidence indicated an ascent, a progressive development. Cumulative category changes, if they occur, would necessarily result in a non-mechanical history, but it may or may not be progressive. Any qualitative category change, by its very nature, had to involve, indeed entailed, mechanical discontinuity. Category change would be inductively unexpected since induction presupposes steady-state uniformity. Imagine adding a new type of part or process to a fixed and complete clockwork. If a categorical discontinuity did occur, it would be unpredictable and mechanically unexplainable as to how it had happened. For the evolutionary theorists the implication was that if there had been an evolution, a category change, it could not have occurred mechanically.

If a system were to change, non-mechanically, non-uniformly, then the present could not be used as the universal key to understanding the past. If systems can evolve, making mechanically discontinuous changes, then any successful inductions must involve idealization and must have an inherently limited applicability. The successes of the Scientific Research Program discovering space-time invariant laws, processes and types of phenomena, must be limited and must be based on local idealizations of relatively isolatable systems.

The accumulating fossil evidence supported the increasingly popular Evolutionary Hypothesis that there had been historical category changes in the things and processes in the biosphere, the geosphere and perhaps the whole cosmos. Most intriguing evidence from the fossil record plausibly supported the further hypothesis that the evolutionary changes had been, in some sense, progressive. But the Evolutionary Hypothesis was born in a period when the enormous successes of Newtonian Mechanics seemed to require that reality was a space-time invariant causal clockwork-like system. How else to explain the wide-ranging successes of the calculable, logico-mathematical predictions of Newtonian physics? There were reasons to suspect that there were inherent limitations of Newtonian mechanical calculations – such as the three-body problem, later rediscovered in the 20th century as Chaos Theory – noted, for instance, by Henri Poincare (1887), but summarily ignored. For the Newtonian practitioners, thinking 'outside of the box' of Mechanics about any eventual limits of the Newtonian Research Program was much less compelling than exploring how far the Newtonian approach might now be extended into new realms.

In his theory of biological evolution, Charles Darwin had introduced the notion of chance-governed variations, chance-governed discontinuities, to account for the occurrence of qualitative category change. Please take note of how strange it was for Darwin to propose a chance-governed process in a period when Newton's rigorously causal-law-governed clockwork model was being hailed as by far the most successful theory in the history of science! And yet, there was a new awareness of 'apparently' chance-governed phenomena in the environment. At the very least Darwin's chance-governed process was needed to account for the, then recent, observations of the chance-like distribution of differences in populations. The measurement and rigorous documentation of the 'regularly irregular' distribution of characteristics in biological populations was an obsession of Darwin's cousin, Sir Francis Galton. Adolphe Quetelet, Galton and many others had begun to document these patterns of the chance-like distributions of variations of all sorts of populations and collections, both animate and inanimate. How could one account for these regular irregularities – 'normal curve' distributions – except as the result of non-uniform chance-governed processes?

The chance-like 'regularly irregular' distributions of the biological variations meant that they must, by their very nature, be generated by a mechanically discontinuous process. These chance-like variations by themselves provide evidence of mechanical discontinuity in the supposedly clockwork-mechanical continuity. Darwin was asked about how we were to make sense of the chance-like nature of these 'variations' occurring within a Newtonian clockwork reality. Most commonly, when pressed, Darwin offered a Promissory Note, suggesting that the evolutionary process didn't 'really' have an irreducible chance-governed component. Rather, he suggested, these were complicated processes that 'appeared' to be chance-governed, but were really just 'observational uncertainties', reflecting the limitations of our current understanding of the mechanical details. But this chance-governed aspect of biological evolution has not been eliminated or even reduced by the considerable observational advances of the subsequent 100+ years. And the mechanical 'cause' of the chance-governed, random variations remains completely mysterious.

The modern neo-Darwinians have not abandoned Darwin's Promissory Note, wherein the 'apparent' chance-governed phenomena might 'eventually' be made sense of in the Mechanical Philosophy. However, they have abandoned the traditional commitment to the idea that biological evolution – the history of life – is progressive. The neo-Darwinians have now embraced the notion that 'evolution' is just 'change' – mechanical rear-rangements. This begins to appear to be a rather absurd position in light of the accumulating, well-documented fossil evidence indicating an historical sequence from a few simple organisms 3.7 billion years ago to the present biosphere. Just change? 'Merely' mechanical rearrangements?

Another long-standing auxiliary hypothesis of the neo-Darwinian lit-any, to account for the observed chance-like variations in biological popu-lations, is that the chance-like variations are due to 'mechanical errors' in the reproductive process. But the notion that a mechanical system can make 'random errors' is just oxymoronic nonsense. Does it make sense to explain counter-evidence to a Newtonian clockwork with an auxiliary hy-pothesis that claims that the system occasionally makes 'random errors'? Errors relative to what standard, relative to what expectation? Imagine in Lakatos's planetary thought experiment if one responded to the failure of the theory to account for variations away from the predicted course of the seventh planet by suggesting that this was due to the fact that the planets, although governed by the universal gravitational laws, sometimes make mistakes. Is this a scientifically credible position? I think not.

You simply can't account for any sort of 'real' category change in a me-chanical system. Gould hammered this point by raising serious uncom-fortable questions about the nature of the sequence of biological evolution, about the path of the history of life on Earth. Gould argued that categori-cal novelty – mechanical discontinuity – could arise in the history of life if and only if the 'apparently' chance-like variations were 'really' chance-governed. Gould wasn't buying Darwin's uniformitarian Promissory Note. Gould insisted that the chance aspect needed to be 'real' if the system was to introduce 'real net differences' – real category changes (viz. qualitative dis-continuities) between past, present and future. If evolution is 'real change' (possibly progressive), you need 'real' mechanical discontinuity. Since me-chanical discontinuity can't be made sense of in terms of mechanics, the discontinuity must at least 'appear' (paradoxically from a mechanical per-spective) to be arbitrary and to arise through a chance-governed process.

To propose that there has been a real historical category change in the *types* of biological phenomena over time means that there has been a conceptual discontinuity such that the concepts (or categories) used to understand the biological phenomena of the past are inadequate to understand, are unable to make sense (mechanically) of the biological phenomena of the present. (To presage a core theme of later chapters, the technological advances of the present cannot be understood in terms of the technologies of the past. Modern technological inventions cannot be logically deduced from the technologies of the past. There is a qualitative difference, a 'revolutionary' conceptual discontinuity. Biological organisms are, per hypothesis, embodied technologies.)

To make the point simple and clear, Gould, in his book, *Wonderful Life: The Burgess Shale and the Nature of History*, offered a powerful thought experiment, asking the following question: if we replayed the tape of the history of life on Earth, what reason do we have to expect the same outcome? Gould's answer was that, since the variations must be chance-governed (viz. mechanically non-uniform) in order to generate category change (qualitative novelty), we have no reason to expect the same sequence or the same outcome. The evolution of life must be 'complex' – a middle-ground mix of mechanical order and, from a mechanical perspective, mechanically discontinuous disorder. The process must involve irreducible mechanical discontinuities, and, the outcome must be an under-determined middle-ground mix of historical uniformity and historical non-uniformity.

Gould emphasizes that there is no reason to expect that the key variations, and in the right sequence, would occur on a replay, even so much as to allow for multi-cellular life to emerge. In any replay, the history of life on Earth might never have moved beyond the primitive bacterial level. Indeed, the bacterial level itself, the first primitive cells, may never have emerged.

At the extreme, from a mechanical perspective, there is no reason to expect life to arise at all. Indeed, it is fundamentally challenging to imagine how life – a thermodynamically non-equilibrium process – could have first arisen in terms of the mechanistic processes of physics and chemistry that presuppose thermodynamic equilibrium. Even now, from within the classical Scientific Research Program, it is unclear how a phenomenon such as life is even possible. How could one make sense of mechanically discontinuous, non-equilibrium life-processes in terms of mechanically continuous clock-work equilibrium processes?

If there has been real historical category change it must 'appear', in terms of any possible mechanical research program perspective, to be 'mechanically irrational', to have an irreducible chance-governed component. This doesn't mean that it is 'really' chance-governed anymore than the perception of the competitive perspective's experience of cooperative behaviors as irrational means that the cooperative behaviors are 'really' irrational. It just means that it doesn't make sense in terms of competitive concepts, processes and reasoning.

Gould's argument is that the history of life (evolution) is inherently unpredictable – per hypothesis, because it is mechanically under-determined. This means that given the history of life that we observe in the fossil record, what we can experimentally verify, there is no possible formally scientific explanation. There is no possible uniformitarian, inductive, mechanical clockwork explanation. If there is real, definitive evidence for mechanically discontinuous category change, it would constitute an answer to Popper's Question applied to the Scientific Hypothesis. Real definitive evidence for a mechanically discontinuous category change (viz. a random mutation) from one type of regularity (or order) to another type – breaking the uniformity, breaking the symmetry, breaking the conceptual continuity – would force one, would be a forcing argument, to conclude that the Scientific Hypothesis and the Scientific Research Program are inherently incomplete: there is no One mechanical order governing all phenomena. The evidence for biological evolution is then, per hypothesis, evidence of the limit of the Scientific Hypothesis and the Scientific Research Program. If the history of life involves category change it means that reality is not completely intelligible in mechanical terms, in terms of One space-time invariant mechanics.

Gould's representation of biological evolution and the resulting structure of the biosphere suggest a middle-ground, a mixture of clockwork and cloud-like processes and structures.

John Barrow, currently one of Hawking's colleagues in the Department of Applied Mathematics and Theoretical Physics (DAMTP) at Cambridge University in England, picked up on Gould's line of reasoning and applied it to the entire cosmos in his book, *The Constants of Nature*. Barrow asked, in effect: 'In light of the 'irreducible probabilistic factor' demonstrated in quantum physics, if we were to replay the tape of the history of the cosmos, what reason do we have to expect the same outcome – in particular, the same outcome in terms of the most fundamental constants of nature?'

Barrow outlines the difference between a 'rigid world' view, compatible with a Mechanical Theory of Everything, where there is only one possible history and the 'flexi-world' view, a sort of middle-ground of order and randomness, like Gould's, where there is no reason to expect the current observed outcome from re-runs of the history of the cosmos.

Barrow says, "Today, as physicists have followed the path towards deeper and more universal theories of the forces of Nature they have moved steadily towards the flexi-world view. There do seem to be constants of Nature that are not absolutely fixed by an all-encompassing Theory of Everything. Some appear there but are allowed to take a whole continuous range of values. Others don't appear explicitly in the Theory of Everything at all but emerge at particular stages in the evolution of the universe by a random process, like the needle balanced on one end that falls in some particular, yet unpredictable, direction. These constants take on values which manifest the way in which the outcomes of the laws of Nature need not possess the symmetries of the laws themselves: they are far more complicated and haphazard" (*The Constants of Nature,* page 181)

John Barrow

Barrow's answer, mirroring Gould's, is that in a cosmological replay we have no reason to expect the same outcome. The chance-governed 'fluctuations' entailed by quantum physics mean that cosmological history is, in some irreducible sense, in some irreducible aspect, under-determined, clockwork-mechanically under-determined and so, unpredictable. Barrow pushed the argument to the core of physics, to the constants of nature, pointing out that, on the basis of the new physics, there is no firm reason to expect even the formation of the elementary particles: protons, neutrons, and so forth. Even if they did form there was no reason to expect that they would have the same properties that allow, for instance, for the formation of atoms and molecules.

And if the history of the cosmos is to some irreducible extent mechanically unpredictable, under-determined, this also means that there could be no mechanical, no scientific, explanation of the organizational structure and processes we now observe, of the actual, currently observed outcome. If you couldn't have predicted the outcome, then you can't explain it, after the fact, with the same theory. There could be no mechanical, causally regular explanation, no clockwork-only explanation and no deterministic One-order explanation. Furthermore, the present regularities cannot be the complete and consistent key to understanding the past, cannot be the key to understanding the history of the cosmos and how it came to be as it is.

If scientific models of the universe are to be judged by their ability to predict, the proponents of such models, those still thinking inside the box of the Scientific Hypothesis, now find themselves in 'the worst possible situation imaginable.' According to Barrow's argument, no possible mechanical model could ever predict a future state completely and consistently because all historical sequences are under-determined. There must always be some uncertainty, regardless of your choice of mechanical models. And yet... scientists clearly can and do predict many phenomena. These 'successes' have been – so far – admittedly limited and local in space and time. But how are we to explain these apparent scientific, mechanical successes in a post-scientific, post-mechanical worldview? The classical Scientific Hypothesis postulated that our predictive successes were due to the existence of universal objective laws and regularities. But this isn't a credible explanation anymore. The history of successful predictions can't be explained in terms of One universal mechanics, in terms of the Scientific Hypothesis. Any post-scientific More General Theory must be able to explain the successes of the various mechanics, as well as their failures. If all classical, strictly mechanical theories are inherently incomplete, then all 'successful mechanical predictions' involve idealizations, and the corresponding mechanical theories are all special cases.

Barrow took his inquiry one step further, noting that the questions that he and Gould were asking were not unique to the historical sciences: geology, biological evolution and cosmology. Barrow notes that scholars that study the history of modern civilizations often consider 'virtual histories', asking 'what if' questions about how Western Civilization might have taken a different historical course. "Virtual history [counterfactual history] tries to predict (reason) what might have happened if some pivotal events had not occurred in the past or had been slightly changed," notes Barrow. For instance, what if Julius Caesar had survived the assassination 'attempt', or Martin Luther had died of cholera as a child; or what if the printing press, or any of a number of other technologies, had not been invented when and where they were?

"This excursion into the philosophy of history aims to show that [professional historians] are engaged in a lively debate that is curiously analogous to that going on within cosmology. ... Virtual natural history is an essential part of modern cosmology," says Barrow (page 194, *The Constants of Nature*).

The Engineering Perspective

Another discipline where historical 'what if' questions are standard fare, one that Barrow and Gould don't mention, is engineering. Engineers are constantly analyzing failures, asking what would have happened, for instance, if the materials in the bridge had been stronger, if the design of this or that component had been different. If you ask an engineer the historical question – how a certain operating structure came to be as it is – he is likely to answer that he can only give you a probable answer, since there were many different practical, causal pathways to each operational situation. The engineers' 'operational situation' here is not conceived of as perfectly defined, in perfect clockwork terms, but as an idealized system conceived in a particular, limited way, serving a particular purpose or type of purpose. How did that bridge come to be as it is? There are always many possible historical paths, the result, the outcome, having been under-determined by the past. How, then, did the universe come to be as it is?

Engineers are also quintessential Participants, always asking what might be possible in the future, asking about possible, more desirable, futures. Engineers explore and experiment to understand what could be, what might be possible, in the future. Engineers seek to bring about qualitatively better designs and operating structures. Engineers, in the broadest sense, are Participant problem solvers – creative value-actualizers.

What I am struggling to clarify here is the difference between a 'random' (viz. just 'change') aspect defining histories and a 'progressive' aspect defining histories. Both would 'appear' equally, to any strictly mechanical worldview, to involve mechanical discontinuities. If biological evolution and human socio-technological evolution are progressive and the process that brings about the progress is not mechanically coherent, then both the process and the outcome will appear to be mechanically under-determined and, equivalently, mechanically discontinuous and, from any mechanical perspective to be, to some irreducible extent, chance-governed.

We need a post-mechanical theory, a More General Theory to discover the 'coherence' of the progressive evolutionary process of innovation and its products.

Just because the inventions and innovations that, per hypothesis, are the marks of progressive biological evolution and human socio-technical history are not predictable, are not deductions from the current biology or the current technology and knowledge, doesn't mean that they are random, the result of chance-governed processes. Just because they are not mechanically predictable doesn't mean there is no historical coherence. More plausibly the historical coherence is to be found, per hypothesis, in some sort of a progressive Participant enterprise, a problem solving narrative where current technology and knowledge under-determine the advances (innovations) and are better understood as enablers of experimental exploration in the process of discovering, developing and implementing novel technologies and knowledge.

The Structural and Functional Implication

Another consequence of Gould's and Barrow's common theme of the mechanical discontinuity of actual history is that the outcome of such a sequence, at any point in time, – for instance, the current structures and processes of the biosphere and the cosmos – will not be mechanically intelligible. The outcome can't be made sense of completely and consistently in mechanical terms. This is simply another way of repeating the general point that the universe is not a mechanical clockwork. Consequently, the structures and operations of the universe are not completely and consistently mechanically intelligible.

A universe, a reality, with an irreducibly progressive history would, per hypothesis, appear to all possible mechanical perspectives to have an irreducible probabilistic aspect, an irreducible discontinuity amid the continuity, an irreducible difference amid the sameness, irreducible (discontinuous) change amid the constancy. Qualitative, mechanically discontinuous change suggests a qualitative emergence – a progressive history.

Quantum Theory and the Confirmation of Historical Development

The initial conclusion of my First Hypothesis research was that, from the mechanical perspective, what appears to be a chance-like aspect of reality might be understood differently, as revealing an irreducible historical aspect to all phenomena. If the historical aspect of reality is 'real' it must appear to be mechanically incoherent – non-law-governed, chance-like. A simple line of reasoning from the embrace of complementarity in the new physics is supportive of this tentative conclusion.

Complementarity in quantum theory entails that the choice of how to observe is a choice between mechanically discontinuous alternatives: the particle and the wave alternatives; (actually in the 'mixed' range between the extremes). Since the observer's choice moves the system (including the observer) asymmetrically down one historical path rather than another, the choice is irreversible. Furthermore, the series of choices is cumulative. Each choice moves the system/observer to a future state that is, to some irreducible extent, mechanically discontinuous from the prior state of the system. The new choice facing the observer is, to some irreducible extent, mechanically discontinuous from the original choice. Choice doesn't diminish or disappear; it simply transforms in a non-mechanical manner. Nonetheless, the future choices are, in some irreducible aspects, qualitatively distinct from the previous choices. The possibility space has transformed, to some irreducible extent, non-mechanically.

What has been under-appreciated is that, in quantum theory, although individual choices are represented as just mechanically arbitrary, they are logico-mathematically different – conceptually discontinuous. One option cannot be made sense of logico-mathematically in terms of the other option. Since the actual choice sequence is qualitatively transforming, a simple logico-mathematical account of the choice sequence is precluded. Whatever one might be able to say about the character, about the coherence, of the choice sequence, it cannot be understood as mechanical; for instance, as a simple logico-mathematically consistent rearrangement by eternal, continuous mechanical processes.

The 'choice' entailed by complementarity is ubiquitous and eternal, but the qualitative character of the choices transforms over space and time. The cumulative sequence is not completely un-determined and it is not completely determined – it is middle-ground, under-determined.

Another way to express these points is to emphasize that quantum theory is a Participant theory. In a Spectator theory the inquirer's choices as to what to believe and how to investigate are detached and have no influence on the nature of reality and the course of events. Observational information and influence flows in only one direction, from 'objective' reality to the idealized, detached observer. Spectator theories are comfortably mechanical, and learning is a hypothesized convergence of the Spectator's theories to an invariant objective target-reality. Participant theories accept observer involvement entailing that the overall system develops with the Participant's choices – and, per hypothesis, possibly develops progressively in some non-mechanical manner.

If the universe has a 'real', irreducible, mechanically discontinuous aspect, that aspect would appear to all possible mechanics as mechanically incoherent, mechanically irrational – as non-law-governed, perhaps as chance-governed. This doesn't mean that it is 'really' chance-governed anymore than the perception of the competitive perspective's experience of cooperative behaviors as irrational means that the cooperative behaviors are 'really' irrational. It just means that it doesn't make sense in terms of competitive concepts, processes and logico-mechanical reasoning. Similarly, the overall irreducible historical aspect of reality may have its own coherence, unintelligible in mechanical terms.

What, then, might be an alternative coherence of a mechanically discontinuous historical sequence? One possible example of a phenomenon that is historically progressive and coherent – by its very nature – is learning. However, learning (say the history of science) according to the Logical Positivist representation of science, is logico-mathematically inductive and conceptually uniform and continuous, paralleling the Scientific Hypothesis's presupposition of a mechanical reality. The Logical Positivist representation of learning is uniform and non-progressive. On the other hand, Popper, Kuhn, Lakatos and Feyerabend argue that any 'real' learning – what we mean by learning, by its very nature – is, at least to some irreducible extent, progressive, conceptually discontinuous, revolutionary and logico-mathematically incommensurable with the earlier learning, with earlier ways of understanding.

29

The Middle-Ground – My Second Hypothesis

In searching for the More General Theory's new way of understanding reality and our place in it, my Second Hypothesis arose from reflection on one of the prominent reactions within the academic community to Thomas Kuhn's presentation of the under-determinism theme in his book, *The Structure of Scientific Revolutions*. Although always in the background, it wasn't until some years later after returning to the U.S. and beginning my university teaching career at Linfield College, in Oregon, in 1980 that I was able to productively pursue the connections between 'progress', 'values' and 'learning'.

A popular response argued that if all possible evidence under-determines the choice of what to believe, under-determines the choice of how to make sense of reality, then some 'other' non-evidential factor must 'complete the determination', must complete the otherwise under-determined choice. In other words, if my choice of theories, my choice of what to believe, is under-determined by evidence, then some additional factor must be introduced to account for what I actually do choose to believe.

Complementarity entails that we are always, necessarily, choosing. And yet, from the point of view of quantum theory, it must be literally impossible to discover the complete mechanical causal determinant of the choice. There can be no scientific – causal mechanical – explanation of the observer's choice. By its very nature the choice appears to be mechanically arbitrary, mechanically incoherent, mechanically unintelligible.

Since the choice is always in the range between two complementary – irreconcilably different, qualitatively distinct – possibilities, the under-determinism theme, in quantum theory, is that the choice, whatever it is, however it is to be understood, represents a bias.

It was this 'qualitative' distinctiveness of the quantum theory choices that was easily interpreted in the broader academic community as involving a *value-bias*. The natural suggestion was that the final determinant of the actual choice must involve an irreducible, qualitatively 'evaluative' aspect.

Here, then, values reappeared – to the consternation of the Positivists who had hoped to close the door once and for all on religion and anything else that couldn't be calculated or reasoned logico-mathematically from simple observations. Quine had been quite explicit in his presentation of under-determinism saying that religious beliefs could not be 'objectively refuted', giving substantial solace to the established religions, but equal solace to any crazy idea that didn't specify its criterion of success.

The exploration of the First Hypothesis strongly indicated that the More General Theory's new way of understanding – the new coherence – involved an irreducible non-mechanical historical aspect. Building on the indications from the First Hypothesis, my Second Hypothesis was that the More General Theory's new way of understanding somehow involved a cumulative, historical series of value-laden choices.

In the philosophy of science community, the feminists and the Marxists were the first, referencing Kuhn, to offer evaluative critiques of the Positivist's representation of science. If one's choice of what to believe necessarily involved a value component, then the current 'supposedly objective' scientific representation of reality was an 'unjustified' male-oriented interpretation (reading) and was dominated by competitive capitalist presuppositions. In physics, they noted, military funding was enormously influential. In biology, the neo-Darwinian 'mechanism' presupposed the free-market economic values favoring the already advantaged, wealthy classes. Both the feminists and the Marxists gave alternative intellectual histories with many, many examples that at least made sense in terms of their perspectives.

From the beginning of the acceptance of under-determinism these controversies could not be resolved. What was the right way? What should we believe?

The classical scientific high ground, postulating one 'unique, objective understanding of reality' quickly eroded. Lacking a More General Theory, the situation drifted into 'post-modernist ambiguity'. The 'post-objectivist', post-modernist movement, focusing on the indefinite, under-determined nature of any formalized standard of 'right choice', pushed toward an extreme openness for the most part, just short of the unlimited, ideological relativism of un-determinism. Extreme relativism argued, in terms of the Parallel Hypothesis, that reality simply had no 'nature' – was un-determined, with no 'real' 'unambiguous' law-like relationships at all. Truth – or what each of us takes to be the truth – would no longer be objective but, rather, 'simply' a matter of each person's presumably arbitrary values.

And yet the choice entailed by complementarity is not unconstrained, is not un-determined. The choice is under-determined. The choice is not completely arbitrary. Complementarity and under-determinism simply require that there is an irreducible, mechanically indeterminate, freely evaluative *component* to all choices.

Nonetheless, the gradual embrace of under-determinism spawned a veritable industry of misguided attempts to provide self-referential 'scientific explanations' of why people believe what they believe, why they choose what they choose. Under-determinism seemed to many to imply that what people decided to believe was finally determined by their values, chosen prior to, or at least independent of, the act of observation. Later choices of how one understood new situations were presumed to be finally determined by a person's current value-biases, established by earlier value choices. The theme of the theory-ladenness of observation, discussed in the First Impressions thought experiment, supported that way of thinking. Once you had made a choice of how to understand a person or situation, you would *tend* to see everything in those terms in the future. However, the phenomena of conceptual conversions, where one switches to the opposite way of understanding, presented counter-evidence. Prior choices don't always, or perhaps ever completely, determine later choices. Conversions break the conceptual (inductive) continuity.

More generally, Kuhn's presentation of under-determinism arose from an examination of the history of science. The model of 'real' learning advocated by Kuhn, Feyerabend, Popper and Lakatos required regular discontinuous, non-inductive movements beyond one's current conceptual understanding; moving beyond one's prior choices, moving beyond one's prior understandings.

The psychology, sociology, anthropology and economics of the scientific enterprise were studied to seek out various influences on different scientists in different periods in different societies. The Marxists were, of course, inclined to see predominately class influences. Self-referential and cross-referential extensions of these studies led to curious enigmas. In so far as psychologists, sociologists, anthropologists and economists are themselves presumed to be empirically based scientists, the new studies implied that we should also ask self-referentially about the psychology of the choice of theories of people working in psychology; we should also ask about the sociology of the choice of theories of people working in sociology; we should also ask about the anthropology of the choice of theories of people working in anthropology.

Further disciplinary crossover studies should ask about the psychology of the choice of theories of anthropologist, the sociology of the choice of theories of psychologists and so forth.

Critics from all sides emphasized that the current, establishment choices – the current dominant scientific and religious understandings of reality, couldn't be empirically justified. Many of these numerous studies were meritorious with credible experimental designs. They have revealed and demonstrated that there are many types of 'influences' on what people choose to believe. Plausible as these accounts sound as determinants of choice, and supported by considerable evidence, they failed to appreciate that, from the point of view of the well-demonstrated new physics and under-determinism, the choice must be free (viz. mechanically under-determined) to some irreducible extent.

But how are we to understand 'free choice'? Only a few commentators really appreciated that the choices of belief systems are inseparable from the choices of the observer/inquirer as required by quantum theory.

These studies to decide, to explain, 'how we decide', always involved a Spectator perspective, parallel to the general Spectator enterprise of trying to understand 'how the world works'. The 'world' is tacitly presupposed to be 'objective' and 'out there'. These social scientific Spectators are attempting to explain the choices of other Spectators, from outside, somehow 'detached' from the actual context of moment to moment decision making of those other Spectators. Insofar as they suppose themselves to be 'scientists' in the Spectator's mechanical tradition, they are uninterested in the question of 'how we *should* decide'.

The Ph.D. thesis of one of my graduate school friends in London was a study of how city councils actually make decisions – in contrast to, he supposed, how they say they make decisions. Perhaps this was a reasonable study. However, having served on several decision-making policy bodies, I suggested to him that the more important question for those serving on such bodies was 'how *should* we make decisions?' My friend, emphasizing that he was a social *scientist*, responded that my question wasn't one that he could, or probably ever would, study. My friend was a Spectator, unconcerned with the 'real world' evaluative context of a Participant decision-maker.

What remained largely underappreciated in the Marxist and feminist debates and in the social science studies was that with the embrace of under-determinism and complementarity, one's choice of values was just as under-determined by all possible evidence as one's choice of facts. Indeed, and perhaps most important, fact choices and value choices now seemed to be inseparable. Since choice, by its very nature, involves a bias, the choice of what to believe about reality *factually* is no longer a value-free, value-independent choice about 'objective reality'. The choice of how to observe, or how to act in a system is no longer separable from the choice of why to observe or act in a system one way rather than another. That was the bombshell.

Despite the ample evidence of how people are influenced by one or another factor, there must always be an irreducible aspect of under-determined freedom in their choice, because that is the nature of the individual's place in reality, as a Participant.

Another way to make the point is to say that if there were an 'objective', observer-independent, causal explanation, then the choice wouldn't be a 'real' choice in the sense required by complementarity and quantum theory. If there were an 'objective', observer-independent, causal explanation of observer/actor choice, we would be forced back to complete determinism – a position already undermined by the embrace of complementarity and quantum uncertainty. No possible deterministic mechanical framework can make sense of 'real choice' – come what may.

As physicist Stanley Jaki once expressed it, 'there is overwhelming evidence for choice – for anyone who hasn't lost touch with common sense.' I always liked Aristotle's poignant reference to everyone's common experience of 'the agony of deliberation.' The existence of institutions of judgment (courts) and the numerous policy bodies in government and business make no sense if reality is fully determined. Real, meaningful inquiry and learning are pointless in a mechanically deterministic reality. Genuine discovery, invention and creativity would be illusions.

Further evidence for the reality of the inherent value bias of choice is seen in the perennial controversies. To a large extent it is the experience of these perennial disagreements that inclines us to search for a post-scientific More General understanding.

No possible deterministic mechanical framework, attempting to 'explain away' real choice, can provide a self-referentially consistent account of how the question of choice ever arose in the first place. There is, of course, the grossly implausible, self-referentially incoherent notion, popular amongst some ideological defenders of determinism, that among the possible deterministic universes we just coincidentally happen to be in one where it 'appears' that there is considerable evidence that we have real choice and that there is real practical value to learning.

The challenge and the prospect for the More General Theory now appears to involve providing a coherent middle-ground unification of the sciences and the humanities, erasing the ideological division between supposedly 'value-neutral facts' and supposedly 'fact-independent values'.

My First Hypothesis as to the coherence of the More General Theory was that it involved an essential, non-mechanical, historical aspect – indeed, perhaps a uniqueness in both time and space that is not mechanically intelligible. My mature Second Hypothesis became that the coherence of the More General Theory must have something to do with the historical narrative of the cumulative value choices. The More General Theory must be able to make sense of the evidence for 'real' choice – understanding it in a new middle-ground way, not completely free (or random) and not completely mechanical (fully determined). All we can say so far in terms of the Second Hypothesis is that the new coherence, the new understanding of the nature of reality and our place in it, must have something to do with a narrative of historically cumulative value choices.

If value and fact choices are inseparable, then, perhaps, as we learn, as we are increasing our *understanding* of facts, we should also somehow be increasing our understanding of what is valuable. And insofar as we embody that learning perhaps we embody that value understanding.

This fact-value merger is a clue to the unresolved issue in classical philosophy of science as to how to make sense of our intuition that one theory can be *better* (of greater value) than another.

The rebels regularly argued that there is no universal logic of inquiry, no universal logic of learning, no universal logic of coming up with successful new theories. There is no logico-mathematical scientific method that enables us to generate new better theories. Consequently, there can be no logico-mathematical explanation of how we make real advances in the history of successful inquiry. This is a version of the theme that 'if your theory can't predict an outcome, then it can't explain the outcome after it has occurred.' If your theory of learning can't generate new advanced theories, then that same theory of learning can't explain how you learned those advances after they have occurred. There can be a story, an historical recording, and a narrative of the exploration and the discovery, but not a deductive, logical explanation. Lakatos regularly commented that different people had different ways to come up with novel hypotheses – new theory choices to explore – possibly leading to discoveries: some drank wine, some went hiking in the mountains, some listened to opera and so forth. Along the same lines, Feyerabend became renowned for emphasizing the absence of any universal 'scientific method' of discovery.

The point is that there is no deductive, logico-mathematical way to produce qualitatively new, successful better theories – better ideas. Assuming that we can learn, one's choice of what to believe is not logico-mathematically determined by what one believed previously. Again, this is intimately connected to the Positivist's problem of explaining in what sense the new better theory is 'better'.

What choices might be justified, might be better? On what basis – indeed, on what *sort* of basis – would one fact-value system of beliefs be judged 'better'?

Having lost the traditional scientific standard – objective reality – for choosing what it is uniquely rational to believe, the default understanding of the final evaluative determinant of choice seemed to be, at least in the free-market-dominated West, whatever served the selfish interests of the individual. On the other hand, in socialist-leaning countries, the rational determinant of the choice was cooperative; whatever served the group interests.

The split is expected by complementarity, reinforcing the hypothesis that value choices and factual belief choices are inseparable and yet independent. Imagining competition and cooperation, per hypothesis, as separable value orientations – like research programs – understands them as being just as under-determined by the evidence as if they were factual choices about the universal nature of reality.

With the loss of objectivity, there has been a shift from the theoretical context toward the practical context corresponding to a shift from the traditional Spectator perspective toward a Participant perspective. What people believe is perhaps primarily a matter of what works for a particular purpose. Purposes – teleology – now appeared to be inherently local and perhaps inherently biased. What works for one purpose doesn't automatically work for another purpose. In any given situation there can now be 'rational disagreements' about how best to understand and deal with a situation. These hypotheses might differ widely with no prospect of finding an 'objectively defensible' common understanding, based on a globally common purpose.

In Participant thinking, one's beliefs are naturally based on practicality and empirical demonstration. Do they work? Would one belief system, one theory, work better than another – at least for now, in this place, at this time, in this circumstance? The move away from absolute objectivity does not force us to an absolute relativism, but it does introduce an element of relativity – the 'free-ness' of choice and a resulting 'rational diversity'. There is an expansion of plausibly rational alternatives. The new precaution is to avoid extrapolating the local success of one's ideas, of one's theory, to a global ideology. All local successes are inherently biased and should not be taken as a representative sampling of the One universal global order.

Hawking, in his recent book, *The Grand Design*, endorses 'model-based realism,' proposing belief in models of reality that are useful – but only in so far as they useful, when, where and how.

The perennial debate between free-market individualism and socialism, regarding how we should live, regarding how to organize society, might now be thought of as a series of local 'rational disagreements' over how to advance society. In the Participant context these are not just complementary Research Programs, about how to understand 'descriptions', but more like complementary, experimental and exploratory Research and Development Programs, exploring how we should live. Thinking beyond the 'one right answer' attitude, each approach might appreciate the value of the other as 'a loyal opposition'. The complementary alternatives are locally competitive and globally compatible. With a Parliamentary Attitude, the perennial controversies are meta-cooperative dialogues involving 'a common faith' that the oppositions can find a creative, constructive middle-ground, a 'win-win' middle way path forward.

From a middle-ground perspective, one can easily imagine that in one setting more cooperation is better and in another setting more competition is better. Complementarity had given firm ground to the speculation that ideas of what is real and what is better are 'essentially contestable concepts.' Any 'practically rational' choice, being under-determined, is constrained, nonetheless; constrained in any given situation, local in space, time and circumstance. It would not be 'practical' to choose to believe just anything.

Interpreted in the Participant's practical context, the choices entailed by complementarity always involve uncertainty. All practical choices, seeking to bring about, to actualize, some better future, are always, to some irreducible extent, *experimental and exploratory*. You can't know the right path for sure before you act; failure is always possible.

The representation of choice in terms of complementary value-framework alternatives – like competitive and cooperative – makes sense in a middle-ground context. Sometimes one chooses to be competitive, sometimes one chooses to be cooperative. In the middle-ground these are compatible contraries. – They are locally competitive with each other in any given situation and yet still rationally, globally compatible alternatives. In Oakeshott's middle-ground society, necessarily local, complementary alternative choices and types of choice may serve over time to balance the system.

Consider the following metaphor: imagine the problem of life as the problem of staying on life's winding path. One ideology says that when in doubt always turn to the right. Another ideology says when in doubt always turn to the left. Each ideology, rigorously pursued, would spin in on itself and, per hypothesis, self-destruct. The perennial controversies are globally unresolvable. The middle way path of the balanced system survives and thrives, not the ideologically competitive or cooperative extremes.

The problem of how to decide on how to move forward on the path forward suggests the need for a Parliamentary Attitude, respecting alternative, particularly complementary, rationalities. A middle-ground perspective would expect that people don't really decide all one way, or once and for all, what to believe or how to act. Even when they do decide, here and now, they can change their minds. They can convert. They can develop creative compromises. Indeed, in the middle ground context all choice involves a compromise of ideology, since all choice involves an irreducible element of the opposite. From a middle-ground perspective, a healthy society – and perhaps the healthy individual – is able to entertain and balance rationally distinct alternatives. Locally, different approaches are competitive, while still being globally compatible, loyal oppositions.

Following along the lines of Oakeshott's conception of society, from a middle-ground perspective, one might expect that if a group of free-marketers were put on an island they would soon separate into free-marketers and socialists, and likewise if a group of socialists were placed on an island they would soon separate into socialists and free-marketers. The expectation is that all working societies embody this sort of dynamically balancing – distributed rationality. It would be self-destructive to convert everyone to one or the other ideology.

Complementarity associated with the new physics and with Popper's Question argues for a formal pairing of opposites. But under-determinism might be taken to suggest that there might be many complementary pairings. There are many complementarities recognized in modern physics. A balanced society might be more like an ecosystem, incorporating multiple, compatible – sometimes cooperative, sometimes competitive – alternative fact-value-systems, each viewing and approaching reality differently, selectively through a series of choices. One might expect a sort of 'ecology of rationally balancing beliefs', rationally balancing ways of choosing, rationally balanced ways of making a living. Contrast this image with the mechanistic, objectivist clockwork image from the Scientific Hypothesis, with One universal, time-space invariant order – One universal, ideological rationality, One *logos* – governing all phenomena.

If the objectivist 'one right answer' representation of reality were correct, then, after 3.7 billion years of evolution, it must be considered quite strange that there is so much disagreement in human society, so much 'rational' diversity in the biosphere.

The middle-ground expects that the perennial controversies are complementary, so that the global questions of truth and justice are 'objectively' undecidable. Embracing complementarity, 'the nature of reality' becomes, in universalist, ideological and scientific senses, 'an essentially contestable concept.' There simply isn't One observer-independent, right way to understand the nature of reality, everywhere for all time. Clearly complementarity, as demonstrated in the new physics, points us to a path outside the box of objectivist mechanics toward some sort of middle-ground.

The investigation of the First Hypothesis as to how to make sense of the middle-ground had pointed to the evidence for an irreducible non-mechanical historical aspect to reality. Plausibly, this view naturally extends to the suggestion that there is, correspondingly, an irreducible non-mechanical spatial aspect to the whole of reality – interpreted simply as saying that the structures and functions of reality are not, completely and consistently, mechanically intelligible, as saying that the structure and function of reality is not a complete and consistent clockwork-like mechanical causal system.

The Second Hypothesis is based on the recognition that the choice entailed by complementarity is a choice in the middle-ground, defined as a range between qualitatively distinct mechanically discontinuous alternatives. Since the choice requires a bias, and the bias is plausibly interpretable as a value bias, the Second Hypothesis is that the More General Theory's new way of understanding reality somehow involves values. The exploration of the Second Hypothesis suggests that the new coherence of the More General Theory, the new post-scientific way of understanding the nature of reality and our place in it, might have something to do with a narrative of historically cumulative problem solving value choices.

30

'Real' History as Progressive
Learning – My Third Hypothesis

My Third Hypothesis as to the More General Theory's new way of understanding reality and our place in it focused on the possible binding coherence of a narrative with historically cumulative value choices. My Third Hypothesis was that the narrative of the historically cumulative value choices was not an arbitrary rambling but somehow constituted a progression. Fairly late in my research I began to see the centrality of inquiry itself – *and learning* – to the understanding of a possible progressive narrative coherence of history.

Descartes's Nightmare

In his famous *Meditations on First Philosophy* Rene Descartes reflected on the history of his own personal inquiry into the nature of reality and the choices he had made about what to believe over time. He reflects on the fact that, as he grew up, he held a number of different beliefs. Each belief system was eventually rejected as false in favor of a new belief system. Descartes does not focus on whether the sequence has been progressive. He primarily concerns himself with finding some foundational belief system that cannot – by its very nature – be open to error and falsehood. Descartes reasons that all empirical, observational methodologies have been shown to be at least occasionally unreliable. Descartes sees himself as facing a potentially unresolvable open-ended uncertainty – precluding a true, completely reliable, understanding of reality.

Descartes's Nightmare, as I refer to it, is that he has just been moving from one arbitrary belief system to another. Beliefs change, but is there any progress? The Nightmare is that he is just wandering aimlessly from one arbitrary belief system to another. Each choice generates a different understanding, a different way to make sense of the universe. Descartes no longer has any confidence that the next choice will be 'better' than the last – closer to the truth. Descartes takes seriously the counter-inductive inference to the earlier Promissory Note – reasoning now that since all theories in the past have turned out to be false, the most reasonable expectation is that the current theory and all future theories will turn out to be false.

The embrace of complementarity entails that choice and uncertainty are necessarily intractable. This entailment precludes any ultimate convergence to a complete and consistent, objective Theory of Everything. If the observer's choice is mechanically and logico-mathematically arbitrary as represented in the standard mechanical interpretation of quantum theory, then, from the Spectator's objectivist point of view, we are continually – eternally – confronted with Descartes's Nightmare. There is movement, change of belief, but no way to represent this as a progression – as arriving at 'better' theories, at a 'better' overall understanding of the nature of reality. And as the rebels' critiques of Logical Positivism established, since all theories have an unlimited number of possible predictions, there is no way to establish quantitatively – by number of successful predictions currently recorded – that one theory is objectively, universally 'better' than another.

My investigation of the Third Hypothesis, as to the possible coherence of the historical sequence as a value choice sequence, began with what I call, The Clue: There is an unexpected experience that nearly all scientific researchers have, namely, that an advance in knowledge while providing closure on some questions, at the same time opens up a vista of new questions – new *types* of questions. These new questions were literally inconceivable in the terms of the pre-advance knowledge and understanding. For instance, after the Copernican Revolution, one could newly ask about circumnavigating the planet and perhaps consider launching an artificial satellite to orbit the Earth. These opportunities and the associated questions were literally *inconceivable* in the earlier, otherwise empirically well-established, Earth-centered framework and Flat Earth Theory.

The emergence of conceptually discontinuous new types of questions is completely unexpected in terms of the Spectator's conceptually continuous convergence model. The Spectator and Logical Positivist representation of inquiry tacitly presupposes One mechanics, One order, One time-space invariant coherence. Consequently, the expectation would be for a logico-mathematically, conceptually continuous, inductivist history of science. The emergence of conceptually novel questions is unexpected, doesn't make sense. Kuhn's careful treatment of the real history of science, with arguments for conceptual revolutions, laid the groundwork for appreciation of The Clue – for the appreciation of novel questions and for an understanding of inquiry as recursively enabling, as qualitatively emergent, as progressive. The answer to one question leads to new questions that were inconceivable in the context of the original way of understanding.

With serious doubts as to whether there is one, objective, right way to understand reality, there is a natural concern with the opposite, relativistic extreme. The worry that any sequence of theory choices might be simply an arbitrary wandering arose, as Descartes's 17th century concern with the Nightmare suggests, well before the conundrums of quantum theory.

The primary critics of the Logical Positivist model – Popper, Quine, Kuhn, Feyerabend and Lakatos – all maintained that earlier and later theories in an advancing, learning sequence were incommensurable – qualitatively, conceptually new, different ways of understanding. 'Real' learning, according to these critics, is revolutionary – logico-mathematically discontinuous. Any new, more advanced way of understanding is always qualitatively new, conceptually new. An advance in science is a conceptual advance, a logico-mathematically discontinuous, conceptual advance. Lakatos argued that any genuinely new, more advanced theory should be expected to identify and explain some novel *type* of phenomenon, should be able to pose and answer some novel *type* of question. To clarify Lakatos's point, the new advanced theory should be able to identify and demonstrate a novel phenomenon that was inconceivable – incoherent – in terms of the previous way of understanding. 'Real' learning, a real advance, moves us to a new coherence, moves us to a new way of making sense of reality and enables new ways of questioning reality, enables new more advanced experiments and explorations.

The Clue – emergent questions – points to the Popper-Quine-Kuhn-Feyerabend-Lakatos developmental theory of learning. Their rebellious new interpretation of the problem of theory-choice is to place it in a qualitatively progressive historical learning sequence, *characterized by qualitatively better and better theories*. This is also a response to Descartes's Nightmare of qualitatively distinct, but disordered, alternative understandings. If there are qualitatively 'revolutionary' conceptual advances in 'real' learning then you 'see' – understand – the universe in progressively emergent new ways as you learn. It is then natural to expect progressive new types of questions – questions inconceivable in the earlier ways of understanding.

The Paradigm Shift from Spectator to Participant

Quantum theory and Popper's Question led us to embrace complementarity. Complementarity led us to the entailment that the 'reality' of choice as an irreducible component of the historical evolution of the universe. In quantum theory nothing *actualizes* without an observer's choice. Quantum theory cannot be understood as referring to an observer-independent, 'objective' reality. Quantum theory cannot be understood within the Spectator's Mechanical Philosophy as capturing or referring to One complete and consistent, observer-independent, universal mechanics.

John Archibald Wheeler

Quantum theory must somehow be understood as a post-scientific Participant Theory.

One of the core founders of quantum theory, Nobel Laureate Wolfgang Pauli expressed it concisely: "In the new pattern of thought we do not assume any longer the *detached observer*, occurring in the idealizations of this classical type of theory, but an observer who by his indeterminable effects creates a new situation, theoretically described as a new state of the observed system. In this way every observation is a singling out of a particular factual result, here and now, from the theoretical possibilities, therefore making obvious the discontinuous aspect of physical phenomena."

(Pauli, Wolfgang, *Writings on Physics and Philosophy*, edited by Charles P. Enz and Karl von Meyenn, (1994), pages 32-33)

John Archibald Wheeler was one of the 20th century's leading cosmologists, renowned for coining the expression 'black hole' and as teacher of Richard Feynman and Kip Thorne. In an article entitled "Genesis as Observership," Wheeler coined the expression "Participatory Anthropic Principle" – agreeing with Pauli, articulating the implications of the new understanding of observation in quantum theory that sees us as Participants embodied in a historically developing quantum universe.

Wheeler remarks: "Modern quantum theory, the overarching principles of 20th Century physics, leads to quite a different view of reality, a view that man, or intelligent life, or communicating observer participators, are the whole means by which the very universe is created: without them, nothing. … The universe does not exist 'out there,' independent of us. We are inescapably involved in bringing about that which appears to be happening. We are not only observers. We are participators. In some strange sense, this is a participatory universe." (Wheeler, John Archibald, Quoted in Denis Brian, *The Voice Of Genius: Conversations with Nobel Scientists and Other Luminaries*, (1995) page127.)

What then is involved in representing the inquirer as a Participant – naturally 'inside' the system? What is involved in representing inquiry as naturally 'inside' the universe, as an aspect of the nature of reality?

The move from the Spectator to a Participant representation of inquiry is a discontinuous Paradigm Shift, is a move 'outside the box' of the Spectator presuppositions to a More General Participant Theory. The detached Spectator representation of the inquirer is no longer viable, no longer able to make sense of the evidence from the new physics. The Spectator representation of both inquiry and the inquirer as 'outside', as 'detached' from, the universe is clearly an idealization. In the More General Participant Theory the Spectator representations of successful inquiry are to be newly understood as limited special cases. In the Participant representation, inquiry is an active enterprise to be understood as an irreducible, integral feature of the nature of reality.

The Paradigm Shift to the More General Participant representation of inquiry can also be characterized as a Problem Shift – a broadening reformulation of the problem of inquiry, a broadening reformulation of the nature of inquiry. The Spectator's problem of inquiry is to discover *how the world works* – how the objective world (out there) works, independent of the Spectator. The Participant's problem of inquiry is to discover *how to work in the world*, how to do things inside, as part of, reality. The Participant inquirer, as a problem solver, is also naturally attempting to bring forth a better understanding, to bring forth greater value into the universe. As an embodied problem solver the Participant is naturally – by his/her very nature – trying to understand *how to make the world work better – how to bring about a more desirable future.*

Reasoning from The Parallel Hypothesis

The Parallel Hypothesis suggested that there is a parallel between one's theory of how one learns about reality and one's theory of the nature of reality. For instance, the Logical Positivist's attempt to represent inquiry as a Spectator enterprise logico-mathematically converging to complete and consistent understanding of an objectively mechanical reality – 'stood to reason'. If reality had been objectively, uniformly mechanical, successful learning, the history of science, should have proceeded by a universally inductive, logico-mathematical scientific method and should also have revealed systematic, logico-mathematical-like advances.

With the Paradigm Shift to a progressively historical Participant representation of successful inquiry, the implication, reasoning from the Parallel Hypothesis, is that the nature of reality itself develops – evolves – historically, progressively through some sort of qualitatively emergent (viz. self-referential, recursively enabling) learning process. As Wheeler outlined, the Participatory evolutionary history of the universe is a progressive history of active Participant 'learning by doing.'

If successful Participant inquiry is emergent, cumulative and progressive it follows that the nature of the Participant's reality must also be, to some irreducible extent, emergent, cumulative and progressive. The Participant doesn't simply want to know how things work here and now. He has a broader, *more general* agenda. The Participant inquirer wants to discover how to change how things work, how to progressively develop the structure and function of the universe – including himself. In one aspect the progression is a recursive process, self-expansively 'learning how to learn' – learning how to do more, to explore novel, progressively expanding, qualitatively emerging vistas.

What we need at this point is a more fully developed Participant representation of inquiry. Fortunately, one has already been partially formulated.

American Pragmatism is a Participant theory of knowledge and learning. Charles Sander Peirce, the co-founder, was one of the first to draw out the implications of a boot-strapping theory of progressive evolutionary learning. John Dewey, a student and colleague of Peirce and William James, in his *Essays in Experimental Logic*, argued that qualitatively new *types* of hypotheses emerge as a result of advances in knowledge (viz. The Clue). Each step in learning increases the ability to learn, opening novel vistas for exploration. As a new way of understanding emerges – that could not have been formulated in the terms of the previous way of understanding – new questions emerge, new opportunities to question emerge.

In the Spectator representation of inquiry, learning is 'just for fun'. For the Spectator knowledge provides no real benefit since everything that will happen is fully determined from the beginning. In the Spectator representation of inquiry, knowledge has no real *value*. By contrast, in the Pragmatic theory of knowledge, meaningful knowledge always has at least some potential, practical benefit. To acquire knowledge of how to work better in the world is to acquire something of value. The history of the acquisition of increasingly beneficial knowledge of how to work better in the world presents an historical narrative of cumulative value manifestation.

For the Pragmatist all 'real' meaningful knowledge can, at least initially, be understood as answers within the context defined by the question – *how* to work in the world. In this sense, it is all about *methods*, about *how* we could live in the future. The broader context of Pragmatic inquiry involves the merger of the factual, 'instrumental' question with the progressive value question about *how* (by what methods) we could live *better*. All answers, all solutions are meaningful, are of value, if and only if they are potentially useful in the Pragmatist's naturally exploratory and experimental Research and Development Program – to bring about a more desirable future.

One way to capture the progressive aspect is to note that an advance in learning is recursively enabling, increasing the potential to learn. When one learns one also learns how to learn better. With the new type of questions – of The Clue – comes new methods of questioning.

The shift from the Spectator to the Participant model of learning is also a shift in the standard of judging success in learning. The Spectator's standard criterion of successful *prediction* of what is presumed to be happening anyway, independent of the inquirer, is no longer a defensible 'gold standard' of learning. The criterion of a real advance in the Pragmatic theory of knowledge is a Participant-type of evidence, the *demonstration* of what one can do with the new knowledge that was not even *conceivable* in the earlier pre-advance understanding.

The Pragmatist wants 'the choice of what constitutes a better theory', to be judged in terms of the potential benefit that the choice provides the learning community, recursively enabling.

Discovery in the Participant-Pragmatist representation of learning always requires an irreducible element of novel exploration and experimentation. Discoveries are, by their very nature, 'conceptually surprising', revealing something that wasn't even conceivable in the prior conceptual framework. Advances in learning are always, to some irreducible extent, discontinuous advances in understanding.

Acquisition of a qualitative new way of understanding is not a logical or systematic process. As Kuhn-Popper-Feyerabend-Lakatos argued, 'real' learning is inherently problematic. The inherent uncertainty and ambiguity revealed by quantum theory make the discovery of new, better 'scientific' theories inherently problematic. What mechanics must see as the arbitrariness of choice is newly understood in the Participant-Pragmatist representation as the necessarily exploratory character of real learning, involving genuine novelty in discovery.

Dewey's thesis is not just that an advance enables us to ask new types of questions. Dewey argues that we also learn how to ask *better* questions and, perhaps, how to better ask questions. For Dewey, an advance in learning is an advance in our ability to learn in the sense of being an advance in our method of inquiry – as if it were an advance in what the Positivist's imagined was a fixed, universal 'scientific' method of inquiry. There can be no universal, invariant 'scientific method'; no universal, invariant logic of discovery because the nature of reality itself changes, the nature of reality develops. As we learn, the nature of reality develops. As we learn, the nature of reality learns. With our advances in understanding, our method of inquiry, the 'logic' (viz. method) of our inquiry, advances. A characteristic of the Participant-Pragmatic representation of inquiry and knowledge, consistent with the Problem Shift to 'how to work in the world', is that it is all about *method*, answers to the defining Participant questions: *how* we should live, *how* might we make the world work better. In the Participant-Pragmatic representation of inquiry and knowledge, new questions and increasingly advanced methods of inquiry emerge in unexpected and unpredictable ways.

The embrace of complementarity entails that the uncertainty never goes away. The uncertainty, embodied in the current questions, transforms and develops. This is equivalent to saying that the opportunity for progressive developmental questioning, exploration, discovery and the progressive development of reality is open-ended.

For the Spectator inquiry is represented such that the problem that learning is trying to solve is the problem of ignorance. Spectator learning is a convergence to knowledge of the nature of a fixed, time-space invariant reality. Spectator learning attempts to eliminate doubt and uncertainty. For the Participant inquiry is represented as emergent and progressive and so doesn't attempt to eliminate uncertainty in the classical scientific sense. The Participant inquirer's problem, his choice, doesn't decline or converge to a fixed solution. Since complementarity, and the uncertainty it entails, is ubiquitous in time and space, the Participant inquirer's opportunity to learn never declines. For the Participant, the choice, the uncertainty, entailed by complementarity, is irreducible. For the Participant the problem of learning, the opportunity to learn, emerges, expands and develops.

The Engineer as Participant

One possible characterization of the Participant-Pragmatist inquirer is as a research and development engineer. Engineers conceive of themselves as practical problem solvers, where problem solving is understood as attempting to move from a current state of affairs to a future more desirable state of affairs. Engineering, as both inquiry and action, is attempting to learn how to bring about a better world through a process of actually trying – 'learning by doing', somewhat blindly, experimentally – to bring about a better world. Participants, understood in this broad sense as engineering problem solvers, are naturally attempting to progressively improve the structure and function of reality, self-reflexively including themselves.

Learning in the Participant-Pragmatist framework is part of an expansion of the ability to act in the world, of the ability to act in more diverse ways, perhaps more efficiently, perhaps more effectively, for whatever purpose. An advance in learning is also an advance in the ability to learn and a qualitative transformation of one's choice, a qualitative transformation of one's freedom to act. An advance in learning is an expansion and a diversifying emergence. Learning and the 'applied' developmental action that it enables are part of a progressive and emergent process. Biological evolution in a post-scientific Participant-Pragmatist model is to be understood as occurring through an embodied process of active, participatory problem solving and progressive learning.

Pragmatist and Systems Theorist C. Wes Churchman explored the question of the nature and evolution of a learning system, as a sort of bootstrapping, problem solving, problem generating process, as a recursive, emergent self-construction (*The Design of Inquiring Systems*). Accordingly, in the evolution of a research and development system, each 'advance' to 'better' theories enables the practical development of the system, stimulating a reorganization of the design of the system; further increasing the system's capacity to choose how to live, to problem solve, to inquire, to learn. New, qualitatively different types of exploration, experimentation and questioning are enabled.

If the Participant-Pragmatist representation of inquiry provides, per hypothesis, a More General understanding of reality and our place in it, then it must be able to subsume and supersede the Spectator representation of inquiry, the Mechanical Philosophy and its associated mechanical knowledge. The initial characterization of the question defining the core of the Participant-Pragmatist Research and Development Program is often posited to be 'merely' instrumental – simply about 'how to work in the world'. And yet this leaves open the question: instrumental to what end? The instrumental question is embedded in the more general defining question of the Participant-Pragmatist Research and Development Program: how to make the world work better? Duke University Professor Henry Petroski, in his book, *The Essential Engineer: Why Science Alone Will Not Solve Our Global Problems*, has argued that (instrumental) scientific knowledge is a tool in the more general engineering enterprise. Petroski suggests that instead of thinking of engineering as 'merely' applied science, a better way to understand the relationship is that traditional scientific inquiry is really engineering research. Scientific inquiry is meaningful, and makes sense, only in the more general engineering framework. Science in this new way of understanding is exploring the 'possibility space' of how the world works here and now, when engaged in this way or that, setting the stage for the value-laden engineering problem solving, trying to bring about a more desirable future. Stanford Aeronautical Engineering Professor Walter Vincenti in his book, *What Engineers Know and How They Know It*, has offered a fundamental challenge to both the classical scientific representation of how successful inquiry works and to the associated scientific theory of knowledge. Vincenti points out that scientific knowledge is an essential tool of engineering problem solving, but emphasizes: "science doesn't tell you how to build an airplane."

Gould and Barrow both note that the mechanical interpretations of the evolution of life and the cosmos enigmatically involve irreducible, non-law-governed, chance-like components. Clearly, if there is an irreducible progressive aspect to the nature of reality, it is not mechanically intelligible. Consequently, this progressive aspect will appear in mechanical models of reality as irreducibly incoherent, as arbitrary, perhaps as chance-governed. In the Participant-Pragmatist interpretation evolution appears more like a recursive learning process, like a boot-strapping – yet constructive – engineering enterprise involving genuine exploratory research and development and discovery of novel innovative methods.

During my first two years at Berkeley, I didn't encounter an engineer or learn anything about engineering. I had never been formally exposed to the concept of engineering in high school. In my junior year at Berkeley, I happened to room with two engineering students. It was only then that I realized that I had never even considered engineering as a possible major or career. Still pursuing a major in Astronomy, I had perhaps tacitly imagined myself as a Spectator. Except in the world of common sense, I had never really thought of myself as a Participant in the development of reality. But, upon reflection, the questions I had always been asking about the nature of the universe and my place, our place in it, were already outside the box of the scientific litany and just never make sense in the Spectator framework.

Pragmatist John Dewey characterized the evolution of the cosmos as a sort of exploratory Research and Development Enterprise to bring about, through exploratory problem solving a more desirable future. In Dewey's expression, the enterprise is better understood as 'the construction of the good.' One of Dewey's early inspirations was the German philosopher, George Hegel, who characterized history as 'the unfolding of an idea… and the idea is freedom.' In the broader picture Hegel might be understood as suggesting that the evolution of life and the cosmos is the emergence of an idea, and the idea is freedom in the Participant-Pragmatist sense of 'increasing ability to act in the world, to problem solve, to bring about an open-ended, more desirable future.

31

Meeting with Students with Disabilities at Science World

Once you start to think outside the box of the Scientific Research Program as it is manifest in the academic setting it becomes clear that it is represented as 'detached' from real world problem solving. The 'official' concern at least is with description and passive prediction, with 'pure' research and knowledge unsullied by real-world practical concerns. The Scientific Research Program is concerned only with *describing* 'how the world works' – in mechanical terms, in terms of repeatable, time-space invariant, causal relationships.

Hawking's 'real world' life as a Participant in society, most dramatically reported here in his interaction with students with disabilities, is not simply about discovering the unchanging, timeless laws that determine the inevitable course of events.

Canadians are different from Americans. Microsoft Canada is correspondingly different from Microsoft USA. Jackie Slemko of Microsoft Canada barely hesitated when I asked her to chose between two options for a possible 'secondary event': a VIP dinner, similar to what Nathan Myhrvold was planning in Seattle, or a session with students with disabilities. Once I had described the meeting with the students in Portland, there was no hesitation. "Yes, let's set up a meeting with students," said Jackie. "That may go over very well right now in British Columbia."

By far the most striking feature of Vancouver's Science World is the building, a huge Buckminster Fuller geodesic dome, more than a half sphere, but not quite a full sphere. Shining silvery metallic, it markedly enhances the Vancouver skyline, particularly at night, when illuminated by powerful spotlights. Coming face-to-face with this incredibly beautiful building, as you come around a corner in Vancouver, is breathtaking. The location, in a relatively flat area of the downtown, makes it visible at a distance from many directions. The medium is the message. It speaks to you: 'this is science, this is technology and engineering and architecture – a glimpse of humanity's potential and the promise of a more desirable future, a more beautiful future.'

Inside, Science World is all lights and gadgets, typical of the international science museum genre. A variety of 'teach yourself about this' interactive displays are built in and around the complex internal environment – everywhere. A full-service museum store greets you on your left on the way in and on your right on the way out. Science World is about both teaching and learning – neither of which makes any sense if we live in a completely deterministic reality.

Jackie let me know that everything about Stephen's meeting at Science World needed to be coordinated through Sid Katz, the current director. I had met with Sid on my earlier reconnoitering visit to Vancouver a few months earlier. Luckily the enthusiasm of the Ministry of Education meant that they would take care of nearly everything: the television production, broadcasting, transportation of the students – lights, camera, action. I met again with Sid to walk through basics like where we would park and Stephen's route from the parking lot to the stage. Joan had emphasized the need for a 'green room', where they could prep Stephen just before he goes on stage.

Sid Katz's biographical profile reveals an energetic creative mind. He was an excellent selection to be the director of a modern science museum. Sid is a polymath – pharmacologist, broadcaster, folksinger and baseball expert. As chief executive officer of Science World, as well as an active researcher in University of British Columbia's Faculty of Pharmaceutical Sciences, Katz had the drive, the charisma and the network to plug science and technology into British Columbians' everyday lives.

Meeting again with Sid, working on last minute details, he let me know that he and his colleagues at Science World had come up with a great idea: to create a new, permanent display to celebrate Hawking's visit to Vancouver and Science World. Specifically, the proposal was to mount a wheelchair and Hawking's name on the wall of the entrance hallway just as it opens into the large central area. Sid was genuinely surprised that I was not excited about the concept.

"I don't think so," I said, trying to be delicate. There is always pain when someone shoots down your fabulous new idea.

"I don't know for sure, Sid, but my sense is that Stephen prefers to be thought of as a physicist who, by the way, happens to have some disabilities, rather than a person with disabilities that became a world-class physicist. I mean really, the point is that he doesn't want to be recognized or remembered for his disability," I said.

"But don't you think that his overcoming his disability would be inspiring to students – not only now but in the future?" Sid offered, perplexed that I didn't get such a simple, yet powerful vision.

"Physicist first, disability a distant second," I said, "That's my understanding of Stephen's attitude."

"But we could mount the wheelchair over a large picture of the Andromeda Galaxy or something. It would look so cool," Sid said struggling for a common ground, "And inspiring. Symbolic of his scientific achievement."

"Well, Sid, why don't we both think about it overnight," I said deflecting the dissonance by delaying the decision. "I can ask Stephen about it," I added, "Maybe I'm being overly sensitive here." We kicked around several alternative ideas, but nothing captured the common imagination.

Sid set up a 'green room' for Stephen. Another larger adjacent room was set up for a press conference scheduled for a few minutes after the completion of the meeting with the students.

Stephen had let me know that he liked to do live press conferences. This was a vital part of Hawking's input as to how the tour and public lectures were to be structured. Since, in his public lectures, Stephen is giving a talk that he has prewritten into his computer system, one might form the mistaken impression that he is just along for the ride on his robotic chariot. Stephen made it clear that he always wanted to have a live audience question and answer session following the presentation of his lecture. The press conferences, the opportunity to respond live to questions from newspaper and television reporters also made everything more human and intimate. I could see the appeal for Stephen, all this being part of his role model agenda in defining a new standard for persons with disabilities in society. He wasn't taking the easy road; he wanted to be engaged and fully professional, disabilities be damned.

By the time Stephen and company arrived, Science World was packed with the students with disabilities, many with their parents. The Canadian Broadcasting Company's camera crews were already well-positioned. Everyone had evidently arrived much earlier. The stage, on the main floor, was at the center of the geodesic dome, in the center of everything. Today, for these students, it was the center of the universe, defining a new frame of reference for them to better understand their role as Participants with disabilities in the development of the universe. The crowd kept growing and finally numbered about 600 – well beyond the standard capacity for the audience area surrounding the central stage. A broad stairway wound up behind the regular seating area up to an encircling balcony, overlooking the main floor, that afforded a reasonable view of the stage. Neither the spiral stairway nor the balcony were official seating areas but were nonetheless double-lined with standing onlookers.

The grandiose setting and the ambience were quite a bit different from Stephen's earlier meetings with students in Portland and Eugene.

After unloading from the van near the front door, Stephen and crew headed to the private green room for a quick pit stop to check his systems. Computer, voice-synthesizer working? Is Stephen seated comfortably? Time for tea? Make time. Nurses rule! "They'll wait," says Joan.

Exiting the greenroom, traversing the floor, Stephen's crew dissolves into the crowd as Stephen drives his chair up a ramp onto the stage. There was fussing and fiddling by the television people to have him in precisely the best position. Brian Clegg, General Manager of Microsoft Canada, opened the festivities, welcoming everyone and expressing appreciation for being able to sponsor Dr. Hawking's visit to Vancouver. Next was Anita Hagen, Minister of Education for British Columbia. She was undoubtedly the one who had been taking the flak for the new policy of mainstreaming students with disabilities. In the audience were many of those who were benefiting from her stand. These were her people. This was her moment. As in Portland, the students with disabilities, most in wheelchairs, were a diverse and motley group. Ages ran the full range, with a notable new presence, unlike in Portland and Eugene, of local university students with disabilities, primarily from University of British Columbia and nearby Simon Fraser University.

"We are going to be broadcasting this event via satellite to nine other sites throughout the Province," Hagen emphasized.

The real introduction of Stephen Hawking is to come from Rick Hansen – well-known in Canada by his popular media handle – 'The Man in Motion'. A few years earlier Hansen announced his plan to wheel around the world to raise awareness about the potential of people with disabilities and to raise funds for spinal cord injury research. As a result of a car crash at the age of 15, Rick sustained a spinal cord injury that paralyzed him from the waist down. His "Man In Motion World Tour" took him through 34 countries and across four continents, an epic journey of more than 40,000 kilometers, prominently reported in the Canadian media. The trip took him over two years to complete and raised more than $26 million for research.

Anita Hagen introduces Rick Hansen: "His courage and determination inspired us to believe in the possibility of a fully accessible and inclusive society and a cure for spinal cord injury.

"Rick was the first student with a physical disability to graduate in Physical Education from the University of British Columbia. He went on to become a world-class athlete, winning 19 international wheelchair marathons, including three world championships and competed for Canada in the 1984 Olympic Games.

"In 1987, Rick was appointed a Companion of the Order of Canada and has re-

Rick and Stephen

ceived several honorary degrees. As a positive role model dedicated to improving the world around him, Rick has a remarkable ability to engage and motivate youth. He regularly shares his message of hope, inspiration and the importance of making a positive difference in the lives of others with young people across Canada."

Like Hawking, Rick is a Participant, and not to be missed, he is also a strong supporter of fundamental, exploratory scientific research.

"Through Rick's leadership, his Foundation over the years has generated over $200 million for spinal cord injury-related programs and initiatives."

Rick Hansen wheels onto the stage where Stephen is already present. Applause and recognition, almost everyone there knows and admires Rick Hansen.

Taking the hand-held microphone from Anita Hagen, Rick begins, "Ladies and gentlemen, it gives me a distinct pleasure to be able to introduce to you your guest speaker and lecturer and our honored guest here this afternoon, Dr. Stephen Hawking. As you know, Dr. Hawking has a specialty in cosmology, or the study of the universe on a grand scale, and I have had the opportunity, in preparation for this introduction, to spend a little time researching this gentleman. And, as I went through researching, I reminded myself that there were a lot of commonalities that we shared. We both have grandfathers that originated from Yorkshire in England. I also studied the fact that in his days as a student, in his early days, he was a bit of a slacker and so was I."

This last comment draws laugher and a chorus of indistinct moans from the audience.

"You know, both of us ended up in marriage and in testing our marriages in close quarters. Dr. Hawking, in the close quarters of university dormitory residence, me in a motor home on 'The Man in Motion Tour'. And I think, as well, we both dealt with a disability part way through our lives.

"Unfortunately, when we come to the field of academic accomplishment in scientific study, this is where Dr. Hawking leaves me in the dust because I was extremely pathetic, and I can imagine the incredible levels of contribution that he has made for our society as a result of his work. His work is well known throughout the world and I think, in introduction, it doesn't really need to be refocused on. What I really would like to focus on this afternoon here is something probably much more universal, the study of the spirit and the human contribution that this man has made during the course of his life. You see, even though Dr. Hawking may not specifically focus on this area of contribution, the fact that he has dealt with a disability, the fact that he has overcome many obstacles to continue to pursue his love and his passion of life, the fact that he has been able to overcome those obstacles and to strive forward to excel and to achieve has made him an amazing symbol of

the strength of the human spirit. He has been an inspiration to me and to millions of people who are looking in their world to be recognized as equals, even though they may have a disability. He shows us that the unbelievable is possible, that we cannot only be recognized in our society as equals if we have a disability, but that we can contribute and we can even lead.

"Ladies and Gentlemen, I hope that you will join with me in giving a great Vancouver welcome and British Columbia welcome to Dr. Stephen Hawking."

There is massive and sustained applause – a standing ovation – from those who can stand. Rick wheels off stage down the ramp, leaving the stage empty except for Stephen.

Stephen begins: "It is very nice to be in Vancouver. I have been here before but not since 1977. It seems just as good as I remember.

"I will be talking about science this evening, but this afternoon I want to tell you about my disability.

"I am quite often asked what do you feel about being disabled with motor neuron disease, or ALS…?"

With minor changes Stephen continues with much the same talk he gave in Portland and Eugene.

At the end of Stephen's talk, Sid Katz joins him on stage to direct the question period. Sid explains, in his own way, "Dr. Hawking doesn't believe in canned answers… He will answer each question directly."

Sid has pre-arranged the first questioner, Chelsea Smith, 11 years old. She also speaks with a voice synthesizer. Sid tells us she is a student from Malaspina School, in Nanimo, on Vancouver Island.

> **Student question:** "Can you see a reason for the existence of people with disabilities while believing there is no god?"
>
> **Stephen's answer:** "I think everyone has some form of disability, none of us is perfect. Some disabilities are just a bit more obvious than others. But I believe the human spirit is capable of triumphing over almost any handicap."

I tell Stephen later that this is a brilliant answer, my favorite, and that we should use it on one of the posters of Stephen that we had been considering as a project. The shortened version: "Everyone has disabilities. Some are just more obvious than others."

What a marvelous wisdom!

> **Student (Curtis, from Surrey B.C.) question:** "I would like to know if you watch TV and if you do, which television programs a lot or maybe just a little."
>
> **Stephen's answer:** "I must confess I don't get much time to watch television. I mainly watch the news but I also look at the science program *Horizon* which is the same as *NOVA*. And of course I watch *Star Trek*, particularly when I am on it."

Another burst of sustained applause, laughter and cheering erupts. Stephen's recent appearance on *Star Trek* is popular and apparently well known to many in this audience.

Sid tells us, "Michael Whitman, age 14… also uses a speech synthesizer, but it isn't portable, so he submitted his question to me in writing." Sid reads it aloud.

> **Student question:** "Given your great dependence for physical care, how do you feel you can relate equally to other people?"

This is another of my all time favorite student questions – no beating around the bush, cutting right to one of those intimate concerns for everyone with severe physical disabilities. How about we discuss this, unexpectedly, on Province-wide live television?

> **Stephen's answer:** "I don't feel that is any problem. It is true I depend on other people to look after me, but we all depend on other people. Very few of us would survive in the jungle on our own. What one needs is the confidence to feel one is also making a contribution. I think we all are."

32

Life in the Universe at the
Vancouver Paramount

The stage crew at the Paramount Theatre wants Dr. Hawking to come by in the afternoon for a sound check. This is a fairly standard, reasonable request particularly appropriate given the novelty of Stephen's presentation coming through a voice synthesizer. Back at the Hotel Vancouver, following the session at Science World, Stephen isn't excited about an extra trip to the Paramount Theatre for a five-minute sound check.

Everyone finally agrees that bringing Stephen to the theatre much ahead of time is an unnecessary inconvenience.

"Really, it should be so simple," said Tim. "They can just plug-in to the voice synthesizer. There is a standard output plug."

Right! High-tech. Very simple. Questions of the compatibility of international impedance standards didn't cross anyone's mind.

I had produced events at the Vancouver Paramount Theatre before. The stage crew was professional – unionized, of course – and fun to work with. We did arrive early enough to run a brief sound check just before the outer lobby doors were opened and the crowd poured into their seats. But it wasn't clear that things were working. There was a buzz in the output – a sort of low-pitched hum. The sound technician assured us that he could dampen that out, so Stephen went to the star's dressing room for a cup of tea.

After several 'welcomes' I introduce Seattle Public Television (KCTS/9) CEO Bernie Craig, who gives another welcome and a brief introduction with a couple of highlights about Hawking.

Having learned of Hawking's penchant for opera including Richard Wagner and being a bit of a Wagner fan myself and with my overactive imagination and sense of showmanship, I had come up with the idea that Stephen's entrance should have musical accompaniment. Given that he is entering on a wheelchair – which I now refer to as his chariot – the obvious choice of music is Wagner's spectacular *Ride of the Valkyries*.

(In Wagner's epic *Ring Cycle*, the Valkyries (special female selectors of the heroes in battle) are entering on flying horses. Non-opera aficionados might recognize the music as it was used, somewhat incongruously, in the dramatic helicopter assault scene in the Vietnam War movie, *Apocalypse Now*.)

The dramatic music starts and builds, and Stephen begins to move from the left wing of the stage. I am blocking his way. "No. No. Wait," I say. Joan, standing next to Stephen, asks, "Why?" "Let the music build," I say. "This is showbiz!" I say. Stephen gives me a few seconds, then he starts to move, in brief – 'out of the way, Bristol.' The music is glorious and, in this setting, up-lifting. As Stephen enters the stage, as soon as the audience can see him, there is a huge, spontaneous and sustained standing ovation. It is an extraordinary moment. There are tears in people's eyes. More applause and cheering. There is something truly significant about what is happening here. It isn't just about Stephen. It is 'un événement sociologique' – a sociological event.

Tim follows Stephen out onto the stage and plugs the sound cable into the outlet on the voice synthesizer. As the dramatic musical theme fades and the applause quiets there is a contrasting silence – except – I hear the hum-buzz. Tim departs the stage joining me in the wings. Stephen takes about a minute to load up his talk to the mainstream of the computer. The audience and the auditorium have become utterly, exceptionally, almost uncomfortably quiet – except for the hum-buzz. Then he asks the crowd, "Can you hear me?" This is standard practice that allows the sound technician to make sure the volume level is appropriate. I hear him fine as does the audience, but I am still hearing the hum-buzz. I am hoping the sound tech can find a way to eliminate this. From the up and down – mostly up – of the volume of the hum-buzz, I can tell that the sound tech is trying. Tim and I are standing just off stage to the stage-right. We exchange glances. I am also standing near the stage manager. "What do you think?" I ask him. He is talking into his headset microphone to the sound tech.

Stephen hasn't actually started yet. He hears the hum-buzz too. If we let this go until Stephen is into the presentation, it is going to be much more disruptive to switch to another system.

I make an executive decision. "Man, this just isn't working! This isn't OK," I say.

"All right," says the stage manager, "it's your decision." He says something into the headset and then hangs the headset on an outpost, picks up a standard microphone on a stand – which I now notice has already been setup and was waiting – and walks out onto the stage. Tim follows him. Stephen and Tim are communicating via the eyes-up, eyes-left system. Tim is undoubtedly asking for Stephen's 'OK' to proceed with the change. I don't know what the three of them are saying to each other but the switchover is proceeding. They unplug the main sound cable from the voice synthesizer, at which point the hum-buzz instantly stops. There are a few hands clapping and an audible sigh from the audience.

The height of the new microphone is adjusted to the level of the output speaker on Stephen's voice synthesizer – positioned about three inches away – and a new sound check ensues. "Can you hear me?" The volume is too low. The audience answers in mass, "No." They had, of course, heard it well enough to conclude that they weren't hearing it well enough. Stephen cycles the computer and again asks, "Can you hear me?" This time the sound tech is pretty close to the right level. The audience answers with a booming, "Yes," and bursts into applause. Why the enthusiastic applause? Contact! There is something special that is happening here – simple but powerful. They have just made contact with the great cosmologist. The enthusiasm has carried over from the standing ovation when he entered. That he is here at all is amazing. They love him, and they love that he is here – that he has the courage… to lead. When voice contact is established – obviously not a trivial thing – there is this palpable excitement. Touch! The doorway has opened to a magic kingdom ruled by a truly unique mind, by a truly remarkable person. Something extraordinary has begun.

The title of the lecture is *Life in the Universe.*

Hawking begins: "In this talk, I would like to speculate a little on the development of life in the universe and, in particular, on the development of intelligent life. I shall take this to include the human race, even though much of its behavior throughout history has been pretty stupid and not calculated to aid the survival of the species."

This last remark garners an enormous laugh from the audience. And we were off and running. The text of a later, but very similar talk is available on Hawking's website. Hawking illustrated one of his concluding themes: 'there are likely to be many civilizations in the galaxy and we will inevitably meet them.' Then he warns about being too friendly with aliens too soon – making the point with two contrasting video clips. The first is from the movie *ET: The Extra-Terrestrial*, where the alien is cute and friendly, family-oriented and the sort of alien one might want to hang out with. Hawking comments that this is a hopeful and comforting image but perhaps not realistic. The second video image is from the movie *Independence Day,* the dramatic story of a distinctly unfriendly alien invasion of Earth.

The audience Q&A was animated, but the questions were fairly routine. One question was a bit rambling but centered on making sense of life and the universe – and our place in the universe – if everything is completely determined as projected according to some Scientific Theory of Everything.

Hawking's answer: "I have noticed even people who claim everything is predestined, and that we can do nothing to change it, look before they cross the road."

The audience loves this answer – explicitly revealing Stephen the Participant.

Andrew Dunn, Stephen's graduate assistant from a few years back, had suggested earlier in the day that we liven up the question session by inserting the following question: "Have you ever run over Mrs. Thatcher's toes with your wheelchair?"

Hawking's Answer: "No. But I almost got Charley once..."

Canadians, being part of the British Commonwealth, easily pickup on the references. Apparently Hawking had been invited to one of the Queen's occasional Garden Parties and had targeted Prince Charles. In Britain there are 'Royalists', those who are enthusiastic supporters of the Monarchy, and there are non-Royalists, those who are not strong supporters. Stephen is in the second group – one of the reasons perhaps that he is not Sir Stephen.

The final question was the obligatory one about the existence of God.

Hawking's crowd-pleasing answer: "I don't know. And it is not for me to say."

Microsoft Canada had arranged a post-lecture VIP reception in the upper foyer/lobby of the Paramount Theatre. It is jammed. Stephen wheels his chair around through the crowd stopping at various junctures as people ask questions and have their picture taken next to Stephen. They love to look over his shoulder at his computer screen, watching as he composes an answer to someone's question.

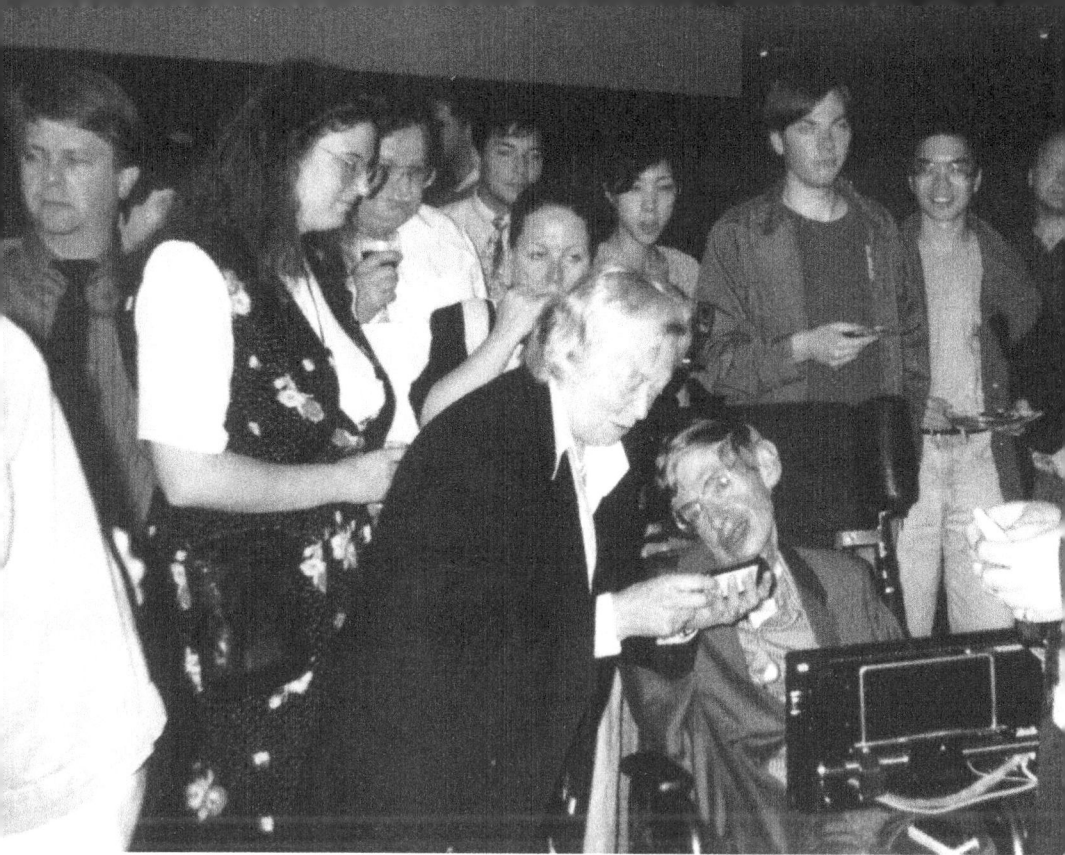

Joan with Stephen at reception

Late Dessert at the Hotel Vancouver

After the reception everyone was tired but still rather keyed up. Stephen had had a large meal just before we departed the Hotel Vancouver for the theatre but hadn't managed to secure any of the goodies at the reception. On the way back to the Hotel, Joan, who had been quietly communicating with Stephen, spoke up, "Stephen would like to go out somewhere." By now it was after 10 pm and it was unlikely that any restaurants would be interested in serving us. We arrived back at the Hotel Vancouver, only a few blocks away, as the 'where to go' discussion continued. None of the locals had any ideas. Then one of the staff at the Hotel suggested that we check out the Hotel's rooftop restaurant. I quickly confirmed that the restaurant would still serve us. Stephen then decided that that was where we should go. Despite the fatigue, dessert goes down well. The lasting effect of the success animates the group – laughter and banter. Everyone has stories – the weird narrow corridor underneath the stage constituting wheelchair 'access' for

Stephen to reach the stage… the sound problem – still not understood, and the full, comical story of Stephen and Charley's toes.

Andrew, who had been sitting in the audience for the lecture, jokes that Stephen – "who never gets tired" – was fighting to stay awake during part of the lecture; big meal beforehand, not much moving around. Andrew and Sue Masey are heading on to Japan in the morning. The rest of us are headed to Seattle by way of Victoria, British Columbia.

All these Vancouver encounters make much more sense in terms of a Participant reading of reality. We were dealing with real world problems. Some of the problems were unexpected, not predicted, perhaps unpredictable, having arisen as if by chance. As Participants, we were all involved in exploring innovative real world solutions to the advancement of students with disabilities and to the social inclusion and further enablement of all people with disabilities. Moment to moment practical problems had called for novel, creative solutions.

Stephen always looks both ways before crossing the street.

Several of us take a late night swim, compliments of the Hotel Vancouver Health Club. The pool is on the top floor – roof – of the Hotel. The pool and surroundings are enclosed by a very large sort of elongated glass bubble, providing spectacular views of most of the city at night. The beauty of the large, illuminated, emerald green pool – highlighted against the overall darkness of the night sky – is breath taking.

– Perquisite of the job.

Part Three

Complementarity in Relativity and the More General Theory

33

My Presentation about Complementarity to Stephen

In line with my modus operandi, hosting speakers in the Institute's lecture series, I had proposed an adventure for Hawking for his Vancouver-Seattle visit. The Vancouver lecture was first, so I suggested that in traveling to Seattle from Vancouver we would take a slightly roundabout route through Victoria. The three-hour drive from Vancouver to Seattle, although pleasant, is less than inspiring. Victoria is located on Vancouver Island, not to be confused with the City of Vancouver, that is located on the Canadian mainland. Like a travel guide I described Victoria to Hawking and his crew. I suggested a leisurely lunch and a visit to the renowned Butchart Gardens, fifty-five acres of stunning, floral show gardens, a National Historic Site of Canada.

Unfortunately, I sold them on the idea before I checked the ferry schedules in detail. We would have to get up rather early to catch the first ferry from Vancouver to Victoria. And this boat arrived in Victoria only three hours before the last hydrofoil from Victoria to Seattle. After exploring options, I broke the news to Hawking and crew. First, I proposed that rather than being pressed for time in Victoria that we skip Victoria and just travel at a more convenient time later in the day by hydrofoil directly from Vancouver to Seattle.

The Joans expressed disappointment with missing the Butchart Gardens. Were there no other options? I went through the details. Sighs of disappointment greeted the apparent conclusion that the rational path was to skip Victoria and travel directly to Seattle. Stephen, of course, being always in charge, was then consulted.

"What do you think Stephen?" Joan asked.

"No." This was another of those slightly enigmatic one-word responses.

"You mean you want to go to Victoria despite the short time?"

Stephen indicated 'yes' with an eye-brows up gesture. There was a sense, perhaps only in my mind, that 'this has been promised, so it shall be.'

"OK, then, that's the plan," I said, swallowing hard with a sense of bewilderment creeping over my face. Over the next few hours I scrambled to figure out what was possible. Boarding large hydro-foil ferries was often a hassle with lines of cars, difficulty securing tickets and so forth. I re-contacted the taxi company I had called previously to secure a taxi-van to take us around Victoria. They knew who Hawking was. "No problem," they responded, "We can meet you at the dock when your ferry arrives from Vancouver, tour the Butchart Gardens, not all fifty-five acres, but enough – and deliver you at the ferry building in time to catch the last hydro-foil to Seattle." They also acquired our tickets, took charge of loading the baggage and notified the ferry company that 'an important person with a disability' would be boarding at the last minute. Canadians are especially accommodating to persons with disabilities. –– I am saved!

The morning after the evening lecture in Vancouver it would be necessary to arrive at the Vancouver Ferry Building by 10:30 am, entailing a somewhat early rise for Stephen but, more of an issue, a shorter than normal time for the nurses to prepare Stephen for the day.

Joan put it succinctly, "It's not the most desirable schedule. We usually like to take the morning at a leisurely pace: bathing, dressing, breakfast, tea and taking care of any necessary medical procedures. Stephen also likes to read in the mornings. But we can move fast when we need to."

I had mentioned to Stephen earlier that I would like to have an opportunity to present an idea or two that he might listen to and critique. I specifically presented this as an uncommon opportunity for a physicist and a philosopher of science to review our separate thinking and our cultural assumptions on one or two fundamental issues. I had proposed a couple of possible times and Stephen decided the best time would be on the three hour ferry trip from Vancouver to Victoria.

Once we had departed Vancouver and were well underway, Tim, Jennifer and Christy are laughing and flirting top deck. I venture downstairs into the ferry's interior where Stephen had set up, to see if this was the right moment to make my little presentation.

Stephen was reading. Joan sat next to him, holding the article in front of him, nearly head-high. I watched for a moment, wondering just how and when to

Christy, Tim and Jennifer

interrupt. On a signal from Stephen, Joan turns the page. I come closer, standing quietly, watching. Joan finally notices me.

"Oh, here you are. Stephen's ready for you. Let's set up over there, so you can have a comfortable seat," says Joan.

Stephen Hawking, all to myself, and I – with a head full of half-baked ideas, moving slowly between small islands heading to Victoria, a modern city retaining its 19th century origins, all this embedded in a geo-history resonant with ancient culture – appropriate background for modern reflections on ancient questions.

I started in a way that seems a little inane with hindsight.

"Look, to begin with…" I said, pausing.

"An apple is not an approximate orange. I mean some people seem to think that because one type of thing can be represented in many of its characteristics as an approximate version of another type – size, shape, weight – that there are no 'real' qualitative differences," I said.

Stephen just gave me a slow blink, an "OK."

367

"Some people seem to think that when you extend a quantitative approximation to some sort of 'eventual' limit you can turn one type of thing into another. They imagine everything to be a calculable, quantitatively continuous, logico-mathematical approximation of everything else.

"But such limit arguments are often misleading. To make the transition from one qualitatively distinct conceptual type to another conceptual type, by means of quantitative approximations, you need to jump over a conceptual asymptote – the final dividing line – thereby surreptitiously erasing the real conceptual difference – illegitimately," I said.

All this was my way of leading up to complementarity.

"Somewhat similarly," I went on, "a particle is not an approximate wave and a wave is not an approximate particle. They're opposite types – one is local and the other non-local in space-time. Some people seem to think that through approximations and limit arguments you can mathematically unify particles and waves, reduce one to the other, so that there is only one type. What I have been thinking about is that if one accepts the complementarity of particle and wave phenomena, this view strongly suggests that the Newtonian and Maxwellian Research Programs must be complementary. And so it has seemed to me that the space-time frameworks of each of these research programs should be understood as complementary."

I waited a few seconds for a comment, but Stephen wasn't doing anything but looking me straight in the eye. After a few seconds, another slow eye-brows up, "OK."

"Einstein's critique of Newtonian space-time was quite explicit, insisting that the Newtonian program had tacitly presupposed absolute simultaneity. Cosmologist John Archibald Wheeler had an illustrative expression: 'time is nature's way of preventing everything from happening at once.' But according to Einstein in Newtonian space-time everything did, indeed, happen at the same time – absolute simultaneity. Einstein's critique pointed out that causal communication between spatially separated places, even with light signals, takes a period of time. And yet the Newtonian clock-work universe had to keep the same time everywhere. The only way to model the Newtonian universe – where time was synchronized everywhere, instantaneously – was to imagine that 'everywhere' happens in one place, as a physical point, a dimensionless point in both space and time. By analogy, Wheeler might have added that 'space is nature's way of preventing everything from happening in the same place.' In Newton's universe, as characterized by Einstein, since everything must happen at the same time [absolute simultaneity], presumably, it must happen in same place – as a point-reality. Einstein remarked that Newtonian reality is a completely localized space-time 'point-reality.'

"Maxwell's universe is the opposite, a complementary type of universe, where everything is entirely distributed, extended over space – *and time.* Really, the distribution is, I think, over a different, opposite *type* of space-time, per hypothesis, over a complementary *type* of space-time. Einstein remarked that Maxwellian reality is a completely non-localized space-time 'field-reality.'"

A crucial step in my argument is that the interpretive space-time framework of Einstein's Special Relativity is, in fact, none other than the framework of Maxwell's field-reality. Curiously, it wasn't obvious for some time for many in the scientific community that Einstein's Relativistic space-time framework was actually the Maxwellian space-time framework. Maxwell hadn't discussed 'relativistic effects'. This late recognition is not to take away from Einstein's accomplishment in developing Special Relativity. Einstein was focused on the relativistic effects of localized particles traveling in a Maxwellian field – as captured in the title of his famous paper leading to Special Relativity: *On the Electrodynamics of Moving Bodies.*

However, since there are no particles in a purely Maxwellian field-reality, Einstein was clearly, actually focused on the curious consequences of the interface of Newtonian particle physics and Maxwellian electro-magnetic wave physics. The crucial step was that Einstein interpreted the relativistic effects of moving particles within a Maxwellian space-time. This later led to the introduction of the famous $E = mc^2$ (viz. energy (E) equals mass (m) times the speed of light (c) squared). Kuhn's objection was that with $E = mc^2$ Einstein has introduced a different concept of mass – relativistic mass. Other critics, arguing along similar lines, have maintained that what he really introduced with 'm' was 'a relativistic mass-like effect.' In other words, particles 'appear' to have an increased mass when accelerated to higher velocities. But is this a real 'objective' increase in Newtonian mass or just an increase in some sort of relativistic mass-like effect? For his part, Einstein is quite generous and explicit in attributing the Relativistic revolution to the groundbreaking work of Maxwell. However, the full significance of Maxwell's presage of Special Relativity's space-time framework has remained underappreciated.

"So I think there is a general understanding today that the defining space-time framework of Einstein's Special Relativity is – at least in a crucial way – already contained in Maxwell's space-time framework."

Hawking was aware of all this and gave me an eyebrows up acknowledgement. I didn't conclude that he was necessarily agreeing, but at least that he was following what I was saying.

I went on, "I started thinking about this idea after switching to philosophy of science in my senior year at Berkeley and reading Thomas Kuhn's arguments in his *Structure of Scientific Revolutions*. In my first few years as an undergrad in physics at Berkeley, I was taught – 'indoctrinated', according to Kuhn – and accepted the argument that the successes of Newtonian physics were correctly explained as special cases – approximations – within Einstein's more general Relativistic Research Program.

"However, Kuhn presented a rather forceful critique of that standard textbook consensus, arguing that Einstein's Special Relativity did not properly subsume (include) and supersede (explain the successes of) Newtonian physics.

"Kuhn's argument was that traditional Newtonian 'rest mass' was not a special case of the 'mass-like property' expressed in $E = mc^2$. His argument wasn't difficult to understand. Clearly, there are easily observed relativistic mass-like effects associated with observations of fast moving objects (viz. particles). However, the standard textbook presentation has always been that as the velocity of the object decreased to zero, the relativistic mass-like effect defined by $E = mc^2$ goes to zero and then the 'relativistic mass' somehow turns into Newtonian rest mass.

"Kuhn's critique has to do with this limit argument, with what happens at the limit of the object's velocity – decreasing, decreasing – eventually coming to rest. The limit argument, by 'jumping the asymptote', illegitimately merges and attempts to erase the difference between the Relativistic concept of mass (or mass-like effect) and the Newtonian concept of mass.

"With this merger, 'rest' becomes a *type* of 'motion'. In Ancient Science rest and motion were always considered to be contraries – qualitative conjugates – complementary in modern terms. For the Ancients arguing that rest was a type of motion, a special case of motion, would be equivalent to saying that constancy was a type of change. In modern terms it is like arguing that particles are a type of wave. It is like arguing that locality is a type of non-locality, and that a point-reality is a type of field-reality.

"I still find it troubling that, whereas in quantum theory, position and momentum are clearly accepted as complementary, somehow Kuhn's obvious argument remained unappreciated.

"Kuhn argued that despite the fact that Newtonian and Relativistic physics use the same word – 'mass' – they don't mean the same thing. "The physical referents of these Einsteinian concepts are by no means identical with those of the Newtonian concepts that bear the same name. Newtonian mass is conserved; Einsteinian is convertible with energy. Only at low velocities may the two be measured in the same way, and even then they must not be conceived to be the same." (Kuhn, Thomas S., page 101, *The Structure of Scientific Revolutions* (1967) Univ. of Chicago, Phoenix Books).)

"Of course, under the Logical Positivist Mechanical Philosophy where everything is logico-mathematically calculable from everything else, there are no real qualitative conceptual differences in the universe, and so there is no problem with the textbook account that erases the qualitative difference; no problem accepting that Newtonian 'rest mass' is a type of 'always in motion' relativistic mass.

"Kuhn's alternative argument was that there was a conceptual discontinuity in the transition from the Newtonian to the Relativistic way of understanding the universe. Kuhn pointed out that if Newtonian physics had been properly subsumed and superseded by a more general Relativistic physics – making them conceptually continuous – you should be able to derive Newtonian physics from Relativistic physics. According to Kuhn, the textbook arguments presuppose what they are supposed to prove: "The argument has not done what it is purported to do. It has not, that is, shown Newton's Laws to be a limiting case of Einstein's. For in the passage to the limit [velocity zero], it is not only the forms of the laws that have changed. Simultaneously we have had to alter the fundamental structural elements of the universe to which they apply is composed." (Kuhn, Thomas S., page 101, *The Structure of Scientific Revolutions* (1967) Univ. of Chicago, Phoenix Books).)

Kuhn's conclusion: "The derivation is spurious."

"So, Stephen, I am thinking that Newtonian physics is not properly subsumed and superseded by Relativistic physics. Like the difference between apples and oranges – more specifically, like the difference between particles and waves – it seems to me that Newtonian space-time and Relativistic space-time are different *types*. And the argument to that effect seems to point to the hypothesis that Newtonian physics and Relativistic physics are complementary – like particle phenomena and wave phenomena."

I pause, hoping for something profound from Stephen. I can't tell if he is composing a remark on his computer. I glance over at Joan across the room. She is puttering, cleaning up, rearranging. As I turn back, Stephen is looking straight at me. I surmise that he is not ready to comment. So I continue.

"Reflecting on this over the years led me to try to more explicitly articulate the difference between the space-time frameworks. I have a couple of thought experiments. I don't think they work formally, but they are, I think, helpful attempts to clarify the difference.

"Consider the following. Imagine expanding the Newtonian point-reality into a spherical bubble – where it is always the same time everywhere in the bubble. There is spatial difference, a volume, but with no time difference. That's an image of a Newtonian spatial volume. Then compare that image to the distributed Relativistic field-reality. In the Relativistic reality there is a different time at each point in space. Notice that there is no Newtonian space – no volume where it is the same time – within the Relativistic space. And likewise there is no Relativistic space – no volume where there are different times – in the Newtonian space. I think this points at the nature of the essential difference – *in type* – of these two space-time frameworks.

"Both these conceptual universes are perfectly symmetric – the same throughout – but in opposite ways. The Newtonian universe – the point-reality – is the same everywhere and always and the Maxwellian/Relativistic universe – the field-reality – is the same in being universally different from the Newtonian, universally distributed – everywhere and always. These two space-time symmetries are opposites – complementary symmetries – opposite types: one perfectly local and the other perfectly non-local.

"The expanded Newtonian bubble can also be understood as a characterization of our normal sense of 'now'. Even with the embrace of the relativistic notion that there is a light-time separation between locations as between Earth and Mars, so that communication isn't instantaneous, it makes sense to talk about what is happening on Mars *now*, even though I can't see or hear it for a few minutes. Likewise, it makes sense to talk about what is, or might be, happening in the Andromeda Galaxy *now,* even though it is 2,538,000 light years away. When we refer to 'the universe' as One – with one age – we seem to be presupposing this 'expanded Newtonian-like bubble now'. It is unclear whether the Newtonian 'now' can be made sense of in a universe that is represented as 'objectively' Relativistic. Similarly our normal Newtonian 'here' seems to presuppose a constancy of place (locality) as time changes. However, in a completely Relativistic framework, a change in time entails a change in space-locality.

"The proper transition to a More General Theory should subsume and supersede *both* Newtonian physics and Maxwellian physics as limited special cases, each with limited validity as in quantum theory, with particle phenomena and wave phenomena. And what I am thinking is that it is at least unclear whether the Newtonian 'now' can be made sense of in universal, 'objective' Relativity."

Except for Joan, who is either arranging things or sitting quietly, everyone else is above, on deck, appreciating the views. I still have Hawking's full attention, so I seize the moment and go on.

34

The Correspondence Principle and the Copernican Extrapolation

I recalled for Hawking a crucial conversation I had had with Feyerabend in Berkeley. "Early in my exposure to the plethora of arguments of Popper, Quine, Kuhn and the various Logical Positivists in the philosophy of science, feeling a little disoriented, I had asked Feyerabend to point me at the heart of the problem. "Where is the core mistake in the modern representation of the new physics?" Feyerabend didn't hesitate. He said, "The Correspondence Principle." As I came to realize, Kuhn and Bohr were emphasizing the same point.

"The Correspondence Principle is about how a new better theory is supposed to subsume (include the truth content) and supersede (explain the successes of) prior theories. The Spherical Earth Theory subsumes and supersedes the Flat Earth Theory, the latter becomes a limited special case within the former. One can understand and explain the successes of the earlier, Flat Earth Theory, in terms of the new superseding Spherical Earth Theory in that the Spherical Earth is so large that, for rather small humans, it looked flat and one could move around over small distances quite successfully, confirming it was flat. The expression, associated with the Correspondence Principle, is that the truth content of the Flat Earth Theory is *subsumed* by and so contained in the 'more general' superseding Spherical Earth Theory. The Flat Earth Theory is properly *superseded* by the Spherical Earth Theory insofar as the Spherical Earth Theory can explain the successes of the Flat Earth Theory as limited special cases, as successes in a limited range of validity. In other words, the success of the Flat Earth Theory is explained as resulting from the 'reasonable' idealizations of small people on a large sphere. The Correspondence Principle requires that for a new theory to be adopted as 'better' – to subsume and supersede another theory – it must be able to explain the successes of the earlier theory, understanding them in a new way, as having involved some sort of idealization that limits the earlier way of understanding.

"That Newtonian particle physics should be contained in, subsumed by, the new Relativistic physics as a special case was required by the official version of the Correspondence Principle. In the Positivist interpretation, the Correspondence Principle required that later better theories should always be 'consistent, conceptually consistent, conceptually continuous, inductive generalizations' over earlier theories. The continuity of these inductive generalizations requires the same *type* of conceptual understanding in the earlier and later superseding theory. In the Positivist representation of inquiry and the history of science, this continuity was essential in order for the advance to be an advance *within* the Scientific Research Program, *within* the Mechanical Philosophy. No conceptual discontinuities, no conceptual revolutions allowed!

"What Feyerabend was pointing to was a problem with this Positivist representation of scientific advance. The Positivist representation 'stood to reason' if you presuppose Galileo's Pythagorean theme that the language of nature is mathematical – with one added clarification. Since the laws of logic govern mathematical reasoning the Positivists reasoned that the laws of logic, such as the Law of Excluded Middle, rule the language of mathematics. This meant that there must be a logico-mathematical consistency between the earlier and later more advanced theories. For instance, as long as you are reasoning from one theory of space, time, mass and energy to another theory of space, time, mass and energy, where the concepts of space, time, mass and energy are the same or at least where one can logico-mathematically reason, and translate, these concepts from one to the other, then the Positivist logico-mathematical representation of the history of science would be confirmed and preserved. However, if advances involved conceptual discontinuities – Kuhnian revolutions – then the Positivist version of the Correspondence Principle and their representation of successive advances in the history of science wouldn't work, wouldn't check out with the facts. The Kuhnian revolutions meant that succession wasn't logico-mathematical and violated their requirement for conceptual continuity.

"In order for there to be the conceptual continuity required by the Positivist version of the Correspondence Principle, keeping it all with the Scientific Research Program, they needed to chose one or the other of the two 'objective' research programs – either the Newtonian or the Maxwellian – and subsume (viz. 'explain') the other as a special case. Either the Newtonian conceptual framework captures the 'real' objective reality and Maxwellian wave phenomena are somehow idealized special cases, or the Maxwellian conceptual framework captures the 'real' objective reality, and the Newtonian particle phenomena are somehow idealized special cases. In order to satisfy the requirements of the Correspondence Principle, in order to retain continuity and remain within the Scientific Research Program, one of these two research programs needed to subsume and supersede the other.

"Since it was clear that the Newtonian framework couldn't explain, let alone 'make sense of', the newly demonstrated relativistic effects the default search was for a more general, post-Newtonian theory that could make sense of the relativistic effects, as well as properly subsume and supersede the successes of Newtonian physics. By 1904, Henri Poincare and Hendrik Lorentz had made major contributions to a theory of relativity. The defining feature of Einstein's 1904 Special Relativity was *the replacement* of the classical space-time transformations of the Newtonian space-time by the Lorentz transformations of Maxwell's electromagnetic space-time framework.

"Stephen, this looks to me like a *lateral conversion* between complementary frameworks, not a conceptually continuous, inductive generalization. The successes of the Newtonian framework simply don't make sense in the new Relativistic (Maxwellian) framework.

Making Sense of an Objective Relativity

"It has bothered me for a long-time as to whether one could even think in terms of a purely Einsteinian Relativistic reality, with no preferred frame of reference," I said.

"Once, when we were all three together, I asked Lakatos and Feyerabend whether they thought it was possible *to conceive* of a universe, to think in terms of a completely Relativistic physics, to think in terms of a purely relativistic language of observation and experimentation. Lakatos launched into a sort of diatribe: 'It would take the normal post-doctoral (post-Ph.D.) physics student at least an additional six years to be able to fully appreciate and to be able to think in purely relativistic terms and perhaps another ten years to be able to design experiments without any need to think first in terms of, or with any reference to, his experimental apparatus in a local Newtonian framework.'

"As he was talking I had the distinct feeling that Lakatos might be putting me on. An occasional smirk seemed to cross his face as he looked occasionally to Feyerabend for confirmation of the number of years. It was as if he were giving me the official litany of the physics community, tongue in cheek, so overstated that I should get the covert message. It was as if I had asked whether the emperor actually had any clothes and the response was to suggest that only the wisest and most studied, and so enlightened, could actually see the emperor's new clothes. Feyerabend never said anything and I didn't read any clear expression on his face, either confirming or challenging what Lakatos was feeding me. On reflection I do believe that Lakatos was putting me on. This sort of thing was a typical part of the serious, but often preposterous, interpersonal game that Lakatos and Feyerabend engaged in.

"It was some years after I had read Kuhn that I realized that Bohr had also argued that Einstein's Relativity did not properly include, did not properly subsume the Newtonian framework. Bohr had argued that it wasn't possible to make sense of the nature of reality and our place in it in terms of a purely 'objective' Relativistic reality. Bohr argued that any account of 'real' observation and inquiry always required reference to a local Newtonian space-time framework that could not be made sense of in a purely Relativistic version of reality."

Hawking, unexpectedly, gives me a somewhat enthusiastic eye-brows up acknowledgement. My sense is that he is quite familiar with Bohr's arguments, perhaps much more so that Kuhn's.

"Bohr's approach was to argue that all 'actual observation' involved both a *local* Newtonian aspect and a *non-local* Maxwellian aspect. In other words, Bohr argued that all 'actual observation' involved both a *non-distributed space-time (viz. point-reality) aspect* and a *distributed space-time (viz. wave-field-reality) aspect.* Bohr emphasized that the local Newtonian aspect was a presupposition of all possible observation – by the very nature of observation – and was not accounted for in Einstein's embrace of the objectivity of the non-local (Maxwellian) Relativistic framework. Einstein's Relativity presupposed that there was no locally preferred frame of reference in the universe – contrary to the Newtonian presupposition of absolute space and time (viz. now represented by Einstein as a point-reality). Bohr argued that in order to actually make an observation you must necessarily *choose* a local frame of reference.

"Bohr's argument, as I understand it – perhaps just what I think it should have been – is that you can't 'observe' anything, any coherent order (viz. like the shape and motions of the solar system), except in relation to, in terms of, some defined, locally coherent frame of reference. And consequently, you can't gain any knowledge except in terms of, in relation to, some defined, locally coherent frame of reference, except in terms of some chosen way of observing. You can't learn anything, make sense of anything, without incorporating that localizing choice of a particular frame of reference. Your choice of a local space-time framework is part of any particular way of observing.

"Bohr's reasoning simply extended his complementarity theme from

Niels Bohr

quantum theory. The Copenhagen Interpretation of Quantum Theory – embracing complementarity as fundamental – should lead naturally to a Copenhagen Interpretation of Relativity. Just as de Broglie had argued that all actual micro-scale observations must have both a particle and a wave aspect, so it seemed natural that on the macro-scale all observations should involve both a local Newtonian and a non-local Maxwellian aspect. One illustrative image, I think from Poincare, was that your desk is not a desk until you choose the appropriate frame of reference – a geometry – in terms of which to observe, and make sense of your 'experience' as a desk.

"Bohr pointed out and emphasized that all Einstein's examples of relativistic effects were 'relative to' some tacitly presupposed, chosen and well-defined local frame of reference. Einstein's examples always started by assuming a local clock, so that there is a sameness of location over changes in local clock-time, and a local ruler to enable one to measure in a flat (Euclidean) geometry. The actual observations of relativistic effects were always demonstrations relative to the local frame of reference defined by these choices (viz. clock, ruler, geometry).

"In other words, Bohr argued that all actual observations have an irreducible local Newtonian component – by their very nature. The proper understanding of relativistic effects – to be able to understand these observations, these measurements – to be able to make sense of them – you must realize that there is always both an irreducible non-local component and an irreducible local component. (All observation is middle-ground.)"

Bohr understood that the general embrace of complementarity had quite broad implications for our understanding of the universe and our place in it. University of California at Berkeley Historian of Science Professor John Heilbron expressed it nicely in a recent lecture to an American Physical Society Conference: "Quantum physics taught Bohr that physicists can only give an adequate description of all the phenomena presented by experiments if they use complementary concepts like wave and particle to refer to the same underlying reality."

Bohr also saw the More General, post-mechanical, post-scientific entailment, as Heilbron notes: "The primary payoff of Bohr's engagement with quantum physics for his wider philosophy was the discovery that multiple truths come... in complementary pairs."

I continued, "There is a powerful connection between the generally accepted 'actualizing' nature of observation in quantum theory and an 'actualizing' nature of observation in Bohr's approach to Relativity. In quantum theory there is no access, no pre-observational observational access, to the range of *potential* future actualities described in Schrodinger's partial wave function characterization of the observational opportunity. It is only when you implement a choice of how to observe (viz. or how to interact) that you 'actualize' a unique future. For Bohr the situation is similar for the description of Einstein's proposed Relativistic field-reality. Since the Relativistic field-reality is 'objectively' (independent of the observer) distributed in both space and time it is not locally accessible for any observer until the observer implements a choice of how to observe – defining (embodying) a local experimental set-up (clock, ruler, geometry).

"To say that something is distributed in space is easy to grasp. For something to be distributed in time is less straightforward. Suffice it to say that in quantum theory until the observer's choice 'collapses' the distributed potential, whatever it is that is described in the pre-choice characterization is 'not accessible'. You can't 'observe it' independently of choosing to 'observe' it – in one way rather than another. For Bohr, the consequence of the general embrace of complementarity is that from any given state of affairs there are multiple possible futures. The actual future is under-determined by any single mechanics – in both quantum theory and Relativity – until a choice of how to observe (viz. or the choice of a (bias) type of interaction) is made. The actual transition from one state of affairs to a future state of affairs requires, necessarily involves, a Participant observer's choice (viz. or a (bias) type of interaction).

"Einstein's characterization of the new Special Relativistic reality in terms of a 'no preferred frame of reference' field-reality is inherently ambiguous, under-defined and, as in quantum theory, inaccessible, unobservable, until the choice of a local observational frame of reference, a local way of observing, is implemented. In this curious way of representing observation, the future emerges by some process that involves an irreducibly non-mechanical, non-pre-determined *choice* amongst the possible observational futures. Each observer's choices are, of course, constrained by where one finds oneself in space and time (viz. in the actual history of the universe) and by how one has learned to observe (viz. by one's experimental technologies).

"Despite Bohr's early vigorous arguments that the embrace of complementarity entailed a sort of Copenhagen Interpretation of Relativity, the modern textbook consensus today is that the framework of Relativity establishes a new, revolutionary, non-local conception of objectivity – subsuming and superseding the traditional Newtonian conception of local objectivity. The proposal for a new 'objective' Relativistic reality, an observer-independent, inquirer-independent reality, attempts to preserve continuity with the detached Spectator representation of inquiry into a fully deterministic objective reality."

Pulling my thoughts together for Hawking, I summarized so far: "In Bohr's interpretation of the Correspondence Principle there had to be a revolutionary, post-scientific, post-objectivist More General Theory in terms of which the successes of both of these complementary Research Programs – the Newtonian and the Maxwellian – could be understood as limited special cases, in terms of which both these physics and both these oppositely defined conceptions of space-time could be understood as essential but limited special cases."

Copernican Principle Extrapolation

Next I offered Hawking a different thought experiment challenging the possibility of making sense of a completely and consistently non-local Relativistic universe.

"Prior to Copernicus the reigning view was that the Earth was the center of the universe establishing 'the one right frame of reference' in terms of which our observations, our experiences in general, were to be understood. With the Copernican Revolution the Sun was newly understood to be the center of the solar system. The Sun then became the new center of the universe, establishing a new preferred frame of reference. The great success of the Copernican hypothesis was realized in Kepler's discovery of a mathematical description of the laws of motion of the planets. Kepler's surprising advance was followed by Newton's remarkable unification of terrestrial and celestial motions with his theory of gravity. But eventually it became clear that the Sun was just one star among many billions of stars. This realization led to further steps away from the notion that our locality, either in the earlier Earth-centered formulation, or in the Solar System centered formulation, represented the real center of the universe and, consequently, the correct frame of reference in terms of which observed motions and structures were to be understood.

"With even better 20th century telescope technology, we found that our star was part of a galaxy of perhaps 100 billion stars organized and with an apparent, regular, solar-system-like motion around a center, defining a single coherent galactic spiral. So it seemed then that perhaps the galactic center should be considered the center of the universe and should be the basis of the preferred frame of reference in terms of which all observed motions and structures were to be properly understood. But then in the 1920s, Edwin Hubble's observations at Mt. Wilson Observatory established that our galaxy was only one galaxy among billions of galaxies. With each of these discoveries there was a shift in thinking as to the correct 'objective' – observer-independent – preferred frame of reference.

"Each shift to a broader frame of reference seemed to properly subsume and supersede the earlier choices of a preferred frame of reference. In the shift to the Sun-centered frame of reference, the retrograde motion of the other planets – as viewed from an Earth-centered frame of reference – had been newly and 'correctly' understood as merely 'apparent' motion, a consequence of the Earth's motion around the Sun as a planet among the other planets. The shift to the Sun-centered frame of reference altered our experience of the geometry of the solar system. Then, once it was realized that the Sun was moving in relation to other stars, Kepler's Laws, describing the highly regular orbits of the planets around the Sun, were recognized as idealized special cases, accurate and meaningful only with the 'arbitrary' idealizing choice of the Sun and the solar system as defining our locally preferred frame of reference. Even more troubling, Newton's theory of gravity dramatically unifying the terrestrial and celestial now appeared to be based on observations linked to this 'arbitrary' idealizing choice of a Sun-centered solar system as the preferred frame of reference. If the 'real' motion of the Sun and the planets has an essential galactic component, then by taking our galaxy's center as the preferred frame, the motion of the planets would be newly understood as spiral-like, following the Sun in its motion in relation to the galactic center. Further observation revealed that our Milky Way Galaxy is also part of a coherent ('mutually bound') super-cluster of galaxies – referred to as the Virgo Supercluster. There are other super-clusters of galaxies as far into the space-time distance as we have been able to observe. Our Virgo Supercluster is in motion relative to these other super-clusters that also are in motion relative to each other.

"Each such shift to a spatially broader picture is now often characterized as an application of the Copernican Principle, resulting in a further and further decentralization of the Earth-centered view as the preferred frame of reference. Up until recently it had seemed that the Scientific Research Program required that only by discovering the One, true, 'objective' frame of reference (viz. observer-independent, inquirer-independent) could we understand the 'real' motions and structures, not just the 'apparent' motions and structures as experienced in terms of different 'arbitrarily chosen' frames of reference."

(Note: It should be mentioned here that the relevance of these larger galactic frames of reference to the proper understanding of all observed motions and structures means that the complete understanding of our observations of the motion of the planets requires reference to non-local motions and structure in the far flung and time-past history of space-time. As Hawking remarked, a complete 'objective' understanding of all motions and structures requires a resolution of the questions about galactic motion that don't make sense in terms of either the Newtonian or Einsteinian theories of gravity requiring us to postulate a Promissory Note about Dark Matter.)

There is a danger in thinking too much about the lack of a preferred frame of reference. Rainer Maria Rilke in his novel, *The Notebooks of Malte Laurids Brigge*, captures our vulnerability to a sort of cosmic vertigo. The main character describes the dilemma and circumstance of one of his neighbors, Nikolai Kuzmich.

"Under his feet, too, there was a sort of motion, not just one motion but several, interwoven in a curious reel. He stiffened with terror: could that be the Earth? Most certainly it was the Earth. The Earth did move. They had said so in school. He staggered about his room as if on the deck of a ship, and had to hold on with both hands. Unfortunately he further recalled something to the effect that the Earth's axis was at an angle. No, he couldn't take all these motions. He felt wretched. Lie down and rest, he had once read somewhere. And ever since, Nikolas Kuzmich had been lying in bed. He lay with his eyes closed. … And that was when he thought up the business of the poems. It was scarcely credible that it should have helped so much. To recite a poem slowly, with an even emphasis on the end rhymes, was to have something stable, as it were, that you could keep your gaze fixed on – your inner gaze, of course." (*Notes*, chapter/section 49).

"Taking the distributed Relativistic space-time framework as 'the One right objective framework' takes the Copernican Principle to the limit, entailing that the 'objective framework' of the universe is the framework where, for actual individual observers, there simply is no preferred frame of reference at all. One might think of this as a preference for 'the non-preferred-frame-of-reference' frame of reference. One difficulty is that in such a completely space and time distributed Relativistic reality it becomes impossible to make an unambiguous 'universal', 'objective' specification of any Newtonian-like locality. You can only specify your location by making arbitrary choices to define a special, limited local frame of reference. Such a local reference frame, for instance, the Sun-centered solar system, cannot be given a 'universal', 'objective' specification in terms of its 'objective' location in either space or time. For Bohr, of course, this means that you can't *observe* 'the universe'. Characterized as completely distributed in space and time, 'the universe' – prior to an observer's choice – is inaccessible, inconceivable, completely and consistently, in One way. Even talking about or referring to 'the universe' and 'the age of the universe' in a completely distributed Relativity is inherently ambiguous. Any chosen reference frame is a limited idealization, a special case.

"Whenever I try to make sense of an 'objective' Relativity, embracing the notion that there really is 'no preferred frame of reference', like Nikolas Kuzmich, I get a little unsteady on my feet.

"The extrapolation to an 'objectively' Relativistic, 'non-preferred-frame-of-reference' frame of reference involves another limit argument jumping over the conceptually discontinuous asymptote between qualitatively distinct opposites. It is like reasoning from the incompleteness of each of a series of locality postulates to the completeness of a new *type* of 'locality', to a universal, non-locality locality. It isn't a valid argument.

"Taking Einstein's Relativistic reality as the new 'one right objective frame of reference' seems to me to be a *conversion* from one type of space-time framework to the complementary type of space-time framework. It is not, as Einstein and the Logical Positivist representation would have it, a conceptually continuous, uniform, inductive generalization over all possible, local Newtonian frameworks.

"Look," I said, pausing to collect my thoughts...

35

Einstein's Real Problem and the
Other History of Science

Coming at it from another angle, I continued, "Another line of evidence and reasoning supporting the hypothesis that Einstein's move to Relativity was a conversion rather than a proper inductive generalization over Newtonian physics comes from an examination of the actual historical record.

"In the middle of the 20th century, there was a dramatic revision of the 'official history' of Einstein's reasoning in coming up with Special Relativity. For several decades, the more or less official story told within the broader scientific community, as well as by historians of science, had been that two experimentalists, Albert Michelson (1852–1931) and Edward Morley (1838–1923) had run an experiment in 1887, at what is now Case Western Reserve University showing that there was no difference in the speed of light, regardless of the direction of passage. The physics of wave transmission was thought to be well-established and understood. Waves were transmitted through (by means of) a medium – water waves in water, sound waves in air. The Michelson-Morley experiments suggested that light and electromagnetic waves, in general, were not transmitted by a medium – were not transmitted by what had been conceived of as some sort of very sparse material 'ether'.

"Later in the 20th century, the official story had become that the young, brilliant Einstein had reasoned, somehow, from this conflict of theory and counter-evidence to come up with Special Relativity. But this official story is now understood, at least in the history and philosophy of science communities, to have been a 'rational reconstruction', an ideological rationalization, developed by reasoning that it must have happened this way because that is what one would expect in terms of the official Logical Positivist theory of how scientific advances occur.

"Embarrassingly, it wasn't until the early 1950s, not long before Einstein's death in 1955, that his biographer, Harvard physicist and historian of science, Gerald Holton, wondered about the accuracy of the textbook story and asked Einstein. In effect, Einstein said that the Michelson-Morley experiment was certainly not what had led him to his problem context and to his solution, which was Special Relativity. Einstein remarked that he was aware of the Michelson-Morley result and had seen it as consistent with his line of theoretical development, but he hadn't viewed it as particularly clarifying or contributing anything fundamentally important to his reasoning. That was a weird sort of wake up call to the historians of science."

Stephen was listening. He gives me an eye-brows up acknowledgment, indicating, I think: 'OK. I know that history. Continue.'

"Historians are supposed to be trained to be aware that it is the winners of a war, typically quite biased, who write the history of what happened. The early textbook accounts of how Relativity entered physics were written by the dominant Positivist-oriented historians of science, firmly committed to viewing advances in science as logically and conceptually continuous. Since their theory of how science worked presupposed that the universe was governed by One logico-mathematical order, advances must be driven by logico-mathematically consistent evidence. Therefore it 'stood to reason' that the reasoning from Newtonian physics to Einstein's Special Relativity should have been – must have been – driven by a logical analysis of novel experimental evidence such as from the Michelson-Morley experiment.

"One uneasy feeling about this formulation, that should have alerted us, is that if the advance is simply a matter of the logico-mathematically consistent counter-evidence and logical reasoning, then why is it that Einstein's contribution is so celebrated? To put it another way, if the succession to Special Relativity is a matter of straightforward logical reasoning, what makes us think that he was so smart – a genius – implying parenthetically that the rest of the physics community was so oppositely dumb and blind? Indeed, the transition didn't happen overnight; the scientific community had a long time to look at the evidence. The Michelson-Morley experiment was performed in 1887 – and Einstein proposed his Special Relativity in 1905. And even after it was officially proposed it was at least a decade before Einstein's Special Relativity was tentatively accepted in the preponderance of the physics community.

"Many prominent scientists of the day, well into the 20th century, never accepted Special Relativity, including Michelson himself, who received the Nobel Prize in 1907 for his experimental work. Einstein's colleague, Max Planck, had suggested that: "A new scientific theory does not triumph by convincing its opponents and making them see the light, but rather because its opponents die, and a new generation grows up that is familiar with it." (((Wissenschaftliche Selbstbiographie. Mit einem Bildnis und der von Max von Laue gehaltenen Traueransprache., Johann Ambrosius Barth Verlag, (Leipzig 1948), p. 22, as translated in Scientific Autobiography and Other Papers, trans. F. Gaynor (New York, 1949), pp.33-34 (as cited in T.S. Kuhn, The Structure of Scientific Revolutions))).

"Again, that's weird. If science is a nice, logical, systematic enterprise, why should it be so difficult for the top establishment scientists to under-stand the next advance? Obviously, Kuhn's alternative representation of ad-vances, as involving revolutionary conceptual discontinuities, as involving logico-mathematical discontinuities, suggested an account more in accord with the actual historical record."

Cal Tech physicist Sean Carroll recently argued for my same formu-lation of Einstein's real problem context: "The original impetus behind special relativity was not a puzzling experimental result (although the Michelson-Morley experiment certainly was that); it was an apparent con-flict between two preexisting theoretical frameworks. On the one hand you had Newtonian mechanics, the gleaming edifice of physics on which all subsequent theories had been based. On the other hand you had James Clerk Maxwell's unification of electricity and magnetism, which came about in the middle of the nineteenth century and had explained an impressive variety of experimental phenomena. The problem was that these two marvelously successful theories didn't fit together. Newtonian mechanics implied that the relative velocity of two objects moving past each other was simply the sum of their two velocities. Maxwellian electromagnetism implied that the speed of light was an exception to this rule." (Carroll, Sean, *From Eternity to Here* (2010), page 82)

"I recall sitting in one of Nick Maxwell's graduate seminars at University College, London, where the presenter, a fellow graduate student, remarked in response to a question about his historical account of some episode in the history of science that he 'didn't need to look at the actual historical record because it was clear from the logic' – his rational reconstruction – 'that the scientific advance must have happened this way.' It was that sort of attitude and cursory, after-the-fact historical scholarship that had led to the historical fantasy – the Michelson-Morley 'just so story' – about the transition from Newtonian physics to Relativity.

"Stephen Jay Gould introduced the use of the 'just so story' metaphor, an allusion to the works of Rudyard Kipling (1835-1936; Nobel Prize in Literature in 1907) into evolutionary biology. Gould pointed out that one could easily make up stories about how things had evolved, stories that seemed to make sense, that tell us that it happened 'just so'. These stories are – absent any real historical inquiry – entirely plausible. 'How the elephant got its trunk' is a famous example, wherein the elephant's original much shorter nose was stretched while trying to escape the grip of a crocodile. The equally derogatory technical jargon in the philosophy of science literature referring to 'just so stories' in the history of science depicts them as 'rational reconstructions', as 'ideological rationalizations'.

"Whenever I would propose my 'insight of the week' to Feyerabend, he would always ask me whether this was supported by the actual – not some imaginary – historical record. I confess that this was rather irksome in the beginning. You are flying high with some great, logically coherent idea, and the professor asks you for historical evidence. I mean, it tends to cramp one's style. But gradually I began to see what had been happening. And the more I looked the more shocked I became. Apparently a large portion of the history of modern science has been systematically misrepresented – rationally reconstructed, ideologically rationalized, according to the Positivist, Spectator representation of science!

"Holton's excellent scholarship served as preamble to Thomas Kuhn's overall challenge to the 'official' histories of science. Kuhn established that there was a significant difference between the real history of scientific advances and the Positivist's rationally reconstructed history of science. Kuhn's arguments generated an enormous literature trying to fit together the 'apparent incompatibility' of what Kuhn documented as the conceptually discontinuous process of actual scientific advance and the official, after-the-fact conceptual continuity of the rationalized 'objectivist' program."

Stephen gives me another eye-brows up: 'Go on, I'm listening.'

"Another supportive critique came from the philosophy of science crowd. They were looking carefully at the logic, the actual reasoning, that had led Einstein to Special Relativity. While at the University of London, I had attended, at London School of Economics, a series of seminars presented by fellow graduate student Elie Zahar. These were part of the weekly 'Popper's Seminar' series. Popper himself was at UCLA for that academic year.

"Elie was evaluating Kuhn's contention that the transition from Newtonian physics to Einsteinian physics was logico-mathematically discontinuous; that you couldn't actually reason from Newtonian physics, plus some 'counter-evidence', to Einstein's Relativity.

"Elie presented the history to us in excruciating detail, citing dates and experiments, the specific publications and the individual players. Particularly in the early years, those who switched from the Newtonian program to Einstein's Relativity were clearly not doing so on the basis of any sort of logical reasoning justified by 'the evidence'. For many years all the relativistic effects that had been experimentally demonstrated had perfectly reasonable interpretations within the Newtonian research enterprise. These defenses were argued by Hendrik Lorentz (1853-1928), the Dutch physicist, who would receive the Nobel Prize in Physics in 1902. A fair analogy is that Ptolemy was able to account for all the experimentally observed planetary retrograde motions in term of epicycles, retaining the Earth-Centered framework. The move to the Copernican Sun-Centered perspective wasn't a logical transition driven by the 'evidence', but was rather a discontinuous shift, presenting a new way of understanding the existing evidence. Indeed, Elie pointed out that in both cases, the actual historical record showed that for decades there was controversy and opposition, often based on well-confirmed experimentally-based arguments that weighed strongly against both the Copernican and the Einsteinian transitions.

"At least among the graduate students we were all a bit stunned by Elie's research and analysis and really weren't quite sure what to make of him. Hadn't the Michelson-Morley experiment forced some sort of logical reasoning that led to Relativity? That was the official story. But – not so quick – maybe that isn't how it actually happened.

"Independent of whatever Einstein had done, Elie also looked in detail for the logic, for the reasoning, that had led the physics community, admittedly rather gradually, to adopt the new Relativistic framework. He just wasn't finding it. "Look harder, Elie!" He did. It just wasn't there. It was in no small part because of Zahar's rigor and scholarship that I began to turn.

"When asked, Einstein, as was characteristic, wanted it both ways. In line with the Logical Positivist notion he imagined that "the supreme task of the physicist is to arrive at those universal elementary laws from which the cosmos can be built up by pure deduction." Yet he was sympathetic to the rebels' insistence that advances are conceptually discontinuous, not arrived at through any sort of logical reasoning: "There is no logical path to these laws; only intuition, resting on sympathetic understanding of experience, can reach them." (Einstein, Albert, 'Motives for Research', a speech delivered at Max Planck's sixtieth birthday celebration, April 1918. Cited in Baggott, Jim, *Farewell to Reality: How Modern Physics Has Betrayed the Search for Scientific Truth* (p. 301). Pegasus Books.)

I continued, "As a freshman and sophomore entering the physics program at Berkeley, I heard the rumors that three or four Berkeley physics professors – the older, established guys – had committed suicide over the years because they just couldn't accept the new physics. To them it was mad and irrational. And yet all the young bucks, the up and coming professors, were converting to the Einsteinian program. The old guys were pushed aside and told that they just couldn't understand it. They were just rumors. I never saw any hard evidence. But I heard the stories several times from different sources. It was at least part of the Berkeley physics cultural undercurrent. I only mention it because if Kuhn is right about the incommensurability of the Newtonian and Einsteinian systems, and yet the physics community was operating under the Scientific Hypothesis – expecting a uniform, conceptually continuous commensurability – then one might expect some potentially tragic misunderstandings and accusations.

"Again, Max Planck's reflection toward the end of his life (1948) that 'a new scientific truth does not triumph by convincing its opponents and making them see the light, but rather because its opponents eventually die, and a new generation grows up that is familiar with it' reinforces the suspicion that at least the greatest scientific advances of the 20th century are not logical, conceptually continuous generalizations over their predecessors. If the advances had been logico-mathematical inductive generalizations there is no reason to expect any difficulty within the community of established physicists in understanding and accepting the new theory. Kuhn's discontinuity theme proposed a basis for understanding the actual (viz. somewhat 'political') nature of the ascendency of the Einsteinian program.

"The real story of the problem context Einstein was exploring had to do with the intersection of the Newtonian and Maxwellian Research Programs; had to do with the failure to integrate, to reduce these two conceptually complementary physics enterprises to One coherent order.

"One of the reasons for the confusion and the failure to appreciate Einstein's real problem context is the way the history of Modern Science has been presented as a single uniform monolithic enterprise. This notion is of course consistent with the Logical Positivist representations that the goal is, as Einstein expressed it: 'The supreme task of the physicist is to arrive at those universal, elementary laws from which the cosmos can be built up by pure deduction.' (Einstein, Albert, "Prinzipien der Forschung: Rede zum 60. Geburtstag von Max Planck" in *Mein Weltbild* pp. 107-110 (1918) in *The Collected Papers of Albert Einstein,* vol. 7, it. 7 (2002)(S.H. transl.)

"However, when you look at the actual historical record, the real history of Modern Science is quite different, and the evidence supports Bohr's generalization of the complementarity theme.

The Real History of Science

"The real history of Modern Science was not the systematic, conceptually continuous progression the Logical Positivist model had expected; not what the Mechanical Philosophy expected. The history of science from the point of view of complementarity looks rather different," I said.

"There were two narratives involved in the actual emergence of Modern Science, and they didn't fit together. – For most students of science, the narrative history of Modern Science begins with Copernicus's move to a Sun-centered universe, followed by Galileo's experiments with falling bodies. Kepler discovers the laws describing the elliptical orbits of the planets followed by Descartes's combination of algebra and geometry, producing new powerful mathematical tools for modeling reality. Christiaan Huygens (1629-1685), widely underappreciated, pointed out the inadequacy of Descartes's straight-line, mechanical 'billiard ball' model in accounting for the centripetal force experienced by a moving body when changing direction. The experimental demonstrations of the centripetal force are one of the definitive demonstrations of inertia. Isaac Newton (1643-1727), who was mathematically uneducated when he first came to Cambridge, taught himself Descartes's mathematics. Building on the work of his predecessors and contemporaries, Newton formulated his Mechanics – his Three Laws of Motion – with the later addition of the unexpected Universal Law of Gravity. In a few sentences, that was the official story of the emergence of Modern Science.

"But there is another narrative of the emergence of Modern Science – the story of electromagnetism. This other storyline is about an almost completely separate research program, investigating very different types of phenomena.

Stephen knew the story, so I only mentioned it briefly. For the reader, let me highlight this 'other' history of modern science – the story of electromagnetism. (It is worth noting that I am ignoring the history of chemistry, biology and other 'scientific' disciplines that deal with phenomena that are, at least arguably, not logico-mathematically reducible to or deducible from physics.)

As with the Newtonian mechanics of motion, many of the most impressive advances in electromagnetism have been made in recent centuries. However, observations and investigations of electrical and magnetic phenomena reach equally as far back as ancient astronomy. Aristotle remarked on Thales's investigation of magnetism at the beginning of the Ancient Scientific Tradition in the sixth century BCE. There are references to the use of magnets in medicine in India around the same period. In fourth century BCE China, lodestones, naturally magnetized pieces of the mineral magnetite, were studied. By the 12th century CE the magnetic compass had been developed improving accuracy in navigation, as well as offering up observations of the unexplained correlation of magnetic and astronomical north. In 1600, when Galileo was a young 36 year old, William Gilbert reported the results of his extensive experiments on magnetism in his enormously influential *On the Magnet and Magnetic Bodies, and on the Great Magnet the Earth*. Gilbert reasoned that the Earth itself was a magnet, proposing this as an explanation of why compass needles align with the Earth's poles. In the history of the electromagnetic research program Gilbert's unification of mineralogy and whole Earth magneto-dynamics ranks with Newton's unification of the gravitational phenomena of apples falling on Earth and the Moon's orbit around Earth.

Records of observations and investigations of electrical phenomena date back even further. Lightning, of course, was nearly ubiquitous, although formal recognition of its relation to other electrical phenomena is more recent. The Egyptians, in 2750 BCE, knew of electric shocks from certain fish, and later Roman authors reported the observation that these shocks could be conducted from object to object. Experiments using amber rubbed on different surfaces to give it a static electric charge pre-date early Greek science.

A series of engineering advances in electrostatic generators enhanced experimental opportunities during the 17th and early 18th centuries. The invention of the Leyden Jar in 1744 allowed for the storage of an electrostatic charge and still more novel experimentation. Two notable discoveries were impressively serendipitous. Luigi Galvani discovered bioelectricity in a chance observation noticing that a spark generator stimulated the twitch of a muscle in a dissected frog's leg in an unrelated experimental preparation that happened to be located a few feet away. While preparing for a lecture on electricity, Hans Orsted chanced to notice that the needle of a compass resting nearby, used for other purposes, deflected when the current in an electric circuit was opened and closed, revealing electromagnetism, the fundamental interrelatedness of electrical and magnetic phenomena.

Technological developments were a common driver as only a few weeks after Alessandro Volta's development of the battery, allowing generation of a steady current, Nicholson and Carlisle discovered chemical electrolysis, splitting water into hydrogen and oxygen. George Ohm, utilizing Volta's battery, determined the relationship between voltage, current and resistance, the basic understanding of the electrical circuit. Charles-Augustin de Coulomb made a number of discoveries concerning the distribution and proportionality of electrical and magnetic forces including that strength of charge declined with distance according to an inverse square law directly parallel to Newton's inverse square law describing the decline with distance of the gravitational force.

Michael Faraday (1791-1867) continued earlier investigations of chemical electrolysis, discerning the discreteness of electric charges. Faraday introduced the concept of the electrical field, discovered polarization of light by a magnetic field and creatively engineered perhaps the single most useful electromagnetic technology of the last two hundred years – the electric motor. In 1831, Faraday also demonstrated that the time-varying magnetic field acted as a source of an electric field and reciprocally, a time-varying electric field acted as a source of a magnetic field. This reciprocity, incorporated in Faraday's Law of Induction, was cited by Einstein as perhaps the single most significant empirical observation stimulating his postulation of Special Relativity.

Through his mathematical articulation of Faraday's concept of the dynamic electromagnetic field, James Clerk Maxwell arrived, in 1864, at his fundamental equations of electromagnetism, noting as a consequence of his treatment that all the different wavelengths of the electromagnetic spectrum transmit at the speed of light in a vacuum. Maxwell then made one of the most dramatic unifications in the history of science, proposing that light itself was an electromagnetic wave. Maxwell's unification easily compares with Newton's unification of terrestrial and celestial gravitational phenomena.

With even this very brief overview of the history of electromagnetic theory it is easy to see that the dominant representation of modern science, as developing from Copernicus and Galileo, focusing on the motion of objects (particles), is seriously imbalanced.

"When these two lines of research are considered as almost completely separate narratives, a very different picture of the history of modern physics emerges," I said.

"Einstein's exploratory reasoning that led to Special Relativity had little or nothing to do with the Michelson-Morley experiment. What had struck Einstein for some years were the considerable differences between the phenomena of Newton's physics, the mechanical motion of objects and the phenomena of Maxwell's physics, the electromagnetic field.

"Per hypothesis, Einstein's problem context was based on the question of the discontinuity, on the question of the logico-mathematical incompatibility of the Newtonian conception of objective reality and the Maxwellian conception of objective reality. Both research programs and their associated concepts of reality had been enormously successful. The problem was how to make sense of these two separate knowledge-bases in one unified, consistent, conceptually continuous framework. The idea that there might be two, discontinuous 'objectivities' is – scientifically, mechanically speaking and from the Positivist logico-mathematical perspective – oxymoronic (viz. two 'one right answers').

"The Newtonian and Maxwellian-Einsteinian research programs are partially independent, yet inter-defined; the former associated with particle phenomena, the latter with wave phenomena. So it seems at least plausible, by analogy with quantum theory, that these overall research programs and their opposite frameworks for representing space-time are complementary," I said.

Finally, having come to the end of my arguments and anecdotes, I restated my hypothesis.

"So I think that the Newtonian space-time framework and the Relativistic space-time framework must be complementary," I said. "What do you think?"

Hawking glanced down at his computer screen and began to compose his response. I just sat quietly, glancing around our lower deck enclave, trying not to stare at him as he composed, yet not to seem to ignore him completely. I guess I was imagining that he might be writing something quite lengthy, given the importance of the question and my forty-minute presentation – a detailed critique perhaps. On the other hand, he hadn't been taking notes.

Hawking's response came in less than a minute. "I think you might be correct," he said and gave me a big, warm smile.

Stephen scanned around searching for Joan who had been watching for a sign from across the other section of the lower deck. Stephen soon began to move, and Joan was by his side in seconds. "Are you two finished then?" she asked politely. Stephen gave her an 'eye-brows up.' "Time for some tea then, don't you think Stephen?" she queried. Another 'eye-brows up.' I had been hoping for a little more of a back and forth dialogue. But that wasn't happening. Acknowledging that and reflecting, I decided that I should be pleased with his answer. It had been simple and direct.

Of course I harbored the lingering concern that Hawking may just have been politely patronizing – giving me some sort of sympathetic but possibly disingenuous encouragement. But I didn't really give that much credence. That wasn't Hawking's way.

The sea has been calm and the day is bright and warm. As we arrive and disembark at the dock in the beautiful central harbor in Victoria, Stephen is immediately swarmed by newspaper and television reporters. I have no idea of how they knew about our travel plans. Their interest is in Stephen's thoughts on what had become a front-page story in British Columbia. A woman in Victoria with ALS has petitioned the government to allow her to have a physician-assisted suicide. "What do you think, Dr. Hawking?"

His answer opened another window on Hawking's Participatory attitude to life. "People should have the right to die if they want to. It is one of the few rights severely ill people have left. But I don't think one should do it."

The Victoria Taxi Company is totally on top of our situation. One taxi van takes care of the 22 bags assuring us that they will be loaded on the hydrofoil before we arrive. Everyone piles into another taxi van. "Will everyone fit?" Tim, Jennifer, Christy, both nurses and I all cram into the back of the taxi-van. Stephen rides up front – 'shotgun' – with his chair securely strapped down. It is close quarters. I query our driver about the time-crunch. He assures me that he has choreographed everything. He will take us around central Victoria and through much of Butchart Gardens, arriving in time for the hydrofoil – the Victoria Clipper. Don't bother asking; he has it all worked out with a precise timeline. Wonderful!

Our driver is very personable and funny and enjoys having us aboard. As we tool around the central city he provides us with a history of the city, as well as a running commentary on the specifics of what we are seeing – the 'very English' Empress Hotel, the famous Parliament Building... Victoria is 'culturally distinct' in British Columbia – very different from Vancouver. There are horse-drawn carriages for the indulgent tourist. I tell stories of visiting here when I was ten, having ginger beer on a picnic in the main city park. The nurses in particular are happy that Stephen insisted on coming to Victoria.

The departure time for The Victoria Clipper is getting close. Our driver reassures me. Soon, we drive right up to the ship to board. They are ready for us – VIP status. Those involved take pride in giving Hawking exceptional attention. Very Canadian.

Once we were settled on board, just out of the harbor, The Victoria Clipper transforms. As the ship accelerates we become elevated on the hydrofoils, the ship's hull no longer touches the water. Speed increases – considerably – compressing what would otherwise be a six-hour crossing to two hours. You can't understand the unexpected, dramatic increase in speed in terms of classical ship technology. You need to understand the experience by reference to the new experimental setup, the new technology, the new way of engaging the ocean – the hydrofoil. I thought about asking Stephen for something more detailed in response to my diatribe but decided that I should just take what I had and be happy.

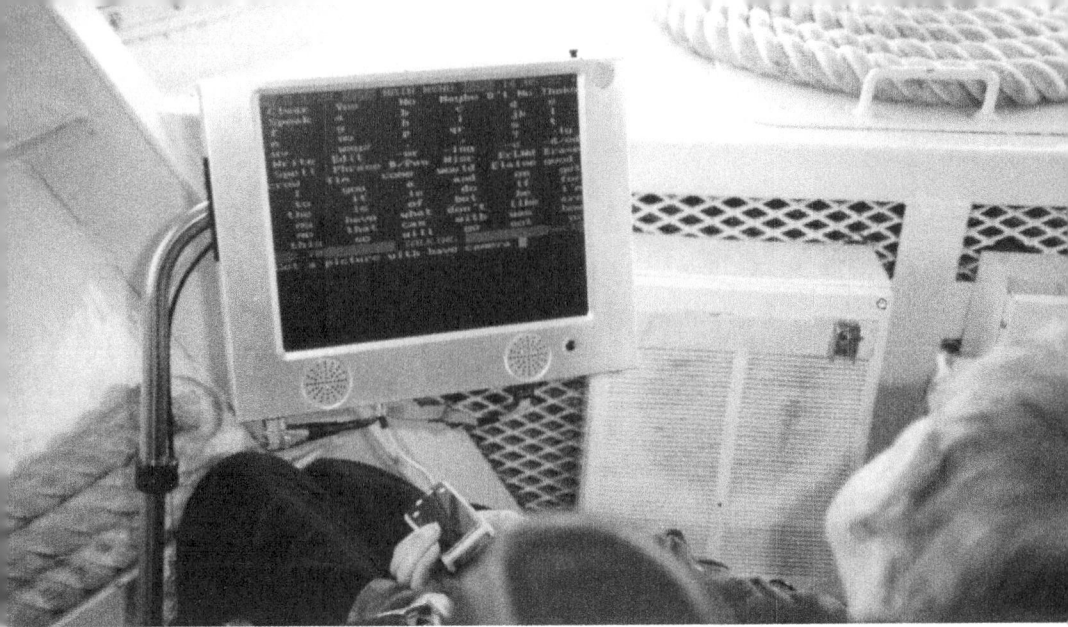

"Take a picture of that."

36

Museum Associates Gallery and Lunch at the Space Needle

Our arrival in Seattle on The Victoria Clipper was uneventful, although a number of the people waiting for their friends or relatives to arrive recognize, or think they recognize, Stephen. Is that actually Stephen Hawking? The only difficulty was for Tim to collect and make sure we had all 22 bags. Everything is transported just a few blocks from the waterfront to the Alexis Hotel.

The Alexis Hotel came highly recommended. It was the number one Seattle hotel – depending on whom you asked. Fortunately, the general manager was an enthusiastic Hawking fan, having bought, if not entirely read, *A Brief History of Time*. We quickly worked out a discount for the rooms in exchange for an ad in the program-magazine and a 'thank you' from the podium when introducing Stephen at the Seattle Opera House on the evening. Stephen had a fabulous suite with a view. The two nurses shared a connecting suite.

Coming out of the Alexis a few months earlier, after negotiating with the manager, I had noticed two doors down the street a storefront with a sign reading 'Museum Associates Gallery.' In the window were a variety of quartz, amethyst and jasper crystals as well as – 'lo and behold' – huge ammonite fossils! I had to visit.

As I walked into the entry, I saw mounted on the wall a framed, fossilized coelophysis, about six feet long (viz. an early bipedal carnivore, a dinosaur from the Triassic, about 200 million years ago). The coelophysis was raised and slightly arched so that it was three-dimensional, coming out of the plane of the frame at the head and tail. It was stunning. Really cool! Just off the entry area was a small room with a desk, a scatter of papers and a telephone. Emerging from another connecting doorway was Richard Berger – a tall, burly guy with a full beard and mustache.

After a cordial greeting and my expressions of admiration and awe, I begin to ask questions. "Are you associated with a museum?" I ask. "No," says Berger, with no elaboration. "But aren't these fossils scientifically significant – I mean rare and worthy of careful study?" I continue. "No," says Berger, with no elaboration. My initial suspicion is that Berger is running some sort of fencing operation dealing in 'hot' fossils. I had never heard of anyone being able to sell a large, fossilized dinosaur. "Where did you get these?" I asked. "Friends," says Berger, with no elaboration. There is a definite air of suspicion between us. I suspect him and, I am thinking, he suspects me of being an undercover FBI agent or something.

"Your collection of large crystals is amazing," I comment. "Where do you find these?" I ask. "They're everywhere," says Berger, adding, "I find them and extract them myself, or occasionally a friend brings a specimen to me." So Berger goes in the field himself, finds and extracts these beautiful minerals and fossils and sells them from a storefront on 1st street in Seattle.

"Where do you find these?" I ask. Berger gives me a coy sort of grin with a slight tilt of his head, his bushy eyebrows rising. I get the message: 'like be serious, why would I tell you, dude!'

So I decide to play along. I figure I can turn him into the Feds later. I just can't imagine that it is legal to sell museum quality fossils on the open market. My scientific sensibilities are offended. Fossils are science not business. The crystals are another matter – at least as long as they are not being taken from subterranean caves within the boundaries of National Parks. It is not too difficult to imagine finding a cache of large minerals on your land somewhere in Wyoming or Iowa. Still, the size and beauty of these specimens inclines me to feel that they should be in museums, not just in private homes to impress guests.

With the initial tension abating, Richard offers to give me a tour of the Museum Associates Gallery – of 'the garden' as he calls it. He tells me to wait for a moment in the entry until he turns on the lights in the garden. Back in a couple of minutes, he invites me in – to what must be one of the most remarkable spaces I have ever seen. It's an exceptionally large room, maybe 250 feet by 250 feet, with a circular path that leads you through a gardenlike setting. There are large plants – ferns, probably artificial, since there are no windows. The foliage is backdrop, emulating ancient tropics perhaps; not the focal point. Most of the lighting is indirect with spots and small floods, darkness above and around, making it impossible to grasp the whole space at once. As I move along the path I am confronted with one gorgeous sight after another: a huge, brilliant, natural copper deposit, an amethyst that must be four feet by three feet – spectacularly lit. At the end of this first passage is a massive relief with a half-dozen dark, fossil fish imprints in a light-colored stone. I want to stop there for an hour and just savor this window into the past, wondering about the thoughts and the world of these fish. Who were they? "How old?" I ask Richard, who is at least fifteen yards behind me. He doesn't answer. He's not accompanying me. He's letting me experience the wonderland by myself. As I turn the first corner there is the first partial overview of the room. The center piece is breathtaking – a six-foot tall, four-foot wide, white quartz crystal – at least eight large spires emanating from a common base – illuminated from deep inside. "Oh, my God!" I keep moving, drinking in the beauty – pyrite crystals; a spiral ammonite with diameter better than three feet; natural sandstone concretions the size of a door; a petrified cross-section of a Sequoia, six feet across; a duckbill dinosaur that lived in Wyoming 65 million years ago.

This is a museum, but more than that. This is a gallery. These are works of art. But the environment that Berger has created is not the Louvre, with pictures hanging on bare walls. The room itself is a work of art. I want to say that it is more than art but I don't have a word for it.

I want to stay here. The room gives one the sense of experiencing the fullness of time – all time. I am surrounded by the past, by my ancestors from deep time. I want to live here – or in a place like this. Then I realize: I do live here – on Earth. I just don't always experience it this way.

I took Stephen Jay Gould to Museum Associates Gallery a few months after this Seattle event with Hawking. Gould, who was then the curator of the Harvard Museum of Natural History, was extremely complimentary of Berger's garden. Gould mentioned that he had often thought that it should be artists that develop the interior design and presentation of museums. Berger's work represented a stunning confirmation.

So Berger and I begin to hit it off. I explain that I am arranging a public lecture for Stephen Hawking. "I must bring Stephen here," I say. "He will love this. A brief history of time right here – well sort of," I muse. "We'll be just next door at the Alexis," I say. I promise to bring Stephen by. "Just be sure to call me before you want to come in, because I am not always here," Berger says. "I will want to show Hawking around personally," he adds. "Right. Will do," I say.

My entrepreneurial mind starts working. "I don't suppose you would like to be a cosponsor of the lecture – along with Microsoft, KCTS/9 and Seattle University," I ask. I am surprised when he expresses skepticism of the proposal. "Terry, you're looking at what is essentially a one-man operation here. I am not in that league," says Berger.

"What about trading that mounted coelophysis in the entry way?" I suggest. Berger laughs, "No way." "Why not?" I ask. "I just can't be giving stuff away for a little publicity," he says. "Advertising like that isn't of any value to me. Everything I do is by referral, word of mouth. So the prospect of being a cosponsor of a large science lecture doesn't even register with me," he says. "Nice try for the coelophysis," he adds, smiling.

"How much to buy it?" I ask, assuming something like $75,000 – $125,000. "$3,500," he says. "No shit! ... excuse my language," I say. "That's almost reachable. Not for me... now. But, you know, once I get out of debt maybe."

"There will be a lot of Microsoft people there. Nathan Myhrvold, one of the VPs – Gates' inner circle – is a friend of Hawking's. He did a postdoc with Stephen before coming to Microsoft," I say.

I keep looking for a win-win relation. I don't recall now exactly how the concept evolved, but the conversation eventually came to an agreement that Berger would help to create an atmosphere by decorating with a variety of items the Seattle Opera House stage and lobby on the evening of Stephen's lecture. There was to be a large, illuminated, white quartz crystal, as well as the large spiral ammonite, on stage. He would mount a dozen other objects for the lobby. We had to have a clear understanding with the Opera House that Berger wasn't selling anything. Standard policy in these venues is that they take 40 percent of the revenue for anything sold on their premises. Berger had no problem with this at all. "I would never make a sale on their site in any case," he said. "I don't operate that way. I am not a cash-and-carry retailer. All I would do is give anyone who is interested my card and invite him to meet with me later at the Gallery," he said.

Berger operates like a normal art gallery that sells paintings and sculptures, except that he is selling nature's creations, admittedly with a serious value-added sense of presentation. It was still win-win. I love this sort of arrangement. It didn't produce any revenue for the Institute, but it was a better event, a better staging and a better atmosphere. And I was able to help Berger without costing me anything.

When I told Stephen about Berger's Gallery, he was immediately interested in visiting. So after lunch at the Alexis one day, we arranged to meet Berger at the Gallery. Being rather naïve, I think I suggested to Stephen that Berger might offer him something as a token. I don't know what I was thinking. It may have been that or it may have been Stephen on his own, but he asked Berger if he would 'donate' something to him, to Stephen. Keep in mind here that the Alexis is giving us complimentary rooms. There was precedent. People like to do things for celebrities, even when it isn't a clear quid pro quo for publicity or recognition.

So it was more than a little embarrassing when Stephen asked Berger for something for free. "Yeah. Right. In your dreams," Berger responded. "I would be happy to sell you something," he added. Stephen took the tour through the gallery garden. It wasn't entirely wheelchair accessible. Berger had placed a ramp in one section, but another section in the back was elevated and normally only accessible by four steep stairs. I of course chided Berger for not being wheelchair accessible. He agreed. "I hadn't really thought about it at the time I created this," he said. "I'll work on it," he added.

Completing the tour of the garden Stephen said: "Pretty impressive." There was a further inconclusive discussion about purchasing something. I don't know for sure if Stephen ever eventually bought anything. But I think so.

When I was at the science store in Science World in Vancouver, my entrepreneurial mind had been wondering why there were so many Einstein posters around and no Stephen Hawking posters. I joked with Stephen, insisting that he was much cuter than Einstein. The spiral shape of the ammonite is reminiscent of the spiral shape of a galaxy. One thing led to another and Christy Richardson knew a professional photographer friend, so we conceived a poster with a backdrop of a spiral galaxy and the large ammonite at Stephen's feet. The favored quotation was Stephen's: "I don't have time to do all the things I can do, so it doesn't seem important to worry about the things I can't do."

Stephen agreed to the project. The photographer decided to photograph Stephen and the ammonite separately. I have copies of both but we never took the next step. Sorry. I did a little Business 101 homework. The prospect was that we might sell 1000 posters over three years, mostly in Science Museums. We might net about $2.00 per poster, that is, if we were lucky and sold them all. When you figured the time and effort to contact all the Science Museums with follow-up, mailing and accounting, it was a sub-minimum wage enterprise. At one point we creatively conceptualized four Hawking posters with memorable quotes. But it just didn't pencil out. So since then the poster project has drifted into second priority land (viz. as soon as I get some extra time...).

High on the priority list for adventures in Seattle was the Space Needle. Stephen had made this clear. The 605-foot tall Space Needle was constructed as the signature, symbolic structure for the 1962 Seattle World's Fair. (Trivia I came across: 'In 1966 eleven-year-old Bill Gates, now Microsoft chairman and co-founder, won a dinner at the Space Needle restaurant offered by his pastor. Gates had to memorize Chapters 5, 6 and 7 of the Gospel of Matthew, better known as the Sermon on the Mount. He recited the sermon flawlessly.') In keeping with the 21st century theme, the final coats of paint on the Space Needle were dubbed Astronaut White for the legs, Orbital Olive for the core, Re-entry Red for the halo and Galaxy Gold for the sunburst and pagoda roof. The top floors house a delightful rotating restaurant and, for those not eating, just above that is a public observation deck. This top section was balanced so perfectly that the restaurant is rotated by only a one-and-one-half horsepower electric motor.

Two of the Space Needle elevators are high speed and can travel at a rate of 10 mph, or 800 feet per minute. Actual travel time from the ground level to the top-house is 43 seconds. Under high wind conditions these high-speed passenger elevators are slowed to 5 mph. On a clear day it is perhaps the most spectacular elevated view for lunch on the planet.

There was some sort of time crunch that day; I don't recall exactly what it was, but I suggested to Stephen that perhaps we should skip the lunch at the Space Needle. Well, if there is uncertainty in the universe it wasn't apparent on this issue. For those who can quickly interpret Stephen's facial expressions and eye-movement signals it was immediately clear. The nurses were readying themselves for departure even before Stephen's voice synthesizer blurted out: "No. Let's go now."

I was aware that the Space Needle would provide a special elevator ride for VIPs. (My father had told me a story of how he and friends, one of whom looked very much like the folksinger Burl Ives, had secured a private ride on a busy day on the celebrity pretense. This little white lie became embarrassing when the look-alike was asked for autographs.) When I made the reservation for our group, I mentioned that this was Stephen Hawking's party and double-checked that they were ready for the wheelchair. I hadn't asked for a special elevator. Nonetheless, when we arrived and got in line, one of the attendants saw us and ushered us around to a waiting elevator. "This isn't really necessary," I say, on the way. "Our pleasure. We are honored to serve Dr. Hawking," she responded. The large, glass, head-to-feet front of the elevator permits a heart stopping view outward and downward on the way up. As we ascend, it's like looking over the edge of a growing cliff.

I can only describe Stephen's behavior at that lunch by analogy. Imagine a ten year old who had been sequestered in a dingy classroom all week, placed on top of a mountain on a clear day – with a digital camera.

"Get out the camera," says Stephen. Tim rummages through the saddlebags hanging off the back of Stephen's chair. "You mean this one. The new one?" Tim asks.

"Yes," says Stephen.

I mean, I love the Space Needle and go there whenever I can when I am in Seattle, but Stephen was so thrilled and enthusiastic – really, like a little kid. Everyone is impressed with the view, particularly on a clear day. And this day was perfectly clear.

"Take a picture of that," Stephen instructs Tim.

So of course we all indulged in teasing Stephen about this. He just smiled away – happy as could be.

The restaurant rotates a full 360 degrees every 43 minutes, so the view is changing every few minutes. "Take a picture of that," Stephen, again, instructs Tim.

The Space Needle is located in the former World's Fair grounds now housing a variety of cultural institutions. We are about a mile north of the formal, downtown Seattle City Center, allowing a perfect view of central city buildings, as well as of the city's hills and inlets.

To the southeast, in the distance, is the enormous Mt. Rainier, the most prominent feature of the broader Seattle skyline – from here it is visible in its full spectacular beauty. Of the volcanic peaks of the Cascades, at 14,410 feet, Mt. Rainier is one of the youngest – a mere half-million years old.

The waters of Puget Sound and its islands appear to the south and west. We can see ferries, like The Victoria Clipper, crossing Elliott Bay, pulling into and out of Seattle's extensive harbor-line. The Olympic Mountains form the distant horizon to the west. An occasional float-plane takes off from the waters of Lake Union, perhaps destined for the San Juan Islands to the north.

The main peak-line of the Cascades, marked by Goat Rocks, is directly to the east. As we rotate, another Cascade peak, Mt. Baker, far to the north, becomes visible.

Space Needle View

Standing right next to the windows, glancing almost directly down, we see below us the Pacific Science Center, Seattle's science museum, as spectacularly beautiful as Vancouver's Science World, but with a decidedly different architecture. Later, as we rotate, the Seattle Opera House, where Stephen's lecture will be, becomes visible below.

The teasing continues. A consensus emerges among the staff and me that Stephen had actually agreed to come to Seattle primarily because he wanted to visit the Space Needle. The lecture just gave him a formal excuse.

"Take a picture of that," Stephen instructs Tim. "Stephen, I am trying to eat my lunch," Tim demurs.

Again, "Take a picture of that," Stephen tells Tim. "Where? You mean over there?" asks Tim. Stephen indicates and directs. Of course there were the obligatory group shots with this or that background view. It was a high point: everyone laughing and teasing and joking – a spectacular setting with fantastic people.

In response to our teasing, Stephen let us know and loving every minute of it: "I am really just a little kid."

37

The Complementarity of Newtonian and Maxwellian Physics

Over the years following my presentation to Hawking on the ferry to Victoria, I have had a number reflections that significantly clarify my hypothesis that the transition to Relativity, represented as defining a new 'objectivity', – is really a *lateral conversion* from one space-time framework to its complement – and not a uniform superseding logico-mathematical inductive generalization.

I had shared with Hawking only briefly, almost parenthetically, one of my main frustrations over the years with the conversion argument. If it were a conversion from one research program to its complementary opposite, then there should have been something missing in the newly favored Special Relativistic program, something lost, something that was previously, properly explained in the Newtonian program. Just as the Newtonian framework was unable to account for the evidence for relativistic effects, with a conversion between complements, there should have been some phenomena well understood and explained in the Newtonian framework that could not be explained, could not be understood from within a completely Relativistic framework.

Having struggled with this question of the reciprocal loss for a couple of decades, I can't say that the answer was obvious, but part of the answer, at least, was the sort of thing that makes you want to thump the side of your head and say – "Of course."

For me the best expression of the answer came from Cal Tech physicist Sean Carroll in his excellent and more than slightly rebellious book,

From Eternity to Here. First of all, Carroll had agreed about Einstein's real problem context: "The original impetus behind Special Relativity was... that these two marvelously successful theories [Newtonian mechanics and Maxwellian electromagnetism] didn't fit together." (Carroll, Sean, *From Eternity to Here: The Quest for the Ultimate Theory of Time*, (2010) page 82)

Carroll then went on to give a straightforward answer to my question of 'what was lost' if, per my hypothesis, it was a lateral conversion. "Like many dramatic changes of worldview, the triumph of Special Relativity came at a cost. In this case the greatest single success of Newtonian physics – his theory of gravity, which accounted for the motions of the planets with exquisite precision – was left out of the happy reconciliation." (Carroll, Sean, *From Eternity to Here* (2010) page 82)

Carroll is clearly suggesting that the move to Special Relativity is not simply an inclusive (subsuming) logico-mathematical generalization over Newtonian physics but involves a conceptual discontinuity, entailing that something was left behind; something was lost in the transition.

Carroll goes on to point out that General Relativity was Einstein's later attempt to rescue Special Relativity from the loss of Newtonian mass by trying to explain gravitational phenomena in a hopefully supersed-

Sean Carroll

ing generalization over the Special Relativistic framework. "Along with electromagnetism, gravity is the most obvious force in the universe, and Einstein was determined to fit it in to the language of relativity. You might imagine that this would involve modifying a few equations here and there to make Newton's equations consistent with invariance under boosts, but attempts along those lines fell frustratingly short." (Carroll,

Sean, page 82, *From Eternity to Here: The Quest for the Ultimate Theory of Time*, Plume (2009))

Once this line of thinking is opened up, one is struck by the fact that other phenomena having to do with inertia, definitively explained in terms of Newtonian mass, are not accounted for in either Special or General Relativity. There are two separable aspects of inertia, definitively explained in terms of Newtonian mass. The first is about straight-line motion. In Newtonian physics 'rest mass' (inertial mass) is that which, by its very nature, requires a force to move it or to otherwise alter its state of motion. This was the standard of the frictionless, billiard ball 'contact causality' of the Newtonian clockwork. The second, apparently separate, aspect of Newtonian inertia had been pointed out earlier by Christian Huygens – the centripetal force exerted by an object in circular motion; for instance, swinging a mass around on a string to which it is attached. Actions associated with the centripetal force were not explainable in terms of the paradigm of Descartes's simple mechanical 'contact causality'. This non-deductive addition to the first concept of inertia is what led to the later, directional vector representation of all motion. In Newtonian physics all motion must be understood as having a direction, by its very nature; you can't talk about the motion of an object without specifying its 'real' direction. But direction defined in relation to what frame of reference?

Newton offered a series of thought experiments about the inertia associated with rotational motion. These thought experiments demonstrated, for Newton, a preferred frame of reference – absolute space – at least in some irreducible sense. Rotational motion, for Newton, was somehow another *type* of motion, not reducible to, not translatable into, simple directional vector motion in one direction. Newton's thought experiments are referred to today in terms of one of the canonical experiments – Newton's Bucket. The experiment considers a bucket of water suspended on a rope, that has been wound-up. The bucket is released and as the rope unwinds the bucket rotates. Newton observed that the surface of the water in the bucket, initially flat, begins to curve – higher at the sides of the bucket, lower in the center. Anyone can repeat this experiment. In another version, Newton considered two stones attached to the ends of a rope. Rotate the system in empty space. There would be a measurable tension in the rope, depending on the rate of rotation. This isn't motion-in-relation-to, isn't motion relative to anything outside the rotating system. There is motion, and there is a measureable tension in the rope, even if there isn't anything else in the universe. This contrasts with all relative, straight-line, 'contact causal' motion that was defined in relation to some other body. Newton argued that these rotational phenomena of inertia could only be made sense of as 'real', not just 'apparent', if they were understood as occurring in relation to the preferred frame of reference of the universe, in relation to absolute space. For Newton, the phenomena of rotational motion were not understandable, could not be made sense of, in terms of the relativity of straight-line motion in relation to something else, in relation to some other object outside the rotating system.

In his book, *The Fabric of the Cosmos: Space, Time and the Texture of Reality*, Brian Greene presents an excellent modern review of these unsettling and still unsettled questions tracing back to the Newton's Bucket thought experiments. On one of his visits to Portland, when addressing a group of advanced high school seniors, I asked Greene, to talk to them about the Newton's Bucket problem instead of talking more about String Theory (the topic of his later, main evening talk). Greene declined, saying, "It is too difficult to explain. They wouldn't understand." I pushed a little, arguing that these 'next generation students' needed to understand that there were fundamental questions about space-time that were unsettled. "Give them something to think about," I said. Greene declined. Instead he offered them a very optimistic promotion of a fully deterministic, and as yet still experimentally undefined, String Theory.

General Relativity was enormously successful in describing and explaining the relativistic effects associated with gravity – relativistic effects that were not expected in Newton's non-relativistic framework. However, that doesn't mean that General Relativity subsumes and explains what Newtonian gravity had previously explained so successfully. My speculation was that just as Newtonian mass is not subsumed and made sense of in terms of relativistic mass ($E = mc^2$), Newtonian gravitational physics is not subsumed and made sense of in the General Relativistic physics. In General Relativity the Newtonian gravitational 'force' becomes what is referred to as a 'fictitious force'. The Newtonian 'force' was 'explained away', replacing it with the geometry of space. Whether General Relativity really explains, really subsumes, the successes of Newtonian gravitational theory is an open question. General Relativity's geometric 'descriptions' have certainly been quite successful in accounting for some phenomena, such as the periodicity of quasars. General Relativity's 'geometry of space-time' also, at least initially, seems to have the characteristics of an objective, observer-independent description of a Newtonian-like absolute space-time. And yet, as John Archibald Wheeler has emphasized, there doesn't seem to be any explanation of why space-time should have evolved into this particular structure – from among a seemingly unlimited range of possible geometries.

In any case both the Newtonian and Einsteinian theories of gravity fail miserably in accounting for the observations of Rubin and Zwicky. Consequently this failure has required the introduction of new 'gravitationally active furniture' – the conceptually discontinuous, increasingly curious – and missing – Dark Matter.

The piecemeal, yet enormously successful demonstration of relativistic gravitational effects – for instance, incorporated in GPS technology – does not justify the extrapolation to a universally, 'objective' General Relativistic account of gravitational phenomena. By analogy, the enormously successful demonstration of a wave-aspect to all particles does not justify the move to a universal wave framework, where one would try to understand particles as a type of wave.

Sean Carroll also emphasizes the importance of an unsettled curiosity: that the experimental demonstration of the inertial properties of Newtonian mass are entirely separable from the experimental demonstration of the gravitational properties of Newtonian mass. This curiosity is particularly troubling in that the Newtonian gravitational mass and Newtonian inertial mass are experimentally measured as having the same value, quantitative, to a high level of precision. Nonetheless, these are two qualitatively different, mechanically discontinuous, ways of observing, of measuring, Newtonian mass. (Kuhn might have suggested that although these two quantitative measures of Newtonian mass are nearly the same, the different experimental set ups measure different aspects of Newtonian mass that are qualitatively, conceptually discontinuous.) The explanations of Newtonian inertial phenomena are experimentally separate from the explanation of Newtonian gravitational phenomena. Consequently, Newtonian inertial phenomena are not explained or even addressed by General Relativity – a theory about gravitational phenomena. What this fact suggests is that the move to General Relativity, explaining the well-demonstrated gravitational relativistic effects, doesn't even (at least clearly) address the question of how to account for the inertial characteristics of Newtonian mass within a Relativistic framework. The experimentally defined concept of inertial mass is not derivable from the experimentally defined concept of gravitational mass – and vice versa. The experimental demonstration of one doesn't even suggest the other.

Confirming that the gravitational and inertial mass are different is the recent proposal to add the Higgs Field to Cosmology. The Higgs Field is a hypothetical invention introduced to account for at least some aspects of the phenomena associated with Newtonian inertial mass. The addition of the Higgs Field to the 'furniture of the universe' is discontinuous in that the inertial mass phenomena associated with the Higgs Field are in no way derivable from or implicit in General Relativity (viz. the latter being concerned with gravitational phenomena).

What is particularly fascinating about this curiosity, and the reason I dwell on it, is that if you have accepted the official litany, the 'indoctrination', that Relativistic physics properly subsumes and supersedes Newtonian physics, then it is not at all clear how even to approach these remaining questions. There is no reason for the student accepting modern physics to imagine that these are even meaningful questions. The subsumption and supersession of Newtonian physics by Relativistic physics is typically presented as 'settled science'.

Kip Thorne and General Relativity

I did have an opportunity to make roughly the same arguments I had made to Hawking to his close friend and colleague, Kip Thorne, Professor of Physics at Cal Tech, in 1997, when I hosted him as a presenter in the Oregon lecture series. With Kip I had the benefit of a much more detailed back and forth dialogue – something I had missed with Stephen. In response to the question about the complementarity of Newtonian and Maxwellian (Einsteinian) space-time frameworks, Thorne, who had been one of John Archibald Wheeler's famous students, answered that it was certainly one of the reasonable, ongoing hypotheses. He said that his book, *Black Holes and Time Warps*, Chapter 11, entitled "What is Reality?" was explicitly designed to address the open question of the 'objectivity' of Relativistic space-time reality.

Somewhat to my surprise, in that chapter, Thorne explicitly honors Kuhn: "The mental processes by which a theoretical physicist works are beautifully described by Thomas Kuhn's concept of a *paradigm*. Kuhn, who received his Ph.D. in physics from Harvard in 1949, and then became an eminent historian and philosopher of science, introduced the concept of a paradigm in his 1962 book, *The Structure of Scientific Revolutions* – one of the most insightful books I have ever read."

Thorne points out that, in practice, there are two paradigms in General Relativity: "The curved space-time viewpoint on General Relativity is one paradigm; the flat space-time viewpoint is another. Each of these paradigms includes three basic elements: a set of mathematically formulated laws of physics; a set of pictures (mental pictures, verbal pictures, drawings on paper) which give us insight into the laws and help us communicate with each other; and a set of exemplars – past calculations and solved problems, either in textbooks or in published scientific articles, which the community of relativity experts agrees were correctly done and were interesting, and which we use as patterns for our future calculations."

Kip Thorne

Thorne is quite explicit about the science being unsettled: "What is the real, genuine truth? Is space-time really flat...or is it really curved? ... They [the community of relativity experts] disagree as to whether measured distance is the "real" distance, but such a disagreement is a matter of philosophy, not physics. Since the two viewpoints agree on the results of all experiments, they are physically equivalent. Which viewpoint tells the "real truth" is irrelevant for experiments; it is a matter for philosophers to debate, not physicists. Moreover, physicists can and do use the two viewpoints interchangeably when trying to deduce the predictions of General Relativity."

I imagine that Kuhn might have argued that since the Newtonian concept of gravitational mass was developed in the framework of flat space-time, it cannot be properly, conceptually, subsumed into the General Relativistic space-time framework of curved space-time. Even though in low gravity situations the measurements should be nearly identical – quantitatively – they are not conceptually the same. Stated simply the Newtonian-like flat space-time mass is measured in terms of force and General Relativity mass is measured in terms of a geometric curvature. Kuhn might have emphasized that what is meant by 'mass' in the two measurements is not the 'same'.

All this brings me once again to reflect on a possible Copenhagen Interpretation of General Relativity. Recalling Nobel Laureate Louis de Broglie's comment: "Two seemingly incompatible conceptions can each represent an aspect of the truth ... They may serve in turn to represent the facts without ever entering into direct conflict." (De Broglie, Louis, "On the Complementarity of Ideas on Self and System," *Dialectica* 2, page 326). Bohr, of course, reasoned that "physicists can only give an adequate description of all the phenomena presented by experiments if they use complementary concepts like wave and particle to refer to the same underlying reality."

Thorne seems to be thinking about a possible complementarity in General Relativity: "Although the two versions of the Einstein field equations are mathematically equivalent [i.e. curved space-time and flat space-time], their verbal pictures differ profoundly. It is extremely useful in relativity research, to have both paradigms at one's fingertips.... Theoretical physicists, as they mature, gradually build up insight into which paradigm will be best for which situation, and they learn to flip their minds back and forth from one paradigm to the other, as needed.... This freedom carries power. That is why physicists were not content with Einstein's curved space-time paradigm, and have developed the flat space-time paradigm as a supplement."

The two paradigms of General Relativity are, so to speak, competitive in different situations and are still, like complements, globally compatible. What Thorne refers to as 'verbal pictures' that 'differ profoundly' is what I have referred to as qualitatively discontinuous, perhaps complementary, conceptions. The continued use of a Newtonian-like flat space-time is, perhaps, not arbitrary or a simple convenient holdover. If the notion of an 'objective' General Relativity had fully subsumed and superseded Newtonian gravity, then there should be no value in retaining the modified, Newtonian-like, flat space-time.

So one must wonder whether Einstein's secondary move – to a General Relativity to resolve the explanatory losses and other failings of the conversion from the Newtonian space-time framework to Special Relativity – has actually achieved the conceptual unification that it was designed to achieve.

As I think Thorne expressed it: "the question is still open."

One of the perquisites of running the Linus Pauling Memorial Lectures Series is that I have the opportunity to sit down and discuss these issues with some of the main players in the physics community. Being a polite group, they accommodate my questioning.

Freeman Dyson is best known in the physics community for his unification of the three competing versions of quantum electrodynamics that had been invented by Feynman, Schwinger and Tomonaga. Many felt that Dyson should have shared the Nobel Prize in 1965 along with Feynman, Schwinger and Tomonaga. Dyson spent most of his career, starting in 1953, at the famous Institute for Advanced Study at Princeton University, where Albert Einstein also spent most of his senior career.

On a recent visit to Oregon to present in the lecture series, we shared stories about Einstein. I told him that when I was an undergraduate at Berkeley I had happened, one year, to live next door to Einstein's son, Hans, who was a Professor of Hydraulic Engineering at the University. When his regular sailing partner was transferred to Stanford, I was in the right place at the right time and became his new sailing partner and crew member. I sailed the San Francisco Bay and surrounds with him for most of two years; more than two dozen outings. When we were sailing, we would talk about everything. On more than one occasion, Hans had assured me that his father "was not a genius." Hans speculated that his father's success had been because, "He had just happened to read the right books in the right order at the right time." With the provocation of that anecdote, Dyson told me that when he arrived at the Institute for Advanced Study, he read two of Einstein's current papers that were circulating pre-publication, and, as a result, Dyson had resolved to avoid Einstein as much as possible from then on. Dyson was not impressed with Albert's intellect.

Dyson is a rebel, as overtly expressed and prescribed in his collection of essays published as a book, *The Scientist as Rebel*. Dyson had always seemed to me to be sympathetic to Bohr's complementarity project. And, following Pauli and Zeilinger and many others, Dyson unabashedly affirmed the ubiquity of choice.

Over a period of several days I had the opportunity to present to Dyson essentially the same arguments I had presented to Hawking about the complementarity of Newtonian space-time and Einstein's Relativistic space-time. "What do you think?" I asked. Dyson, who had been facing forward, looking out the front window of my parked car at the time, swung around with a sense of surprise on his face. I think the surprise was that I had even asked such a rebellious question. He didn't even pause with his answer. "Yes, definitely," he said, and for emphasis added, "Absolutely."

These conversations with and comments by Kip Thorne, Hans Einstein and Freeman Dyson do not rise to the level of arguments. They are peripheral anecdotes, included here as rather blatant 'appeals to authority'. However, on my intellectual journey I took their comments at least as encouragements.

A Nobel Laureate physicist friend, now at Stanford, in a recent conversation, after having listened to one of my 'conversion' diatribes, talked to me, almost in a whispering mode, about the unfortunate intellectual dominance and momentum of the 'Einstein Cult' in modern physics. Challenging the litany, making proposals that involved 'thinking outside the box', was not a smart, career-building strategy in modern physics.

I am thinking that I was lucky to have switched to philosophy of science.

38

Visit with Students with Disabilities
at Seattle University

I arranged to meet with the Seattle Public Schools folks in order to provide complimentary tickets to Hawking's public lecture for k12 students and teachers. It was pretty much standard procedure to give away about a fifth of the 3000 seats to high schools and provide another ten percent at highly discounted prices to local institutions, like Seattle University, which had become an official cosponsor.

I had worked with the Seattle Schools before. In fact, I had been making it a point to hit every school district in the surrounding Puget Sound area for a couple of years now. When I mentioned that Stephen Hawking had agreed to give a public presentation in Seattle I could see the wheels start turning. The next thing I knew I was set to meet the Special Education Coordinator for the Greater Seattle Area, along with a woman named Sheryl Burgstahler, who had some connection with assisting students with disabilities. Sheryl called me and we spoke briefly before the meeting – a couple of days hence.

I was totally unprepared for the encounter. The meeting was in an old school building; one of those 'retired school buildings' that the City still owned, too run down now to be used as a school, but adequate to house the odd administrator of this or that educational support program. Most of the building was closed off. We met in what had been the library. I arrived first and met the local Special Education Coordinator – nice guy. We chatted a little and joked about his 'luxurious surroundings'. We sat at a large, round, oak table that might have accommodated eight people – undoubtedly a furniture holdover from the earlier library era.

Sheryl arrives and I eventually suggest the possibility of a meeting with a group of students with disabilities similar to what we did in Portland in 1992. Sheryl loves this idea, and we start brainstorming possible scenarios.

This is not just Sheryl Burgstahler, but Dr. Sheryl Burgstahler with a Ph.D. and a faculty position at University of Washington. Formally, Dr. Sheryl Burgstahler is Affiliate Professor in the College of Education at the University of Washington in Seattle. Her teaching and research focus is on the successful transition of students with disabilities to college and careers and on the application of universal design to technology, learning activities, physical spaces and student services. She founded and directs the DO-IT (Disabilities, Opportunities, Internetworking, and Technology) Center and the Access Technology Center. These two centers 'promote (1) the use of mainstream and assistive technology and other interventions to support the success of students with disabilities in postsecondary education and careers

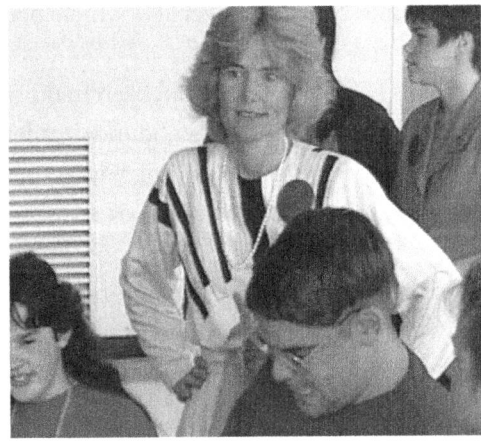

Sheryl Burgstahler

and (2) the development of facilities, computer labs, academic and administrative software, websites, multimedia, and distance learning programs that are welcoming and accessible to individuals with disabilities.'

She is smart and organized and starts talking about how we can put together this meeting with the students. "I have already proposed the idea to Seattle University, so if it is all right that is where we can hold it. They really want it there. They seem to think that it has a strong resonance with their Jesuit mission or something," I say.

Beyond the location all the logistical arrangements of getting the kids there and so forth have now become the responsibility of Sheryl and the folks at Seattle University.

The DO-IT Program was established just the previous year, in 1992, with initial funding from the National Science Foundation and administered by the University of Washington. The program primarily targets high school students with disabilities who want to pursue careers in science, math or engineering. Participants spend two weeks each summer at the University of Washington, attending lectures and labs to get a feel for college life. They also meet with faculty and other students to learn and experience how new technology is enabling students with disabilities to pursue degrees and careers in fields once thought beyond their reach.

'Use of the internet had become a core component of the program. DO-IT scholars use their computers, from home or elsewhere, to communicate with one another and with special mentors from around the world. A sense of community develops from these online relationships and they serve as a source of encouragement to the students in meeting the common challenges in pursuit of their goals.'

From the Alexis to Seattle University we traveled the couple of miles to visit with a group of 24 students with disabilities – about half from Sheryl Burgstahler's DO-IT program and the other half Seattle University students with disabilities.

The van is greeted by a group of dignitaries from Seattle University. The President, dressed in Jesuit garb, and Provost are extremely pleased that 'Saint Stephen' has dignified their humble university with his divine presence. Standing next to where Stephen is exiting the van, Joan leans over to me, smiling and teasing, and says, in a lightly feigned cockney slang, "You better warn this lot off all that talk. We have to be with him at lunch later, you know."

"Yeah, could be big trouble here. These are Jesuits, you know. Stephen might start thinking that he's getting the word from the Pope," I say. Stephen hears us and smiles. The Seattle University people vaguely overhear us, but don't get it. We all smile and chuckle in a good-hearted manner so as not to offend anyone. Stephen rolls back and forth in his seat as his chair lurches backward off the van's elevator. Everyone is smiling as we are shown the path to the gymnasium where the main gathering has been waiting for at least half an hour.

There is an elevator from the ground floor to the gymnasium level. Stephen, one nurse and Tim ride up. Everyone else takes the stairs. As we enter the gym, the Seattle University folks indicate to Stephen where they want him to be located and by what route, wandering through all the chairs and people milling around. Stephen at that point becomes notably more independent, zipping forward, accelerating, following the Provost into the center of the gathering. Tim and the nurses dissolve into the background behind the crowd of newspaper, radio and television journalists; no longer part of the focused excitement.

I ask Joan, "Aren't you going with him?" She doesn't answer immediately but continues towards the back of the gathering with me following her.

"No," she says, "He doesn't need us."

"But…," I pause, not having any definite formulation of my concern.

"He'll be fine," says Tim confidently.

There are more press people than students – double. Cameras everywhere. It feels weird. Is this for the students or for the media? Is this just a public relations event for Seattle University? I am asked by the Provost about whether the press can ask questions after the students. We've already been over this. Stephen always likes to be interactive with the press. "Just wanted to confirm that it's OK, that Stephen agrees," she says. "Yes, at the end, of course. He wants to have press questions, at the end," I say. Gradually I accept that the Provost actually has everything organized and the press managed. She has briefed the press on etiquette and rules of engagement. Clearly, the students are the primary focus, the meaningful center around which all this other madness swirls. It's all about Stephen and the students.

The Provost does a formal introduction, welcoming everyone and laying out the agenda: Stephen will talk to the students first, followed by a question-answer dialogue with the students, and then the press will be able to ask Stephen questions.

Stephen begins: "Can you hear me?"

There is, as always, a sense of exhilaration – and mild relief – in the audience when they first hear Stephen speak. Contact! Touch! "OK, this is really going to be cool," says one of the students I am standing behind.

Stephen begins his presentation: "I am quite often asked what do you feel about being disabled with motor neuron disease or ALS? ..."

Stephen's talk ends and the Provost picks up the microphone and sets the stage for the question period. She tells everyone that Stephen will be constructing each answer, word by word, in his computer. So it will take him three to five minutes to construct an answer.

"So, please feel free to talk amongst yourselves in the mean time. Professor Hawking will let us know when he is ready to answer," she says.

Student question: "If I weigh 121 pounds, and my Mom weighs about 150, and my Dad weighs 200, who would be the heaviest in weight in water?"

"You didn't know that this was going to be a quiz, did you?" the Provost comments.

The whole audience is extremely quiet while Stephen is constructing his answer. Despite the Provost's permission 'to talk amongst yourselves,' the silence is awkward.

I'm thinking, "Somebody tell a joke." But Stephen is quick to answer.

Stephen's answer: "Everyone floats in water so they weigh nothing."

There is a slow explosion of laughter followed by a moan of "Oh yeah!" recognition. Common sense. Right? "I knew that!" – right after he said it. After this icebreaker, the atmosphere is much more relaxed, and people chit-chat while Stephen is constructing his answers.

Student question: "How did you like working on *Star Trek – The Next Generation*? Did you enjoy working with Data? I don't know his actor name. I saw you on *Star Trek* last week and I really enjoyed that part. But how did you like working with the *Star Trek* crew?"

Stephen's answer: "It was great fun. But not to be taken seriously."

Stephen is in the center of a semi-circle of students. He makes it personal, moving his chair back and forth and closer, orienting directly to face each student asking a question. – Respect.

> **Student question** (Kevin Berg, a freshman at Seattle University): "How does it feel to be labeled the smartest person in the world?"
> **Stephen's answer:** "It is very embarrassing. It is rubbish, just media-hype. They just want a hero, and I fill their role model of a disabled genius. At least I am disabled, but I'm no genius."

One of the reporters looking over Stephen's shoulder as he constructed the answer later wrote: "Hawking rapidly picked out words. He spelled out "media" and "hype" letter-by-letter, since they were not part of his computer's library of 3,000 words."

> **Student question:** "I read in your bio-sketch that you were born January 8th. I think that makes you a Capricorn. Do you equate astrology with cosmology in any way? And do you believe in reincarnation?"
> **Stephen's answer:** "Rubbish."

Stephen's answer was so quick that almost everyone missed it underneath the ongoing chatter. Everyone was expecting him to take longer so the immediate conversational noise level was high when he did answer. Not a hugely nuanced response, but there you have it. The Provost finally gathers that Stephen had answered as much as he was going to and moves on. The assembly was startled and there were whispers back and forth for a minute or two – "What was his answer?"

> **Student question:** "Dr. Hawking, welcome to Seattle. Good to have you here, sir. I was wondering, how do you think we can encourage more people with disabilities, no matter what degree they are, to become involved in higher education and how can we encourage

the universities to provide reasonable accommodations, including housing and to meet the needs of the students and the professors with disabilities?"

Stephen's answer: "Education is a winner for the physically handicapped. Nowadays, muscle-power is obsolete. Machines can provide that. What we need is mind power, and disabled people are as good at that as anyone else."

The Provost identifies Aleysa Reed, an 11 year-old Seattle girl with cerebral palsy, as having a question. Aleysa is the youngest student and it's her birthday. She manages a wave and says "Hi." Her mother, Sally Reed, needs to read the question.

Student question: "She, I believe, is planning a potential career in TV, so she was wondering how you go about getting a role in a TV show. And the other question was what do you do for your legal signature because she is running into problems with having to have a legal signature."

Stephen's answer: "I got on Star Trek by accident. I just happened to be visiting the studio for some other reason."

The story is that Star Trek producers learned that Stephen was visiting a nearby studio. So they spontaneously asked if he would appear in an episode. Stephen agreed and a script was hastily prepared. Joan later showed the students how Stephen executes his legal signature – a thumbprint – using an inkpad, with an assist from one of the nurses. This is also how he signs books.

Post Intelligencer newspaper writer Tom Paulson talked later to Aleysa's mother and reported: 'Sally Reed said her daughter became enamored with Hawking years ago after seeing him featured on a public television show.' "The thing that most impressed her was 'here was this brilliant man who needed help eating, just like she needed help'," Sally Reed said.

The next student also needs assistance in asking his question.

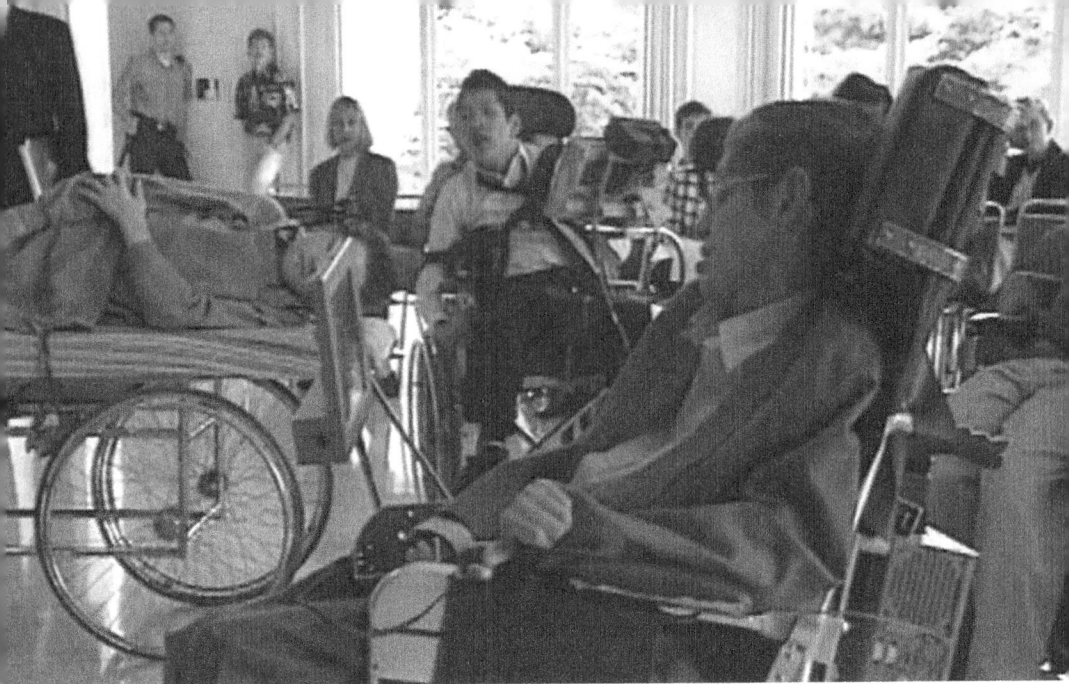

Turning for students to see

Student question (offered by his parent): "Fortunately, he told me the question he wanted to ask over the modem yesterday. He would like to know if you would be interested in joining a group that he is organizing called Worldwide Friends of the Disabled Over the Modem, and if you have a modem address, or if he could give you Worldwide Friends of the Disabled Over the Modem's modem address, so that you could contact them.

Stephen's answer: "I have an electronic mail address, which my assistant will give you."

Although I wasn't privy to any of the exchanges, I am aware that Stephen engaged with members of this group in a lively communication that continued for at least a number of years.

Student question: "Some of us missed the *Star Trek* episode, so can you tell us a little about the role that you played? And I am curious as to how they sort of explained your disability because in Star Trek time they could just hit you with a laser and you could walk or something."

Stephen's answer: "Data called up an image of me, along with Newton and Einstein, for a game of poker. So I didn't belong to *Star Trek* times. I won the game, but I'm not sure how to cash my winnings."

Raucous laughter erupts following this last comment.

Student question: "I'm a recent college graduate, and upon looking for a job, a lot of people have suggested to me that if I identify myself as a disabled person that I would be more likely to be considered for a job due to such legislation as the Americans with Disabilities Act – that is forcing companies to fill quotas and such things. I'm curious as to what you think of the legislation, ADA – if you think it's a good thing, if it's helping disabled people and also what advice do you have for recent college graduates."

Stephen's answer: "It is far too easy for companies to discriminate against the disabled – like they used to discriminate on the basis of race and sex. I just wish we had similar laws in Britain."

Student question: "I have four questions. I guess I will have to choose. I guess I'll ask about the program that you use, when you choose words for it – do you – does it like split it down into categories – because I can't imagine you looking at every word in the English language."

Stephen's answer: "The words are listed in alphabetical order. I select the first letter and then get several pages of words [beginning with that letter]. I keep the vocabulary down to about 3000 words and spell out other words letter by letter. So I have had more time for research."

Following this question and answer it was suggested that the students might gather around behind Stephen so that they could watch his computer screen while he composed his answer. Stephen moved closer into the midst of the students and turned his chair around to make his screen visible to most of them; others shifted around to get a better view.

Student question: (Ewan Day, Redmond High School student): "Dr. Hawking, I am not so sure about how good you are at making predictions, but since this is a fireside chat sort of thing I'm going to go ahead and ask anyway. If you could see fifty to one hundred years into the future what do you see for the advancement of human civilization culturally or socially?"

Stephen's answer: "I think it is possible that we will destroy ourselves. If not, it is possible that most human beings will be made redundant by intelligent computers."

There is a stunned silence. An intellectual and emotional disjunction just occurred! Like – 'I didn't hear what I just heard.' Mr. Optimism just gave us an utterly pessimistic vision of the future. The only way I can describe the effect it had is that it was like having – in a mild way – the wind knocked out of you. You are sort of gasping for a breath. You could tell from facial expression and the body language of the student who asked the question that this was not the sort of answer he expected. His enthusiastic smile just drained from his face. He was looking for some sort of upbeat visionary scenario – consistent with Hawking's life-affirming persona. Weird answer. I am thinking, "Right, so everything is possibly hopeless but, in any case, pretty much meaningless." No one quite knew what to do with this answer. But it clearly penetrated and sobered the audience.

Here was a rare conflagration. Hawking the physicist as Spectator overriding Hawking the Participant.

Student question: "Dr. Hawking, professionally how much discrimination have you encountered and how have you dealt with it?"

Stephen's answer: "Not much. But I think that I have been very lucky. I was lucky to be in theoretical physics and to be quite good at it."

Student question: "What made you really want to go on? You mentioned that in the hospital you saw a young patient die of leukemia, and therefore you said you wanted to go on. But before then what really made you want to go on? In my instance for example there are some days in which I don't want to continue. I mean there are days when I wake up and say, 'why in the heck am I doing this?' And I am wondering if there is some sort of driving force inside you that is saying yeah I can do this. I am just wondering where you are with that."

Stephen's answer: "I don't really know the answer. I have always wanted to do well and felt I could. And I always feel it is better to do something than just be passive and inactive."

The student gave Hawking an 'OK' sign with his hand and a giant smile.

Student question: "How do you keep from being depressed?"
Stephen's answer: "I realized that the rest of the world does not want to know whether you are bitter or depressed. You have to be positive if you want sympathy."
Student question: "I am wondering if you know the meaning of life."
Stephen's answer: "I wish I did so I could stop searching."
Student question: "Do you believe in clairvoyant dreams?"
Stephen's answer: "If I did, I'd be the Emperor of Rome."

The chairperson of Persons with Disabilities at Seattle University Club steps forward: "On behalf of the Seattle University community I want to present you with this pin and to thank you for coming and spending your time with us here today."

Finally the moment arrived for the questions from the press.

Press Question: "Your vision of the future seemed rather grim. And I was wondering what you think we should be doing differently."
Stephen's Answer: "I think it is important that we understand where we are going and what we can do to control it. In a democracy

we have to educate people. Our priority should be to spread an understanding of science and its implications for society."

Press Question: "Professor Hawking, you said that when you learned you were diagnosed with ALS you asked yourself why did this happen. And cosmology is the science of trying to figure out why everything happens. Do you think in any way ALS helped you focus on looking into cosmological theory?"

Stephen's Answer: "I am working on the question of whether information is lost when a black hole forms and evaporates; if it is, it means that the universe is a lot less predictable than we thought it was. Other physicists don't like the idea, but they are beginning to agree with me."

Press Question: "Dr. Hawking, it is obvious what these young people take away from interacting with you. I wonder what you take away from your time talking with them?"

Stephen's Answer: "I take away a sense that they are taking advantage of the best of their situation."

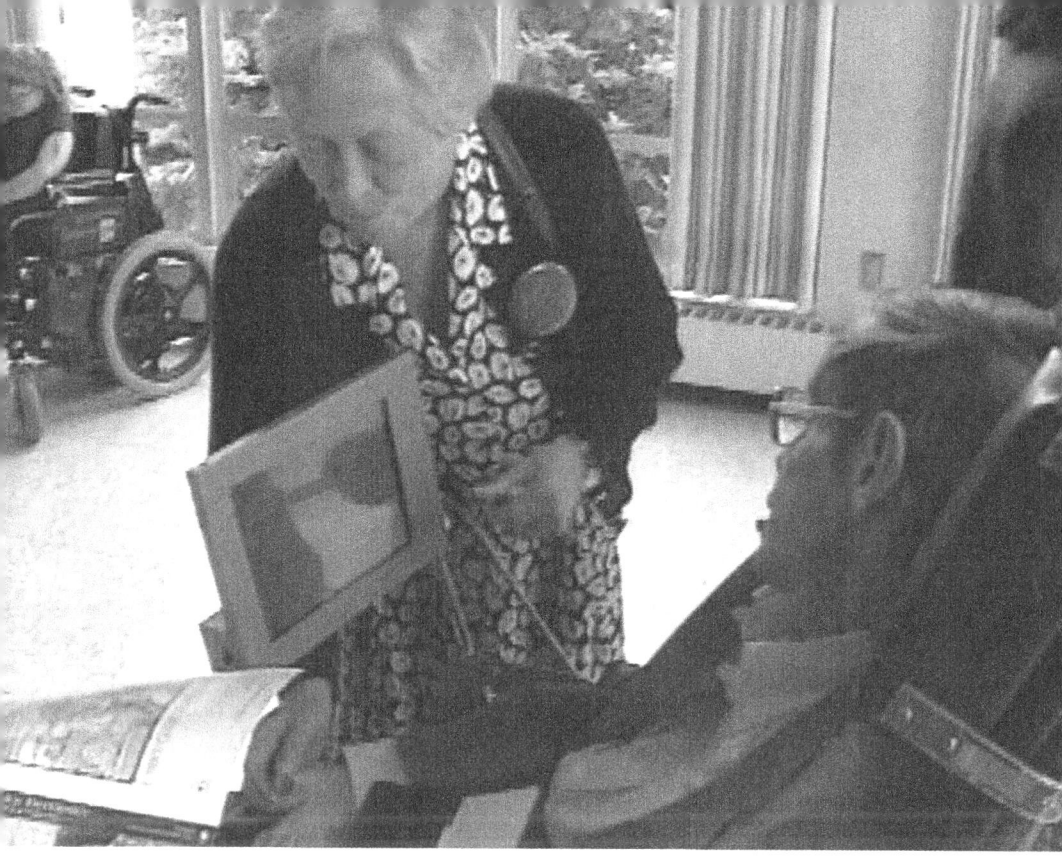

Joan with inkpad for signatures

As the gathering is breaking up, a Seattle Times reporter interviewed one of the students, Mitch Weddle, 16, who has scoliosis. Mitch is horizontal, lying on a gurney, encased in a sort of spring-loaded cage to prevent any motion. He spent the past five months at Shriner's Hospital, in Spokane, recovering from two surgeries that fused his spine and vertebra together. Mitch spent six hours on the road to come to hear Hawking.

"An opportunity to see a man like this is very rare. I had to take advantage," the teenager says.

At the end of the meeting, the Joans passed out Stephen's e-mail address to the students so they could send him messages. Weddle proudly showed off the scrap of paper to his parents.

Asked whether Hawking was worth the six-hour trip, Weddle answered, "He was worth every minute. An inspiration to me is what he is."

39

Anthropic Cosmology and
Symmetry-Breaking

Cosmologist John Barrow, in his book, *The Constants of Nature*, had emphasized that the 'irreducible probabilistic aspect' of new physics entails that if we could re-run the history of the cosmos there is no reason to expect the same outcome. Barrow's semi-probabilistic (constrained) re-running of cosmic history involves the interplay of two orders – a classical cause-effect causal order and its complementary opposite. Max Born had represented the complementary opposite of the Maxwellian field-reality as a chance-governed, probabilistically distributed, Newtonian charged particle-like reality. Following Born's interpretation, Barrow's argument is based on modeling the evolution of the cosmos in terms of the interplay of an order and an irreducible, complementary disorder. In Barrow's approach, the actual outcome in terms of the most fundamental 'constants of nature' is mechanically under-determined.

One of the main motivations for seeking a post-mechanical, post-scientific More General Theory is that, with the embrace of complementarity, all possible uniform mechanical theories – consistent with the Scientific Hypothesis – under-determine both the historical sequence and actual outcome of the nature of reality. And if your Research Program can't possibly predict the history and the actual structural and functional outcome, then you can't explain the history and actual outcome in terms of these same types of theories – come what may.

Hawking and an increasing number of his cosmology colleagues have begun to actively search for some additional, possibly post-scientific, 'principle' that would allow us to make sense of – *this* – the actually observed outcome of cosmic evolution. The most commonly discussed option, the Anthropic Principle, seems to be an exploration of common ground between the Spectator and Participant representations of reality. In 1973, Australian cosmologist Brandon Carter stepped 'outside the box' questioning the extreme 'objectivist' interpretation of the Relativistic cosmos. Carter pointed out that the reasoning to the extreme limit of the Copernican Principle – completely decentralizing the Earth – implied that 'we' Earthly human observers had absolutely no 'privileged' position in the cosmos whatsoever. Carter noted that for human observers to exist at all the history of the universe must have been reasonably fine-tuned – privileging that outcome. Carter's focus was then on how to account for the empirical evidence of a biased ('privileged') fine-tuning of the constants of nature necessary for the existence of human observers.

Carter's baseline argument was that without an extremely precise historical selection of values for a couple of dozen constants of nature there wouldn't be a universe with observers like us. Carter proposed the Weak Anthropic Principle wherein: "the observed values of all physical and cosmological quantities are not equally probable but they take on the values restricted by the requirement that there exist sites where carbon-based life can evolve and by the requirement that the Universe be old enough for it to have already done so." (*The Anthropic Cosmological Principle* by John Barrow and Frank Tipler, p. 16). Note that for Carter, "location" refers to our location in time as well as space. The evolution of carbon, for instance, required many billions of years.

Carter then goes on to formulate the Strong Anthropic Principle, positing that the fine-tuning that makes possible human observers might not simply be a curious random coincidence, but that it might 'somehow' be an inevitable outcome – by the very nature of reality. Carter's Strong Anthropic Principle posits that: "the Universe must have those properties which allow life to develop within it at some stage in its history." (*The Anthropic Cosmological Principle*, p. 21).

The Anthropic Research Program is an attempt to seriously explore 'outside the box', outside the paradigm of the 'objective', observer-independent Scientific Research Program since it involves, at least tacitly, an effort, self-referentially, to account for the presence of human (Participant) inquirers. The Anthropic Research Program is a cautious tentative step toward a post-scientific More General Theory. The effort is still fundamentally descriptive – still as if we are somehow Spectators – leaving untouched questions that face the Participant as to how we should live, as to how we might bring forth a more desirable future. Early Anthropic reasoning didn't attempt to 'explain' the actual nature of the universe in novel post-scientific terms. Anthropic reasoning has been primarily a search for what the nature of the universe must be, must have been, for the current outcome, for *us* to have been possible (Weak) or inevitable (Strong).

Although it is not the main emphasis, Carter's Anthropic reasoning accepts, based on the evidence, that the cosmos has a *'real' history* and that that history has been *progressive*. It took 13.7B years of a stage-by-stage development to generate modern human observers, to generate the modern human inquirers who are asking these questions.

It may surprise some that Hawking has been at the forefront of the exploration for an Anthropic Principle: "I shall adopt the no boundary proposal, and shall argue that the Anthropic Principle is essential, if one is to pick out a solution to represent our universe, from the whole zoo of solutions allowed by M theory. … There may be some principle [of physics] that we haven't yet thought of, that restricts the possible models to a small sub class, but it seems unlikely. Thus I believe that we have to invoke the Anthropic Principle." Hawking's no boundary proposal erases the boundary, drops the relevance of Carter's original distinction between the Weak and Strong formulations, focusing us on finding *the* Anthropic Principle.

Hawking unabashedly broadened the range of phenomena that might potentially be accounted for in terms of the hypothesized Anthropic Principle: "If one is going to have to appeal to the Anthropic Principle, one may as well use it also for the other fine tuning problems of the hot big bang. … Super symmetry breaking is an anthropic requirement. One could not build intelligent beings from massless particles. They would fly apart." (Reference: http://www.hawking.org.uk/quantum-cosmology-m-theory-and-the-anthropic-principle.html)

Proponents of the Anthropic Research Program have never offered an actual Anthropic Principle – have never actually specified any actual anthropic aspect of the nature of reality. The Anthropic Research Program is simply the hypothesis that there should be 'something like' an Anthropic Principle as part of the nature of reality to account for the actual outcome of human observers. The primary result of the Anthropic Research Program so far has been the expansion of increasingly specific, detailed observational evidence that the cosmos, from the beginning and at each novel stage, must have had an extremely precise, selective 'unfolding' to be able to account for the eventual outcome of human observers.

Freeman Dyson captured the sense of the hypothetical Strong Anthropic Principle, commenting: "As we look out into the Universe and identify the many 'accidents' of physics and astronomy that have worked together to our benefit, it almost seems as if the Universe must in some sense have known that we were coming."

Steven Weinberg's First Three Minutes

Nobel Laureate Physicist Dr. Steven Weinberg, in his book, *The First Three Minutes: A Modern View of the Origin of the Universe*, describes the sequence of events in the unfolding of the cosmos in the first three minutes after the start of the Big Bang beginning. Reasoning backward – retrodicting – on the basis of Hubble's evidence that galaxies are receding at velocities proportional to their distance, the consensus inference has been that there must have been a time when everything was at one point; when every where and every when was one point-reality.

Weinberg reasons that as we move back in space-time the smaller and smaller universe would have been hotter and denser. At extremely high temperatures molecules would have been torn apart. There would have been an earlier period when even atoms could not have existed. Still further back, to even hotter and denser conditions, Weinberg reasons that electrons and protons could not have existed.

Then, reasoning forward from the very earliest instants of the Big Bang, Weinberg describes the unfolding of the cosmos in terms of the unexpected emergence of novel elements and relationships through a process of discontinuous 'spontaneous symmetry-breaking events.' Each emergent stage is characterized in terms of a type of symmetry, with specific types of elements and relationships, structures and functions. The subsequent stage emerges with novel types of elements and relationships.

Since mechanical systems are defined by their symmetry, spontaneous symmetry-breaking events are – by their very nature – mechanically discontinuous. Since mechanical systems are defined by their *type* of symmetry, spontaneous symmetry-breaking events are – by their very nature – qualitatively emergent. The transitions are from one type mechanical system to another by a process that involves an irreducible non-mechanical component. In this series of symmetry-breaking events the mechanical symmetry of each stage does not uniquely determine – under-determines – the qualitative characteristics – the elements and relationships and the symmetry of next stage. In Weinberg's description of the emergence of the early cosmos each spontaneous symmetry-breaking event moves us from one *type* of 'objective' mechanical universe to another. At each stage the future of the cosmos is under-determined by the current stage. Weinberg is just providing a plausible account of the emergence of the actual outcome, not suggesting that it was a uniquely inevitable outcome. Weinberg's account is consistent with Barrow's theme that *from a Mechanical perspective*, where one would expect a uniquely determined sequence governed by One space-time invariant order; instead, there is a mechanically discontinuous symmetry-breaking aspect to what has actually emerged. In Weinberg's account there are irreducible mechanical discontinuities, irreducible logico-mathematical discontinuities in the historical emergence of the cosmos.

If the symmetry-breaking events are discontinuities in the order governing the universe, then the history of the universe is not mechanically reversible. Yet the formulations consistent with the Scientific Hypothesis of modern physics still demand, still presuppose, mechanical reversibility.

If the symmetry-breaking events are – by their very nature – discontinuous, then the non-reversible emergence of the order governing the universe is cumulative. If the symmetry-breaking events are discontinuities, then the non-reversible emergence of the order governing the universe is 'really' historical – meaning that it is not a history governed by One, timeless, mechanical order. It is not just that the 'real' (middle-ground) order is changing discontinuously from one mechanical order to another but that the 'real' (middle-ground) order is constructively, cumulatively emerging. The evidence for the historical formation of stars and galaxies, for the historically progressive nucleo-synthesis of the chemical elements, strongly suggests a history where the possibility space is expanding – qualitatively – where things are possible in the later stages that were not either possible in, or inevitable from, the prior eras. The evidence suggests that the emergence of the cosmos is a series of mechanically discontinuous steps leading to a progressive, qualitative expansion of the possibility space; leading to a progressive, qualitative expansion of the Participant's opportunity space.

All possible Scientific Cosmologies are defined by a symmetry principle – the mechanical order governing all phenomena must be universal, must be space-time invariant. The key point about spontaneous symmetry-breaking is that there is *no possible account of spontaneous symmetry-breaking events* – by their very nature – in a universe that is governed by One, symmetric, time-space invariant Mechanical order. The evidence cited by Weinberg for spontaneous symmetry-breaking events constitutes an answer to Popper's Question, applied to the Scientific Hypothesis, as to what type of evidence would force one to conclude that the Scientific Hypothesis and the associated Mechanical Philosophy are false – in the sense of being inherently incomplete accounts of 'all phenomena in the universe'. Spontaneous, symmetry-breaking events cannot be explained or made sense of within the Scientific Research Program – in terms of any possible future scientific theory – come what may.

The evidence for mechanically discontinuous, spontaneous, symmetry-breaking events should be understood as evidence supportive of the More General Theory and the hypothesis that the emergence of the cosmos occurs through some sort of Middle Way process. And yet we are still only struggling toward better articulation of the More General Theory's new way of understanding the nature of reality and our place in it.

40

Nathan's Dinner for Stephen

The life trajectory of Nathan Myhrvold, more than any other individual I can think of, exemplifies the 20th to 21st centuries transition from the Spectator to the Participant understanding of ourselves and our role in the universe. Since the early formulations of the new physics there has been a gradual, largely silent, trend toward a Participant understanding of humans and the nature of the universe. The trend is embodied in lives and choices.

Nearly a decade earlier Myhrvold had received his Ph.D. from Princeton University in theoretical and mathematical physics with a thesis entitled "Vistas in Curved Space Quantum Field Theory." Myhrvold's work on gravity and quantum mechanics had caught Hawking's eye. Hawking accepted Myhrvold's application for a post-doctoral position. At the time, Myhrvold seemed destined for a stellar career in physics. For about a year, Myhrvold worked with Hawking. For diversion he took a summertime leave of absence to work with his younger brother, Cameron, and some friends on a windowing operating system for personal computers. Through a series of happenstances, Myhrvold never returned to his study with Hawking.

The operating system, called Mondrian, was gaining favorable notice in the IBM-and-compatible world until IBM itself issued a commercially doomed program called TopView. Fortuitously, Microsoft stepped in. Chairman Bill Gates had gained IBM's begrudging backing for Microsoft's new Windows program by promising TopView compatibility. Purchasing the Myhrvold-led company, Dynamical Systems Research was the quickest, easiest way to a solution. Myhrvold and his crew moved from Berkeley, California to Redmond, Washington and went on to help Microsoft enhance Windows, the best-selling computer operating system next to Microsoft's MS-DOS.

Nathan was now Vice President for Advanced Technology at Microsoft. In later years, while still at Microsoft, he identified himself simply as part of the Office of the President. Nathan co-authored with Bill Gates their 1996 best-seller, *The Road Ahead,* about the progressive, emerging technological revolution. Nathan had opted to become a Participant. For the technological engineering community, many of them formally trained as 'scientists', the draw to the Participant Research and Development Program hadn't been about the money, although that always helps. For engineering enterprises, generally the excitement – the passion – is about being at the leading edge bringing forth a better, a more desirable future. Many (perhaps most) of the scientist/engineers that participated in the Apollo project, landing humans on the Moon in 1969, worked themselves into early graves for sub-industry wages. Why? The enthusiasm was for the opportunity to engage in the many, novel, creative engineering challenges and opportunities. As Myhrvold and Gates express in *The Road Ahead*, there was a heady sense of participation in bringing about a new era in human history – a new era in the history of the cosmos.

In an interview a few weeks before the Seattle event, Nathan Myhrvold reflected on his earlier period studying with Hawking. One prominent recollection was about Hawking's great sense of humor that continued despite his incredible disabilities. "Before Hawking used a synthesizer, his slurred speech was difficult to understand. But it didn't stop him from kidding around. There were times when he was telling a joke that he had to repeat the punch line five or ten times even for people who knew him well," Myhrvold said. "And he'd patiently go ahead and do so until we got it."

"I'm eager to show Stephen why I'm not an assistant professor someplace," Myhrvold said with a laugh. 'The earlier alternative path is a notion the physicist in him thinks about occasionally,' he admits.

"The stuff that I do here probably is going to touch a lot more people's lives more directly than anything I would do in contributing to a quantum theory of gravity or some theory of cosmology," Myhrvold argued. "On the other hand, I think that you can also make the argument that whatever I did in physics would potentially be far more timeless than something here."

After debating with himself a little more, he concluded: "It's a tossup."

Among the numerous physicists Myhrvold has known, Hawking is the most generous with sharing credit and helping others. "Once Hawking asked if a graduate student could attend an invitation-only conference the topic of which was crucial to the student's thesis. They said only one person could attend, assuming Stephen would," Myhrvold recalled. Instead, Hawking sent a note with his student saying, "I'm sorry you didn't have room for me."

"If you wanted to have a universal metric by which you could compare civilizations on different planets, one of them would be how much did they figure out about the universe they lived in," said Myhrvold. "Hawking's pursuit," he said, "is a testimony to our species and our civilization."

Nathan, the polymath, has added gourmet chef to his repertoire. He's been taking classes, and has, in fact, become an apprentice to one of Seattle's finest chefs – Thierry Rautureau. So Nathan organized a private dinner party for Stephen at Rautureau's restaurant – Rover's.

'Thierry Rautureau is one of Seattle's most famous chefs and his imaginative take on traditional French cuisine is stunningly presented in this restaurant. Set in a small house with private gardens, Rover's has a decidedly French country style, with secluded courtyard seating and a simple, but cheery, décor. The menu is fixed-price, the sauces legendary and the seafood, such as lobster in Perigord truffle sauce, highly recommended. Also featured are game dishes, such as venison with green peppercorn and Armagnac. Every night, the kitchen produces three tasting menus of six or nine small courses, one of which is vegetarian. There is a superb wine list and garden dining in summer. Closed Sunday and Monday. No lunch.'

The invitations went out to about 30 Microsoft brass and friends and their spouses. Stephen and the nurses and Tim were automatically invited.

I had wanted to attend, and I figured that since I was hosting and transporting Stephen in Seattle, it was not unreasonable to be considered part of Stephen's crew. So I had asked Stephen if I could be included as part of his crew. He said "yes" and communicated that to Nathan, who agreed.

We arrived at Rover's a little late. As Stephen was unloading from the van, I glanced at the entrance to Rover's to see Bill Gates lean out the door, feet inside and hanging onto the doorjamb, making an angle of about thirty degrees from vertical; just poking his head and body out to see how we were doing, smiling, looking cheery. Nathan didn't rush out to greet us, undoubtedly dealing with his guests inside. In fact not much is happening as we move toward the entrance – no greeters.

Stephen and the Author

I keep glancing around for the wheelchair ramp. No ramp here in front. As we reached the bottom of the steps Tim ventured up into the restaurant to inquire as to the location of the wheelchair entrance. "Around the back perhaps?" We're looking at a steep set of stairs – maybe eight steps, covering an elevation difference of about five feet. A couple of people come out of the restaurant with Tim. I ask the obvious. "Where is the wheelchair entrance?" Blank looks. "Damn! Rover's isn't wheelchair accessible." I'm thinking, "Nathan the genius screwed up!" How could you invite Stephen Hawking to dinner at a restaurant and not consider whether it was wheelchair accessible. "Give me a break." Stephen's main stowage, his twin battery packs, Brahms and Puccini, are working this evening with Beethoven and Mozart charging up at the hotel. Each battery weighs around thirty-five pounds and looks like two large automobile batteries connected end to end. Stephen and his chair – with batteries – weigh well over two hundred pounds.

To use a computer culture expression, there was no obvious 'work-around'. Carrying Stephen and the individual parts, even if practical, would have been undignified. Stephen doesn't carry easily. Complicated. The batteries fit tightly underneath Stephen with their power leading to a myriad of secondary functions: the chair's power, the computer, the voice synthesizer and more. Tim immediately nixes the idea of lightening the load by removing the batteries. There was a lot of shuffling of feet and 'kicking the tires' as various people size up the situation.

Subtlety is abandoned. The strategy is a frontal power assault, pushing and lifting Stephen in the chair up the stairs one by one. Unfortunately, what was not generally appreciated by the geniuses at the beginning of this assault was the real weight of Stephen, chair and batteries. "Can't be too heavy. We'll just carry it up the stairs." Tim offered that indeed, it might be 'too heavy.' The initial attempt to carry the load all at once was soon replaced with the step-by-step strategy. The chair of course never actually gets off the ground. This becomes a you push, we'll pull effort. More bodies join in as the magnitude of the task is comprehended. The nurses are increasingly anxious – suspecting, along with other bystanders, that this frontal assault may end very badly. Stephen lurches back and forth in his seat as the collective muscle tries to maneuver the chair up the stairs. These stairs are especially steep, doubling the difficulty. Stephen is now leaning backwards at nearly a 45 degree angle. After several false starts and a couple of pauses to rethink the whole idea of a frontal power assault, four strong guys, two pushing and lifting from the back and two pulling and lifting from the front, manage to scale the stairs to the landing. Cheers celebrate the conquest of this Everest.

Throughout this debacle, I am standing back with Christy, Jennifer and our van driver quietly mumbling a rude commentary. "The next challenge is going to be getting him down the stairs," I comment. "I wonder if we could come up with a ramp between now and the end of the dinner?" Christy Richardson is a problem solver. Having articulated the challenge, she locks on. "We'll find one," she says. Since no one thought to invite Christy and Jennifer to the dinner, they, along with the van driver spend the next two hours scouring Seattle for a ramp adequate to get Stephen down the steps at Rover's.

Inside there is much laughter and comradery – and a parting of ways. Stephen and one nurse are at the head table with Nathan and Bill Gates and a dozen others, while the rest of the crew are scattered among a half-dozen other tables seated with Microsoft brass and a few VIP guests. Nathan calls everyone's attention and welcomes the collective, giving a short speech that includes a little more background on his time in Cambridge with Stephen.

Brother Cameron, in Berkeley, had asked Nathan for help. So Nathan asked Stephen for a leave of absence of six months, after which time he would return to complete the postdoc. Stephen agreed. However, after six months, Nathan asked for another six month leave. Again, granted. Cameron's and Nathan's company was soon purchased by Microsoft and Nathan, now a Vice President at Microsoft, never returned to complete his postdoc.

Myhrvold had also brought a serious agenda to Microsoft: new markets, possible acquisitions and the plans for the company's evolving research laboratory. The young, erstwhile physicist stands just outside the spotlight but close to the helm. "Other than me, Nathan has more impact on our long-term strategy than anyone else," Gates had commented in an interview.

Nathan introduces the chef, Thierry Rautureau, who outlines the multiple course menu for the evening.

Stephen then indicated he wanted to say something. He first thanked Nathan for his generosity in organizing and hosting this dinner. And then said, "The real reason I have come to Seattle is to take Nathan back to Cambridge to finish his postdoc." This drew immense laughter. And the dinner began.

The fact that I didn't get to sit with Stephen was a real bummer. There was continuous raucous laughter from the head table throughout the evening. Stephen was being the comedian, with Nathan playing the provocateur and straight man. But it was impossible to make out the specifics of the conversations from where I was sitting. The tenor at all the other tables was subdued by comparison. There were eight of us at my table. The good news, actually great news, was that sitting across from me was Craig Mundie, with his wife on my right.

Everyone around the table shared a brief who's who and what do you do. All were Microsoft brass – Director of this division, Vice President of that. One guy was touted by the others as 'the star of the scaled fonts' – a major plus for all Microsoft products.

It seemed only natural that I should ask about 'the next great thing' that would be coming out of Microsoft. I figured I was on the inside and perhaps I might be privy to some of the current insider scoop. We were at a social event – a dinner, we were all buddies – "So what's happening?" Everything went instantly cold as my question drew a blank from around the table. I noticed a couple of quick glances back and forth – brief eye contact. Still no answer. So I press a little, "You know, what's hot?" Silence. Then one guy down the table (viz. I am sitting at one end, he at the other) starts with this sort of press release litany about Microsoft and its service orientation… yada, yada, yada. That wasn't what I was asking.

There was simply no talk about what I assumed would be the normal 'shop talk' with a lot of computer software geeks. My interpretation of the evening at the time was unsympathetic – 'dinner with a pack of competitive wolves'. There were palpably focused glances when I would ask what seemed to me to be quite reasonable questions. Mind you, I didn't actually care about what 'the next great thing' coming out of Microsoft was. I was just being polite and trying to start a conversation – making it easy by beginning on their turf. Something funny was going on. But I didn't get it.

I misunderstood. It was actually at least a year later before it dawned on me. This was Microsoft. You know – MICROSOFT!!! – a major, publicly traded corporation at the heart of an explosive growth industry. It was the flagship of the fleet. And to recall a World War II expression – 'loose lips sink ships.' I felt downright stupid when I figured it out. These guys had been carefully schooled – *very* carefully schooled – about what a high-ranking executive can talk about at a dinner with a stranger. They were focused all right. And it was coordinated. They were all on the same page. I just didn't come close to understanding the situation. I wasn't merely on another page. I was reading a different book. It was a cultural discontinuity.

I mean I could have said, "Trust me, I won't say anything. I won't tell anybody." And I would have been sincere. But with my infamous inability to keep a secret, I probably would have told five people by noon the next day whatever they had even suggested might be coming next from Microsoft. "Yes, Oscar, you know I had dinner with a bunch of Microsoft executives last evening. Very pleasant." "Oh yeah, what did they tell you was coming next from Microsoft?" "Oscar! You know I couldn't tell you something like that." "Terry, come on!" "Ok, Ok! But promise you won't tell anyone." "Yes, of course."

Within twenty-four hours several hundred people would have heard – including several stockbrokers. "Yeah, got it from this guy who had dinner with a bunch of Microsoft executives last night." Buy Microsoft! Sell Microsoft! The reality of this nightmare had been seriously emphasized to these executives.

Craig Mundie had come to Microsoft following Perestroika, the political and economic restructuring in the former Soviet Union initiated by Mikhail Gorbachev from about 1986. Mundie had only joined Microsoft the year before, in 1992, the same year that Alliant, the company he co-founded and where he was CEO, had shut down. Alliant Computer Systems developed massively parallel supercomputers. The main market for these computers were 'the spooks' – the Central Intelligence Agency (CIA). When the cold war evaporated so did the market for supercomputers. Craig's initial responsibility at Microsoft was to create and run the Consumer Platforms Division, which was charged with developing non-PC platforms and service offerings such as the Windows CE operating system, software for the handheld, Pocket and Auto PCs, and early telephony products. Mundie later headed Microsoft's digital TV efforts. He acquired and managed for Microsoft the WebTV Networks subsidiary.

I had an extraordinarily intriguing and enjoyable conversation with Craig. Because of his assignment to run the Consumer Platforms Division, unlike the others, he was able to talk to me. Based on what he was telling me – his long-term vision of the future of technology and society (non-specific to Microsoft) – I invited him to participate in the Science, Technology and Society series the next year.

When I hosted Craig in Portland the next year, he was just completing the purchase of WebTV. His presentation was about the interface of the emerging computer and the established television industries. Craig was trying to create a partnership between computing and television. One memorable comment, arising from his frustrations, was that 'the television industry was no longer run by engineers, but by accountants and lawyers.' Engineers understand reality in a different way from how lawyers and accountants understand reality. They have different languages that don't translate easily. Creative communication about how to bring about a more desirable future was difficult – was not happening.

A friend of mine once said that one of the nicest compliments you could ever give someone was to say that you would like to work for them – under their leadership. That was the feeling I had for Craig Mundie. Unfortunately he wasn't heading a philosophical initiative. Different paths. But I was glad that he was where he was – a key player in America's corporate infrastructure, trying to bring forth a more desirable future.

The dinner has ended and we now face the challenge of descending Everest. Christy and Jennifer have come up with a ramp. They are intensely proud of this ramp and tell me of their valiant efforts running all over Seattle to find it. They are particularly insistent that I express heartfelt thanks to the van driver, who apparently made an enormous sacrifice of his otherwise committed time in order to ferry the dynamic duo around to find a ramp. The ramp looks solid. It's about eight feet long. It certainly reaches from the top-landing of the stairs to terra firma. A back-of-the-envelope application of the Pythagorean Theorem tells you the problem here. It's the angle. City Code for a wheelchair ramp is 9 feet of distance (hypotenuse) for every one foot of elevation (the vertical side). At this standard the ramp should be forty feet long. With an eight foot long ramp there is a strong likelihood that Stephen and/or the chair are going to pitch forward. Confronting this reality, along with a paucity of alternatives, the Everest expedition decides, nonetheless, to descend by this treacherous 'ramp route'. Wisdom dictates that Stephen should face upward so as to be descending backwards. What happened was not pretty or elegant, but Stephen and chair reach the bottom of the stairs in one piece.

In one period or aspect of their lives, both Nathan and Stephen have thought of themselves as engaged in the Spectator enterprise. Nathan and an increasing number of his 'scientific' contemporaries have made the formal transition to the Participant's Engineering Research and Development enterprise. Nathan's remaining slight ambivalence – 'it's a toss up' – reflects an unresolved question about the relation between Spectator and Participant enterprises. Stephen, as Lucasian Professor at Cambridge, is institutionally, professionally, well-rooted in the Spectator enterprise. Yet it may be that his role as a Participant in advocating for people with disabilities – to bring about a better future for everyone – will be his most significant contribution to humanity as well as cosmological history.

41

The Participatory Anthropic Principle and the Tao Te Ching

Reasoning from the Spectator interpretation of the new physics, there seems to be no way to account for the currently observed (or any particular) outcome of the history of the cosmos. The present is under-determined by the past, and the future is under-determined by the present. The Anthropic Research Program, seeking to think outside the box, is a search for a principle, somehow embodied in the nature of reality, that would serve to 'select' the future, to select the outcome we actually observe. So far, the Anthropic Research Program has only served to gather increasingly diverse evidence for just how enormously unlikely this history and this outcome of the cosmos – with human inquirers – seems to be.

The Anthropic Research Program stemming from Brandon Carter's original Weak and Strong formulations has remained 'descriptive'; still trying to make sense of the universe as 'objective', as observer-independent; still embracing the Spectator's detached perspective. The Anthropic Research Program isn't concerned with self-reflexively accounting for Participant inquiry as an irreducible aspect of reality.

In what I referred to as the Parallel Hypothesis there is a link – a parallel – between one's theory of inquiry and one's theory of the nature of reality. The Logical Positivist's thesis that learning should be logico-mathematical – uniform and conceptually continuous – 'stood to reason' if one embraced the Scientific Hypothesis that all the phenomena of the universe are governed by One time-space invariant (mechanical) order. An understanding of the details of a mechanical reality should be revealed through a systematic, universal, uniform 'scientific method'.

On the other hand, if how we actually learn, as Kuhn and the other rebels argued, has an irreducible discontinuous – non-uniform, conceptually discontinuous, revolutionary character, then the natural expectation is that a careful study of the 'real' history of cosmos would reveal irreducible discontinuities. It stands to reason. Weinberg's symmetry-breaking account of the history of the cosmos can be understood as supportive of the Parallel Hypothesis implication of the nature of reality – reasoned from the rebel's account of the actual history of learning. Reasoning on the basis of the Parallel Hypothesis, the presumption seems to be that the cosmos develops – paralleling 'real' inquiry and cumulative learning – through some sort of genuinely exploratory, progressive, discontinuously unfolding, experimental, learning process.

Carter's original Anthropic Research Program reasoning began by considering how the universe must have developed in order to produce human observers at some point. Human observers happen quite late – after perhaps 13.7 billion years. However, the Parallel Hypothesis implication linking the revolutionary character of learning with Weinberg's symmetry-breaking discontinuities would require some sort of observer-like process from pretty much the beginning of the Big Bang. Radical as this may seem at first, the embrace of complementarity entails that the evolution of the cosmos, the transition from one stage to the next, is mechanically under-determined and requires some sort of selection process in the transition from any one state of affairs to the next state of affairs. Some sort of choice-like selection process must be involved, moment to moment, to actualize any one future from among all the possible futures inherent in any present state of affairs. The key point is that the embrace of complementarity entails that any such choice-like process must be *ubiquitous* – always and everywhere.

The embrace of complementarity seems to suggest that the historical evolution of the cosmos is 'somehow' the result of the progressive (viz. non-symmetric) interplay of complementary orders. The tacit entailment is that the evolution of the cosmos doesn't occur uniformly – doesn't preserve symmetry. The observational evidence cannot be understood in terms of the 'evolution' of One 'objective' universal, time-space invariant mechanical order. The mechanically non-symmetric interplay of the complementary orders generates a cumulative, historical, time evolution – generating the irreversible arrow of time.

Thinking again in terms of the Parallel Hypothesis, the Spectator's 'objectivist' universe embodies the Law of Excluded Middle – there is One order governing reality and 'one right way' to understand reality. On the other hand, the Participant embrace of complementarity suggests a sort of inversion of the Law of Excluded Middle. With the embrace of complementarity, the evolution of the cosmos must occur non-symmetrically, non-uniformly through the Middle Way – between the extremes. Both limited complementary physics are limited irreducible components. The extreme one-way 'objective' formulations of both of these pre-destining mechanics are excluded. Following de Broglie's core contribution to the quantum principle that all 'real' systems have both an irreducible wave and an irreducible particle aspect, for the Participant, all actualized stages, all actualized systems in the evolution of the cosmos should have irreducible aspects of both complements. All actualized stages, all actualized systems should be Middle-Ground, embodying some aspect of both complementary orders.

From the point of view of those still trying to remain in the Scientific Research Program, the evolution of the cosmos is not the result of the interplay of two coherent complementary orders but, enigmatically, the result of One order (rationality) and a complementarity disorder (irrationality). In any adequate More General Theory the evolution of the cosmos must be understandable as the interplay of two limited complementary coherences, not one coherence and one incoherence.

Nobel Laureate quantum physicist Niels Bohr was certainly aware of the natural connection between this 'two complementary coherences model' and the ancient Eastern Taoist Philosophy of yin and yang. Honored in his native Denmark by being inducted into the Order of the Elephant – an honor traditionally reserved for Danish Royalty – Bohr produced a new design for his family Coat of Arms. As the central image Bohr selected the Taoist interpenetrating yin-yang diagram and accompanied this with the motto: 'contraries are complementary.' Physicist Fritjof Capra, in his book, *The Tao of Physics: An Exploration of the Parallels between Modern Physics and Eastern Mysticism*, suggested similarly that the modern Western discovery of complementarity in the new physics is the re-discovery of the ancient Eastern insights about complementary processes embodied at all levels in the nature of reality.

The central concern of the Taoist Philosophy is usually translated as a Participant-like concern with ' how we should live', with 'the way', 'the path', 'the route', or more loosely as about the 'principle' or 'doctrine'. In all its uses, the Tao itself is considered to have ineffable qualities that prevent it from being pre-conceived, pre-defined or expressed in words.

"The Tao that can be spoken is not the eternal Tao.
The name that can be named is not the eternal name.
The nameless is the origin of Heaven and Earth.
The named is the mother of myriad things.
Thus, constantly without desire, one observes its essence.
Constantly with desire, one observes its manifestations.
These two emerge together but differ in name.
The unity is said to be the mystery.
Mystery of mysteries, the door to all wonders.

(*Translation by Derek Lin. Tao Te Ching: Annotated & Explained*, published by SkyLight Paths in 2006 (www.Taoism.net)

As a thought experiment think of the Tao as concerned with the path forward, the path of the evolution of the cosmos, perhaps the ongoing evolution of a more desirable cosmos. Since the future is always in doubt, the enterprise to stay on, or to develop, the path is inherently problematic. Any universal, uniform ideology – what can be spoken – is like saying, for instance, 'whenever in doubt, always turn to the right.' The complementary ideology says, 'whenever in doubt, always turn to the left.' Each of these logico-mathematically consistent, uniform ideological paths spiral in on themselves and self-destruct – what I have come to refer to as 'The Closing Phenomenon'. The message of the Tao seems to be that the path forward is some sort of non-ideological, Middle Way. Could it be that the Principle that the Anthropic Research Program is seeking is, from a Participant perspective, the Middle Way?

The Participatory Anthropic Principle

For those of us schooled in the new physics and embracing Bohr's principle of complementarity, what is striking and disappointing in Carter's Anthropic Research Program is the absence of any explicit consideration of the irreducible element, 'choice'. Where is the discussion of the Participant's problem of choice? In Bohr's Copenhagen Interpretation the embrace of complementarity, in both quantum theory and Relativity forces us to seek some sort of post-scientific, post-mechanical, More General, Participatory Theory.

The hard core 'mechanical objectivists', in order to avoid talking about the scientifically enigmatic 'choice', have moved to the non-selection model of multiverses, where all possible futures are manifest – just not 'here' in our observable universe. Their non-explanation explanation opts for a probabilistic account of the actual outcome we observe: with an unlimited plethora of possibilities actualizing it is 'certain' that this one 'we' just happen to be in was mechanically necessary and inevitable. Jim Baggott in his recent book, *Farewell to Reality: How Modern Physics Has Betrayed the Search for Scientific Truth,* appropriately characterizes this non-sense trend as a sort of 'fairy tale physics.'

One prominent physicist, John Archibald Wheeler, took the bold next step in the Anthropic Research Program toward understanding choice within a superseding, More General, Participatory Theory. Wheeler had worked with Bohr as a student and later, as a professor at Princeton, was the teacher of such luminaries as Richard Feynman and Hawking's close friend, Kip Thorne. Steeped in General Relativity Theory, Wheeler had coined the term 'black hole.' Wheeler outlines a Participant Cosmology, a sort of Copenhagen Interpretation of Cosmology, by making the seemingly obvious move to understand the selection principle – the Anthropic Principle – that accounts for the outcome of current cosmos in terms of the Participant observer's choice. Wheeler invented the expression "Participatory Anthropic Principle", reasoning that reality is created – actualized – by the choices made by Participant observers embodied in the nature of the universe. The current formulations of the new physics – in both quantum mechanics and Relativity – begin their description of the universe as the possibility space of a not-yet-actualized future possibility space. In quantum mechanics, this initial description is referred to as 'the wave function of the universe' – or the wave function of any isolated subsystem of the universe, as the case may be. This description is of a sort of 'virtual' possibility space that is inaccessible prior to being accessed, prior to being observed – prior to being 'actualized'. In both quantum theory and Relativity, the 'the collapse of the wave function', the collapse of the possibility space to actualize one of the possible futures, requires a Participant observer's choice (value bias) of how to observe. Wheeler reasons that there can be no actual universe, no actualized universe, and no developing history of the universe, without actualizing selection, without inquirers – without some sort of Participant-likeness, whose observational and action choices select and actualize the ongoing future.

In his famous article, "Genesis as Observership", Wheeler reports, in a footnote, that a German scholar had pointed out that his view is not new. Wheeler confirms that his Participatory Anthropic Principle model of the nature of reality seems to have been formulated previously by the German philosopher-scientist Friedrich Schelling. Schelling was a student friend and colleague of George Hegel. Schelling had rejected Immanuel Kant's model of an ultimately unknowable reality 'out there' – the observer-independent thing-in-itself (the noumena) – that manifests in observed phenomena in the mind of the 'detached' observer (as phenomena). Schelling argued that this objective-subjective or realist-idealist division was untenable. The solution was to place the subjective in the objective, to place the idea in reality, to place the observer, Mind, inside reality as a Participant.

The Participant's selective (value biased) actions in observation, inquiry and development were to be understood as inherent, irreducible aspects of the nature of reality.

Schelling simply accepts that mind – that which makes 'intelligent' choices – is everywhere and always in the universe. (Just how 'choice' is to be understood, aside from serving to characterize a selection process, still remains to be clarified. Schelling's 'intelligent' choice here might be tentatively understood as itself problematic and developing – as 'problem solving' – as experimentally exploring, seeking to learn how to bring about a more desirable future.) This embrace of mind as an everywhere and always aspect of the nature of reality makes sense in a Participatory physics. Whereas complementarity is ubiquitous – everywhere and every when – the problematic choice, and per hypothesis, mind, is ubiquitous.

According to Feyerabend and Lakatos, George Hegel, Schelling's close associate, was, covertly, a favorite of the 20th century philosophy of science rebels.

The evolutionary philosophies of Hegel and Schelling were important in the development of American Pragmatism, particularly in the work of Charles Sanders Peirce. John Dewey, whose thought became perhaps the most advanced of the American Pragmatists, began with a strong orientation to Hegel's work. In his later work Dewey developed a similar but arguably more sophisticated Participatory evolutionary worldview. Having transitioned from philosophy of science toward a philosophy of engineering, I have argued elsewhere that American Pragmatism is 'the missing philosophy of engineering.'

Wheeler's Participatory Anthropic Principle hypothesis clearly arose from his embrace of Bohr's complementarity. But where did Schelling's view a century earlier 'merging observer and observed' come from? In hindsight it is not surprising that Schelling's first publication was a commentary on Plato's dialogue *Timaeus*. In that dialogue, the main character, Timaeus, is asked: How did the universe come to be as it is? The account given by Timaeus should be taken not as literal but as metaphorical: 'not as a process by which an intelligent Craftsman put the world together at some time in the past, but as a statement of the principles that underlie the universe at all times of its existence.'

Timaeus articulates three themes worth mentioning that illuminate what a Participant Cosmology might involve:

First: Timaeus characterizes the Participatory Mind embodied in the universe as the Demiurge. This is translated in modern terms as the Master Craftsman: *demi-* public + *urge-* worker (from the Greek *ergon-* work). The Demiurge is alternatively referred to as the Architekton; *arch-* main, first, beginning + *tekton-* craftsman, builder. Both 'demiurge' and 'architekton' also suggest a modern translation as architect-engineer, understood in the broadest sense as the problem solver (value actualizer) who designs and builds. In the *Timaeus*, the ongoing business of the Master Craftsman is the bringing about, the development, of the universe as it is.

In the *Timaeus* there are two levels of Master Craftsman. The first is the Perfect One that creates the second, the Imperfect One. The Imperfect One is the one that develops and that is the Soul of the World. This second Imperfect Master Craftsman is the one that is progressively actualizing the universe. The 'imperfection' points at the 'ignorance' of the Master Craftsman about how to proceed; he/she lacks a script, he/she lacks a pre-conceived plan. Human inquirers are somehow 'part' of the second Imperfect One.

Timaeus provides an account of what sounds like a creative Middle Way process resulting in the Middle Ground (viz. intermediate) nature of the Soul of the World and its processes. The first, Perfect Master Craftsman, in creating the second, Imperfect Master Craftsman, "combined three elements: two varieties of *Sameness* (one indivisible and another divisible), two varieties of *Difference* (again, one indivisible and another divisible), and two types of *Being* (or *Existence,* one indivisible and another divisible). From this emerged three compound substances: intermediate (or mixed) Being, intermediate Sameness, and intermediate Difference. From this compound one final substance resulted – the World Soul. From the three, made into a single mixture, he re-divided the whole mixture into as many parts [viz.

that's us] as his task required, each part remaining a mixture of the Same, the Different and Being." (35a-b), *Timaeus*, translation Donald J. Zeyl)

One of Plato's main themes is that human nature (and the nature of all parts of reality) has the same dynamic middle ground mixture of complementary aspects. The human intellect has the same nature as the Intellect of the Soul of the World. Reality is intelligible – to the limited extent it is – because it has the same nature as the Master Craftsman's intellect. (Recall that with the embrace of complementarity reality is not 'mechanically intelligible'.) The Parallel Hypothesis argues that the nature of learning – how (we) the intellect learns – should be, per hypothesis, the same as the nature of reality. The nature of the Participant learner is, per hypothesis, the same as the nature of reality. The intellect is always imperfect because learning is an unfolding, an under-determined, progressive emergence into an open-ended future.

Second: When asked, 'According to what sort of plan has the Master Craftsman (and its many parts) brought about the world as it is,' Timaeus answers that it is – by its very nature – not a 'conceivable' plan (viz. meaning that it is not pre-conceivable). In Taoist terms the Master Craftsman's plan is the path, the way, that cannot be spoken; that cannot be put into words, cannot be ideologically conceived – as if it were logico-mathematically timeless. Any universal, time-space invariant plan that can be specified is not the real plan, not the eternal plan, not the Tao.

There is, per hypothesis, an eternal plan in the mind of the first Perfect Master Craftsman, a plan that 'emerges' or 'unfolds' imperfectly as it progressively actualizes in the spatio-temporal reality of the second Imperfect Master Craftsman.

Timaeus says that although both the path and the outcome are unique, he can only give a *probable* recounting of how it came to be as it is. There must always be an irreducible uncertainty. The present can never provide a perfect key to understanding the history of the past. There were many possible paths to the current outcome, but the details of the actual path cannot be logico-mathematically, rationally, reconstructed. The discontinuous nature of the unfolding of the plan precludes a simple uniform, rationalistic, mechanical recounting. There must always be what 'appears' to be – from a rationalistic, mechanically deterministic perspective – an irreducible chance-governed (probabilistic) aspect.

Third: However, Timaeus assures us that one thing is certain – that the work of the actualizing Imperfect Master Craftsman – by its very nature – always and everywhere moves (or at least is attempting to move) to the good, 'to the extent that this is possible.' What is certain, then, is that there is a direction and the direction is toward the actualization of what is valuable – the direction is toward the ongoing actualization of a more desirable future.

That the direction is toward a more desirable future makes sense, per hypothesis, if the mind of the Imperfect Master Craftsman is that of the Participant Engineer. Since engineers are by their very nature problem solvers, and problem solving is attempting to move from a current state of affairs to a future more desirable state of affairs, then the universe progressively actualized by the Participant Engineer is always attempting, through exploration and experimental learning, to move to the more desirable future – always toward the good. It is important to emphasize that the 'problems' to be solved are not pre-conceivable and so perhaps better characterized as open, non-pre-conceived (emergent) opportunities to be actualized. The 'problems' or opportunities cannot be pre-conceived any more that the plan or the good – they emerge in unexpectable ways.

The very nature of the plan, the Taoist path – and the understanding of the nature of the good – 'unfolds'. Timaeus notes that his present account of the nature of the universe is revisable. This is, I think, per hypothesis, because the plan and the nature of reality that embodies it is qualitatively emergent (is progressively revealed) with each advance. Think of the account one might have given two billion years ago and how it would have had to be revised by the nature of what has emerged since then.

Wheeler's Participatory Cosmology only makes a partial shift toward what is articulated in Schelling and in the *Timaeus*. Wheeler's formulation of a self-actualizing reality avoids discussing the nature of the Participant's problem of how to best choose – the value question. Whether the cosmos implied by Wheeler's Participatory Anthropic Principle would naturally move toward the good just doesn't come up. In the *Timaeus*, as in Schelling and Hegel, there is a more general Paradigm Shift and a problem shift, wherein a new, more general type of question and a new, more general type of inquiry emerges and makes sense. The new, more general type of inquiry that emerges is unavoidably self-reflexive. The new type of inquiry is concerned with the Participant's place in the universe, is concerned with our role in the universe. The shift is in the representation of inquiry: from the Spectator's detached descriptive enterprise trying to understand and explain how the universe works, to the Participant's embodied enterprise with an irreducible concern with the prescriptive, with trying to understand how we should live. For the Spectator the problem is to discover 'how the world works' – presupposing that it works the same way everywhere and always. For the Participant Engineer the question is, in the initial, limited version, 'how to work in the world'. More fully, by virtue of his inherent character as a value-biased problem solver, the question becomes 'how to make the world work better' – 'how to bring about a more desirable future'. Because the Participant is embodied in reality the agenda of bringing about a better future is crucially, self-reflexively inclusive. The Participant can have no prior knowledge of what he is seeking to learn. The Participant has no prior knowledge about what constitutes a more desirable future. The Participant's engineering enterprise is necessarily exploratory and experimental. The bootstrapping effort self-reflexively involves bringing about a more desirable future self – an improving intellect, a better method of inquiry. Pragmatist C. Wes Churchman captures the larger socio-economic agenda in his book, *The Design of Inquiring Systems*. Inquiry is not just about asking 'detached' abstract questions. It involves the building – the creative, recursively enabling, exploratory construction – of a progressive research and development enterprise.

The sense of learning as exploratory and experimental is, I think, captured metaphorically in Ancient Mythology. As the story goes, at a time when the gods were creating mortal beings, Prometheus and Epimetheus were ordered to equip them for living in the world. Epimetheus took on the task of distributing all the skills – so that the birds knew how to fly, fish knew how to swim and so forth. Prometheus upon inspecting his brother's work, realizes that he forgot to provide humans with the skills, with the embodied knowledge of how to live. Consequently, humans lack *foresight* – the embodied knowledge of how to live. One moral of the much larger story is that humans survive and advance in the world, by learning through *hindsight*. Humans try to choose the best way to live, but, lacking foresight (viz. knowledge of how to choose) they make mistakes. Human action is necessarily exploratory and experimental – and, in this sense, learning can only occur through hindsight.

42

The Parallel Hypothesis and Royce's Criterion of Self-Referential Coherence

The Parallel Hypothesis is based on the recognition of the link between one's theory of the nature of reality and one's representation of how learning about that reality does, or at least should, occur. Initially the hypothesis captures the link between one's philosophy of science (inquiry) and one's philosophy of nature. The hypothesis captures the link between one's representation of inquiry and one's presuppositions about how reality works. The Logical Positivist's early Spectator representation of inquiry, suggesting that scientific method was, or at least should be, logico-mathematical – a sort of systematic, mechanical reasoning – made perfectly good sense, given their presupposition that reality was governed by One objective Mechanical order. Their 'mechanics of learning' was naturally supposed to parallel the mechanics governing all the phenomena in the universe.

Kuhn and the others in philosophy of science who rebelled against the Positivist model argued that actual learning was not mechanical – evidenced both in the actual practices of inquiry and in the nature of the result (knowledge). Kuhn and the other rebels argued that advances in the history of learning, as well as the process itself, revealed essential discontinuities, essential logico-mathematical (mechanical) discontinuities. Reasoning on the basis of the Parallel Hypothesis, the rebel position on how 'real' learning works entails a parallel thesis about the nature of reality – namely, that the nature of reality develops in a similar manner, by a mechanically discontinuous learning process. According to the rebel's Participant cosmology, the history of the universe should be a progressive development – advancing by a process that parallels 'real' learning. Empirical investigations of the history of the universe should reveal an irreducible accumulation of historically irreversible, mechanical and logico-mathematical discontinuities.

The rebel's embrace of complementarity leads to a Participant perspective wherein the inquirer – the learner – and the learning process are literally embodied aspects of the universe – irreducible, ubiquitous components of the nature of reality. As the Participant learns, so does the universe. As the Participant learns, the nature of reality develops, the nature of reality emerges – unfolds – in a non-uniform, mechanically discontinuous way. The Spectator's representation of a separation between the inquirer and the universe breaks down completely. The Spectator's tacit presupposition of a qualitative difference between the learning inquirer and the fixed universe, the fixed mechanical nature of reality, breaks down completely.

The Parallel Hypothesis suggests that the inquirer and reality have the same *type* of nature. In the Participant perspective the inquirer and reality have the same *type* of progressive developmental nature. Since the Participant inquirer is to be understood as embodied in reality, the nature of the inquirer and the nature of reality are no longer simply parallel – they merge, having the same character, the same nature. The activity of the Participant inquirer is the process of the evolution of the nature of reality, is the process of the historical development of the universe. The Participant perspective offers a new way of understanding ourselves and our place in the universe. The Participant perspective offers a new, post-scientific type of answer to the question of the nature of reality and our place and role in it.

The Anthropic Research Program began by trying to account for human observers. It soon became clear that the selection processes necessary to bring about human observers must have started much, much earlier in the history of the cosmos.

If complementarity is ubiquitous, the selective choices that bring about this particular, unique universe must also be ubiquitous. Wheeler argued that any actual – actualized – universe requires choices to have been made – from the beginning. Wheeler, recapitulating Schelling and Hegel, moves to a Participant cosmogony and cosmology by erasing modern science's separation between subjective chooser and objective reality, by erasing the boundary between idealism and realism. Similarly, Plato, in *Timaeus*, offered an early, largely metaphorical representation of a Participant worldview, wherein the imperfect intelligent 'parts' (viz. 'us') are characterized as craftsmen – engineers – Participants in the enterprise seeking to bring about a more desirable future.

American Pragmatist Josiah Royce offered an important argument supportive of the essential role of learning in any self-referentially coherent worldview. Royce reasoned that any proposed complete theory of everything must be able to account for itself – both for its own existence as part of the universe (i.e. the existence of the theory itself) and for its having been learned. For instance, if physicists propose a Theory of Everything, that theory must be able to account for physicists in the universe and must be able to account for how they learned that complete theory of everything. Royce initially discusses this in terms of what he calls 'the problem of problems.' He reasons that any theory of everything must be able to account for the evidence of, and for our real experience of, problems and problem solving. Two types of problems come to mind: first, the problem of ignorance and, second, the problem of evil. For Royce, in accord with the other rebels, learning involves problem solving. Royce reasons that any complete theory of reality must be able – self-referentially – to account for the problem solving involved in learning as an irreducible aspect of the nature of reality.

I have come to refer to this view as Royce's Criterion of Self-Referential Coherence – in effect, that any theory of everything must be able to *make sense* of itself. The critical starting point, of course, is that neither 'real' learning nor 'real' problem solving *makes sense* in a fully deterministic universe. I have suggested that American Pragmatism is an early formulation of what might now be referred to in the 21st century as a self-referentially coherent Philosophy of Engineering. The engineer is to be understood broadly as 'a problem solver' and, as engineering textbooks express it, the problem of engineering is the problem of design, the problem of how to design a more desirable future. How should we design the irrigation of our fields? How should we design our houses? How should we design our neighborhoods? How should we design our cities? How should we design our economies (tariffs or not)? How should we design our political system? – to preserve our economy? How might we design a more desirable future? The U.S. Constitution is the outline of a modern, historically edified, experimental design in an ongoing, participatory exploration seeking better answers to the question of how to bring about a more desirable future, seeking better solutions as to how to learn more about how we could and should live – better. The U.S. Constitution is the outline of what C. Wes Churchman referred to as *the design of an inquiring system*. The U.S. Constitution is not only historically edified, but designed to be recursively edified, revised, modified and developed on the basis of the results of the ongoing experiment. The agenda for Participant Engineering is open-ended problem solving, open-ended improvement of the design of reality. The improvement cannot be understood as concerned simply with a reality 'out there'. The improvements, the advances in understanding, the advances in research and development, must self-referentially include us. Nobel Laureate Economist Joseph Stiglitz builds on Churchman's theme, in his recent book, *Creating a Learning Society: A New Approach to Growth, Development, and Social Progress.*

In the rebel's formulation of 'real' learning as problem solving, wherein problem solving is always seeking to bring about a better, more desirable future, Royce's two types of problem – 'ignorance' and 'the problem of evil' – are really one problem. The problem of ignorance (viz. how the world works, and/or how to work in the world) and the problem of evil (and good) (viz. how to make the world work better) are inseparable. In the Spectator representation all knowledge is about objective facts and processes 'out there'. The question of values 'out there' is meaningless, inconceivable from within the Scientific Research Program and the Mechanical Philosophy. In the Participant representation, inquiry is an embodied problem solving enterprise, by its very nature, seeking to bring about a more desirable, a more valuable, future – both in understanding and in the actual structure and function of the world. Participant learning, both the process and the results, cannot be made sense of independent of values, independent of seeking 'better' theories, independent of the question of what constitutes a more desirable future. Since the Participant's 'problem' of how to make the world work better is never completely pre-defined in specifics, solutions must be discovered through some sort of creative, experimental exploration. Since the Participant's 'problem' of how to make the world better is never completely pre-defined in specifics, the process of finding solutions might be better understood as the discovery of solutions and, in discovery, bringing forth something of value. Advances, learning, and solutions might be better understood as value actualizations.

The Participant inquiry is at least about '*how* to work in the world' – about '*how* one might change the world'. What is discovered, what is knowledge, in the Participant representation is these '*methods*'. The broader, defining question of Participant's inquiry is, per hypothesis, about '*how* we should live?' – about the best, or at least better and better ways to live. Again the solutions, the answers – the knowledge – are all about 'methods', about *how*, about practical ways of being and becoming in the world. This characterization of the results of meaningful inquiry, of the nature of 'answers', of the nature of 'solutions', is central to the Participant theory of knowledge – easily associated with American Pragmatism. The Participant's initial, limited exploration of his opportunity space seeks to discover existing 'methods', processes and relationships in reality (self included). What has been previously represented as 'pure' – pre-application – scientific research is newly understood in the More General context as 'pure' exploratory and experimental engineering research. Pure, pre-application research only makes sense within a more general framework defined by the Participant's presupposition of potential future beneficial applications. The American Pragmatists have regularly emphasized that all meaningful knowledge is potentially beneficial. In other words, 'knowledge' is real and meaningful if and only if it is potentially beneficial in some future application.

In the full characterization of the Participant, the 'knowledge', the 'methods' of concern are not just about changing the world. The methods of concern, the sought after discoveries, are those methods that potentially enable the engineer in bringing about a more valuable future.

In the Pragmatic-engineering theory of knowledge, Royce's two types of problems, often thought of in the 20th century as separate, merge into a single new type of question, subsuming and superseding the separated formulations. The new Participant representation of inquiry is More General, both subsuming and superseding these two traditional representations of inquiry: factual (scientific) inquiry and value (religious) inquiry. In the new Participant representation of inquiry, the new type of 'merged' question arises, makes sense, in an unfolding, existentially self-developing reality. Both the 'factual' and the 'value' structures and functions of the universe develop progressively with successful problem solving. The Participant pragmatist-engineer is concerned to bring about – to construct – a better factual reality, that is at the same time a better value/moral reality. Per hypothesis, these 'realities' are inseparable. One way to grasp the relationship is in terms of the way innovations and technological advances provoke novel value issues, novel moral questions about how they should be incorporated into how we live. Obvious modern examples include abortion, environmental issues, privacy in the new electronic era, politico-social-economic issues in a post-agrarian society, and so forth.

The *self-reflexive question* of why we are ignorant of how to live doesn't make sense in a fully deterministic representation of reality. That there should be *self-reflexive questions* about how to live doesn't make sense in a fully deterministic representation of reality. The question of how such questions might have arisen in the first place – might have arisen at any stage at all – doesn't make sense in representations of either a space-time invariant mechanical order or a space-time invariant value system (viz. invariant moral law).

Although much more needs to be explored in articulating the More General Theory, it seems at least plausible that the Participant representation of inquiry and learning is self-referentially coherent – makes sense of itself. The defining question of the Participant paradigm of inquiry is – how should we live? – newly understood as inquiry inherently concerned (self-referentially) with inquiry. The Participant is concomitantly concerned with learning how we should live and with bringing about better ways, better methods of living. Learning how we should live involves trying to bring about a reality that recursively enables us to continually improve our ability to learn still better ways to live. Learning how we should live involves manifesting a better future wherein we can learn more, wherein we can embody more learning. It is along these lines that the Participant enterprise might be characterized as a self-referentially, recursively enabling Research and Development enterprise.

The Participant representation inquiry is perhaps better understood as concerned with innovative methods, with finding and developing better ways of doing things, better ways of living. Learning is practical and occurs through practical, experimental, exploratory engagement. As such, the embodied method of living and the embodied method of learning merge and develop together.

In the Participant paradigm the separation of the sciences and the humanities disappears, is no longer tenable, is no longer defensible.

Middle Ground Realism

The Participant Paradigm allows for a new type of realism – just not an 'only one way' observer-independent, observation-independent, 'objective' realism. The Participant is embodied in a middle ground possibility space, in a middle ground opportunity space. The Participant Paradigm suggests that we should be able to make sense – at least to some limited extent – of a middle ground reality. Any sort of detailed account of the approach suggested by the shift to the Participant Paradigm is well beyond the scope of this book. However, a few preliminary characterizations might be useful here.

Consider what might be taken as a corollary of the Parallel Hypothesis: for every theory of knowledge (epistemology) – as to the nature of what we know – there is a parallel theory of the make-up of reality (ontology) – as to the components of reality and their structure and function. Since the rebels', and the Participant, theory of 'real' learning, and the consequent knowledge (epistemology), is about 'methods', the parallel reality (ontology) is about the structure and function of evolutionary learning processes.

The inherent limit of the Scientific Hypothesis means that all mechanical, clockwork-like processes – throughout the universe, come what may – are better understood as limited special cases. The universe is not One, space-time invariant, causal clockwork. What, then, can be said in terms of the new Participant Paradigm about the superseding coherence? How is the superseding reality to be understood? If the limited mechanical ways of understanding are idealizing special cases, they should be newly intelligible as special cases, within some new, superseding, More General understanding of reality, within some new, superseding, More General post-mechanical coherence.

In his dialogue, *Timaeus,* Plato considers the question of the intelligibility of reality – here and now – the elements, structures and functions of the emerging universe. Timaeus suggests the eye as an exemplar of how the Imperfect Master Craftsman (viz. that's us) brings about systems that recursively and synergistically facilitate further learning and development. As we learn, we learn how to learn better, and we develop tools and systems that enable new ways of exploring and experimenting.

In the Spectator representation learning has an overall convergent, presumably narrowing, agenda – discovery of the hypothesized fixed, objective truths. In the More General Participant representation of middle-ground reality, natural learning and questioning are both exploring and expanding the possibility space of choices. The middle ground opportunity space of choices emerges and expands. Inquiry in the Participant's Research and Development Program is about the discovery and implementation of innovative methods – ways of living.

University of Victoria (B.C.) Professor of Biology Robert G. B. Reid captures the sense of the new, Participant-engineering understanding of biological evolution as a cumulative, exploratory learning and development enterprise in his book, *Biological Emergences: Evolution by Natural*

Experiment. Biological evolution is perhaps, per hypothesis, better understood as a recursively enabling, experimental, exploratory progressive engineering enterprise.

As we learn, we learn how to develop better tools for learning, including our societies. Exploratory engineering is about the creative development and constructive implementation of innovative methods, is about the creative development and constructive implementation of innovative ways of concomitantly living and learning.

Increasing Complexity as a Euphemism for 'Real' Progress

One of the more popular modern approaches to the representation of both biological and cosmological evolution as *progressive* has been in terms of 'self-organizing complexity'. What has been lacking is any account of how the hypothesized 'self-organizing' process *selects a future, more complex state from the opportunity space*. Furthermore, there is still no clear definition of either the nature of 'complexity', or the hypothesized 'increasing complexity'. The Participant Engineering Paradigm hypothesizes that the 'self-organizing' process is perhaps understandable as Participant engineering problem solving – as an exploratory recursively enabling process of autodidactic innovation. 'Increasing complexity' could perhaps be better understood as the progressively intelligent (viz. problem solving) development of an exploratory engineering enterprise.

Innovative learning and implementation *selects* a future from the opportunity space both by developing the future and by bringing a more desirable, enabling future. What I referred to as 'The Clue' is that although 'real' learning is, in part, locally convergent, answering some immediate local question, its additional importance and benefit is to expand the field of potential inquiry and exploration by enabling revolutionary, new *types* of questions, by enabling innovative, new *methods* of exploratory inquiry. 'Real' learning and constructive, creative implementation expands the opportunity space, expands the range of our ability to do things – expanding our freedom in qualitative new ways. Increasing 'freedom' in the Participant perspective should be understood as involving the increasing ability to make the world better and concomitantly as involving the increasing ability to learn how to make the world better.

Hegel seems to have captured this sense of history as an autodidactic unfolding, enabling an embodied intelligence: "History is the unfolding of an idea, and the idea is freedom."

The Participant Engineering Research and Development Program

These characterizations of the Participant Engineering Research and Development Paradigm offer a new way of understanding the nature of reality and our place in it. They offer some much needed substance to the searching 'outside the box' initiatives of both the Anthropic Research Program and Complexity Research Program. The Participant Engineering model, per hypothesis, offers a way to make sense of the evidence of a historically progressive cosmos that is recursively selective in its creative constructive agenda to bring about a more desirable, a better, future.

The embrace of complementarity, however, entails that progress doesn't arise through any sort of uniform, systematic, pre-specifiable process. Consequently, in the Participant perspective advances in learning – inventions and innovations – always arise through an experimental and genuinely exploratory process and are always to some irreducible extent *unpredictable and unexpected*. Innovations are conceptually discontinuous and 'appear' to be chance-governed from a systematic, logico-mathematical or mechanical perspective.

Stanford University Professor of Engineering Walter Vincenti, in his book, *What Engineers Know and How They Know It*, laid down a challenge to the dominant Logical Positivist, scientifically oriented, theory of knowledge. Vincenti notes that: "Engineering knowledge, though pursued at great effort and expense in schools of engineering, receives little attention from scholars in other disciplines. Most such people, when they pay heed to engineering at all, tend to think of it as applied science... Modern engineers are seen as taking over their knowledge from scientists and, by some occasionally dramatic but probably intellectually uninteresting process, using this knowledge to fashion material artifacts... From this point of view, studying the epistemology of science should automatically subsume the knowledge content of engineering... Engineers know from experience that this view is untrue."

Vincenti argues that from an engineering perspective "technology appears, not as derivative from science, but as an autonomous body of knowledge... This view of technology, and hence of engineering, as other than science, accords with statements sometimes made by engineers, such as the following by a British engineer to the Royal Aeronautical Society, in 1922: 'Aeroplanes are not designed by science, but by art inspite of some pretense and humbug to the contrary. I do not mean to suggest for one moment that engineering can do without science; on the contrary, it stands on scientific foundations, but there is a big gap between scientific research and the engineering product which has to be bridged by the art of the engineer.'"

And then Vincenti adds, "The creative, constructive knowledge of the engineer is the knowledge needed to implement that art."

Duke University Professor of Engineering Henry Petroski, in his recent book, *The Essential Engineer: Why Science Alone Will Not Solve Our Global Problems*, has argued powerfully for a rethinking of the history of science, for a rethinking of the history of 'real' inquiry. Petroski boldly argues that the 'history of science' representation of 'real' inquiry is better understood as a limited idealization within the more general, superseding framework of the history of exploratory engineering inquiry and innovation. Petroski proposes to understand the Spectator representation of inquiry in a new way as a limited special case within the More General Participant Engineering representation of inquiry and innovative development.

In Mechanical models of reality all interactions are zero-sum: preserving symmetry, preserving equilibrium, conserving matter/energy. With the embrace of complementarity, in the superseding Participant Engineering Research and Development Paradigm, the evolution of reality occurs

through a non-symmetric, non-zero-sum interplay of complementary orders, resulting in the progressive time-cumulative *production* of a real and
substantial reality. John Barrow points out, in his book, *The Book of Nothing*,
that the defining symmetry presuppositions of the Scientific Research
Program entail that 'if you add up everything in a mechanical universe –
the sum is zero, entailing that the universe doesn't exist.' In the Participant
Engineering Research and Development Paradigm action generates a net
reality – all work (in the engineering sense) has an irreducible net, non-zero-
sum, partially non-symmetric *product*.

A crucial challenge to the Participant Perspective is to demonstrate that
all actual interactions are to some irreducible extent non-zero-sum. In part,
this is what the evidence for a 'real' history, a discontinuous symmetry-
breaking history begins to establish. Interactions in the Participant model,
being non-mechanical, are non-symmetric, not adding up to zero, always
producing a non-symmetric residual. But what is the evidence that this in-
stantiated history is, has been and might continue to be *progressive* in a
value-laden sense? David Warsh, in his book, *Knowledge and the Wealth of
Nations*, tells the story of Paul Romer's Participant Paradigm Shift in economics. Romer's famous 1990 article, "Endogenous Technological Change"
argues for a new understanding of economic activity that supersedes the
classical zero-sum, scientific, supply-demand Steady State equilibrium
models. Innovations are no longer chance-governed, 'exogenous' (as they are
when viewed from the outside) inputs to 'normal' zero-sum, supply-demand
economic activity.

In Romer's 'New Growth Economics' economic activity is understood more generally as an engineering enterprise, as a creatively emergent
Research and Development enterprise. In classical scientific economics there
was no explanation, no way to explain 'real' growth, qualitatively progressive growth. Supply and demand processes were, in effect, the basis of the
classical Steady State model of economic activity. New Growth Economics
subsumes and supersedes the limited successes of classical zero-sum equilibrium economics. (Think of the properties of classical equilibrium economics as analogous to the homeostatic properties of the human body. Our

homeostatic processes and properties enable, but under-determine human inquiry and problem solving activity. A more general understanding of human activity as creatively constructive subsumes and supersedes the enabling zero-sum homeostasis.) In New Growth Economics the economic actor is no longer simply a 'rational optimizer in a zero-sum game'. In the new understanding the economic actor is the engineer; broadly understood as problem solver, naturally questioning, seeking to discover and develop 'new, better, *ideas*', novel, innovative ways. In the New Growth Economics perspective the economic history of human civilization is a conceptually emergent history of recursively enabling technological innovations, recursively enabling innovative methods, recursively enabling advances in how we live and learn. Engineering knowledge, in the sense of the discovery and implementation of progressive innovative methods, is the real wealth of nations.

The idea that 'real', successful inquiry (R&D) involves the creative development of non-zero-sum relationships is fundamental to the Participant understanding. Oxford University Professor Richard Dawkins in his popular book, *The Selfish Gene*, defines selfishness and selflessness in harsh zero-sum terms. Selfishness is where I win and you lose. Winning – gaining something – *necessarily* requires that someone else loses. Selflessness is where I lose and you win. Losing *necessarily* requires that I disadvantage myself so that someone else can gain advantage in relation to me. For Dawkins, in what he pre-supposes to be the universal, zero-sum struggle for existence, selfish behavior is rational and selfless behavior irrational. Dawkins goes to some length to try to convince his readers that all real actions and relationships – by their very nature – are selfish and that the world is completely and consistently competitive. There actually are no 'real' selfless actions and relationships – there can't be. Those who propose that there are selfless actions and relationships are simply confused and misunderstand the phenomena. In his more recent book, *The Extended Phenotype: The Long Reach of the Gene*, Dawkins appears to move toward a more nuanced, middle ground position wherein one of the best ways to be competitive (to produce more off-spring) is to cooperate, to engage in cooperative actions and form cooperative relationships. The Participant understanding subsumes and supersedes all (symmetry-presupposing) Mechanical Systems and their consequent zero-sum analysis. In the Participant model successful inquiry and innovation are progressive – positive sum – bringing about a better world, as well as further enabling better worlds. In order for this to be possible, successful inquiry and innovation must involve the development of creative win-win relationships. Win-win relationships are middle-ground, subsuming and superseding the prior demonstrated value of opposite types of value. Creative, win-win relationships occur through the middle way. All real advances necessarily involve an irreducible element of each (viz. in analogy to de Broglie's insight that all observational opportunities and all actualities have irreducible aspects of both particle and wave. If win-win relationships were not possible then the progressive development of New Growth Economics could not have happened.

Celebrated business philosopher Stephen Covey, in his popular book, *The Seven Habits of Highly Effective People*, provides an excellent account of the notion of win-win relationships in both economic and social relationships. Covey counsels that when entering into a relationship, eschew win-lose and lose-win relationships. If the relationship is not win-win – where both parties benefit – then just say 'no deal'. Win-win relationships are progressive and mutually enabling.

Master Engineer George Bugliarello (Chancellor, Polytechnic University of New York) has argued that engineering students should be taught that modern engineering is a natural progressive extension of processes of biological evolution. Bugliarello's tacit hypothesis is that biological evolution is better understood as a recursively enabling, boot-strapping, engineering enterprise. His Participant thesis is that biological evolution is better understood as a creatively constructive value actualizing enterprise. This requires that the process of evolution involves the discovery and implementation of win-win relationships.

Science writer Dorian Sagan and ecologist Eric Schneider, in their book, *Into the Cool: Energy Flow, Thermodynamics and Life* outline a developmental engineering model of the history of life. Their history of 'the engine of the biosphere' is the history of a cumulative, systemic, recursively enabling, innovative engineering design process. The evolution of the biosphere can only be progressive if it is possible to form win-win relationships.

When evolution is viewed as progressive (viz. not 'merely' random change), the evidence for progress is the same in engineering, economics and in biological evolution: 'an increase in the performance of work' and concomitantly the parameter of development is 'increasing (viz. better designed) capacity to perform work'. In the Participant engineering context 'work' is coherent (viz. over space and time) action – 'to do'. So the argument is that there has been an increase in 'doing' and concomitantly in the capacity to do things – to learn. Life, according to this new way of understanding, evolves expansively to be able to do more things – to do more *types* of things. Understood this way, the 'increase in the capacity to perform work' is an increase in the ability to do things – an increase in freedom. An increase in the actual performance of coherent work – Participant Research and Development – is an increase in the exercise of freedom, is an increase in the exercise of natural exploratory experimentation.

43

'Stephen Sells Physics Better than Madonna Sells Sex'

On the evening of the main lecture in Seattle we parked near the Pacific Science Center and then trekked a quarter mile across the former World's Fair grounds, under the Space Needle and toward the Opera House. The event is sold out and as we approach from the rear entrance we encounter three college age people with a large sign, seeking tickets. I wish I had more freebies to give them, but I don't. We wish them well: "Good luck! Perhaps you can find some high school students willing to sell you their free ticket." We arrive only an hour early, yet in plenty of time for the nurses to fuss over Stephen's appearance and serve him tea.

Richard Berger has populated the lobby of the Seattle Opera House with well-lit and well-displayed crystals and fossils. On the stage is an immense spiral ammonite. The atmosphere is exciting and deep-time historical – appropriate to talking about the cosmos.

Nathan is scheduled to introduce Stephen tonight, and after tea he meets us in our 'Green Room' waiting area. Nathan and Stephen exchange a few niceties. For the first time I get an extended, up close and personal, first impression of Nathan. He is dressed in a blue sports coat with a curious tie, the sort of thing you'd expect of a software engineer – a nerd. Nathan is a new type of corporate executive, a discontinuous contrast to the traditional business professional with his expensive suit and coordinated tie. Nathan strikes me as slightly 'cherubic'; his light colored hair is curly and fine. His presence is somewhat awkward and insecure. He shuffles. He is attentive to Stephen. Nathan is not projecting a 'power' presence. – Just my first impressions.

Show time. I greet and welcome the crowd and introduce Bernie Craig, the CEO of KCTS Public Television in Seattle. Bernie says his piece and introduces Nathan.

When Nathan stepped onto the stage to introduce Stephen, there is a change, not completely incongruous with my initial im-

Stephen and Nathan

pressions; and yet, now, he seems fully self-confident. In front of 3000 people he speaks as he had the night before in front of 60 friends and colleagues. He personifies an air of confident openness.

Nathan begins: "In physics, cosmology, similar areas to Stephen's … I was enormously inspired by his work and that helped me to go on in that particular area. A couple of years later, I had the privilege to work with Stephen. He gave me the first job I got after graduate school and I was a postdoctoral fellow working with him. And now, given a variety of the strange twists and turns life can take you, I'm here introducing him.

"I think there are three things about Stephen that I would like to try to leave you with. The first is, of course, his work. Stephen has given us insights into the fundamental nature of space and time, the laws that make the universe work, the laws that might make any conceivable universe work. The notion that we, at our current stage of development, can make very powerful and profound statements about how the universe came to be, how events that occurred fifteen to twenty billion years ago occurred, is still amazing to me, even though I used to work in that field.

"Second, I think I'd like to mention that Stephen is really one of the great physicists on a human scale as well. He has a tremendous sense of humor, fortunate for me, and Stephen inspires a loyalty in the people who work with him that's really unmatched by any of the other great physicists I have known, or even people in other walks of life.

"Stephen is just a tremendous individual.

"I think the third thing that is fascinating about Stephen Hawking is that, besides being a great individual, besides doing incredible physics, he's taken a third step that really very few scientists ever take and that's gone to the effort of making his work accessible to all of us – to popularize it. His book, *A Brief History of Time*, has sold about 5 ½ million copies worldwide. I was trying to put that into perspective in looking at a bunch of book statistics and discovered that Madonna's book, *Sex*, sold about a million copies. Sooo…"

Nathan pauses for some sustained laughter from the audience.

"… I think that the final thought is we're going to listen tonight to a man who can sell physics better than Madonna can sell sex, despite the fact that Madonna's book and topic are a little easier to grasp.

"Anyway, with that I'll let Stephen come on and tell us about Black Holes and Baby Universes."

Nathan disappears into the wings.

A few seconds later the dramatic music begins – *The Ride of the Valkyries* – and Stephen – encouraged to have, and having accepted, the appropriate pause – moves onto the stage. The audience explodes with cheers and applause. Individuals in the audience rise. In a sort of spontaneous wave the audience of 3000 individuals becomes a singular, standing ovation.

The theatre staff has secured a colorful Persian-type rug near the front edge of the stage. Stephen maneuvers perfectly to its center. Since his position is fairly close to the audience, maybe eight feet from the front edge of the stage, the theatre staff had insisted on constructing a small 'oops-barrier' – a six foot long, three-by-three inch square piece of wood nailed to the stage between the rug and the precipitous edge. In case Stephen made a slip in maneuvering his chair, they didn't want him joining the guests in the front row.

The large screen, surrounded by black curtaining, is only a couple of dozen feet behind Stephen; the bottom of the screen eight feet above him. Tim Hunt is behind the screen from where the PowerPoint slides are to be projected. Tim has a copy of Stephen's script marked with the places where he is to change the slides. Hawking often rewrites his talks at the last minute but – usually – nothing so radical as to alter where the slides are to be advanced.

Joan Grant moves to take her seat in the audience, while Joan Godwin remains just off stage with me, in case Stephen needs her. I learn that there are subtle 'signs' that Joan keeps an eye out for that would indicate that Stephen is having a problem of some sort calling for a little nursing intervention.

Impressions of the moment differ. In contrast to the triumphal entrance, one local reporter commented the next day: "Diminutive and solitary on a stage, his rumpled body looks like an inflatable doll with the air let out – one of the world's most magnificent intellects indentured to one of the world's least useful physiques."

Once Stephen is situated the applause begins to subside, terminating in a profound almost 'deafening' silence that lasts for perhaps a minute. Stephen is cycling up his computer to generate the talk – a software shift from the prior task of moving onto the stage. As the seconds tick by, there is a vague question animating the silence – just hanging there suspended – as to whether there might be a problem.

Finally Stephen booms out: "Can you hear me?"

There is an explosion of cheers and applause that quickly subsides in anticipation.

"This may be the first time you have been addressed by a real computer. You may have seen science fiction films, like 2001, in which there were computers that spoke. But these films are really cheats. The computer parts were spoken by humans."

There is laughter and – in this tech savvy Seattle audience – a recognition.

Stephen continues: "The reason was that computerized speech synthe-sizers were not sufficiently good to be used in films or TV programs. But this speech synthesizer is a great improvement. It varies the intonation and gives me a voice that sounds almost human instead of psychedelic. The only trouble is that it gives me an accent that has been variously described as American, Scandinavian, or Irish.

"Now for our lecture......... Black Holes and Baby Universes.

"I will talk about black holes, which aren't as black as they are painted. Instead I shall show that they can shine white hot and they can be the proud parents of little baby universes.... "

The full upgraded and edited version of this talk was later published in his book, *Black Holes and Baby Universes and Other Essays* (September 1994).

Beginning in the mid-1960s, Hawking began studying black holes. It was John Archibald Wheeler, an American physicist, who coined the science-fiction-friendly term, 'black hole'. "It was," Wheeler wrote later, a "terminologically trivial but psychologically powerful" description.

Hawking noted: "The importance in science of a good name should not be underestimated."

After fellow British physicist Roger Penrose in the mid-1960s, came up with the modern theory of black holes as resulting from the collapse of older stars to a point of singularity with such a strong gravitational pull that noth-ing could escape, Hawking showed that by mathematically reversing this event, one could model the expansion of the universe.

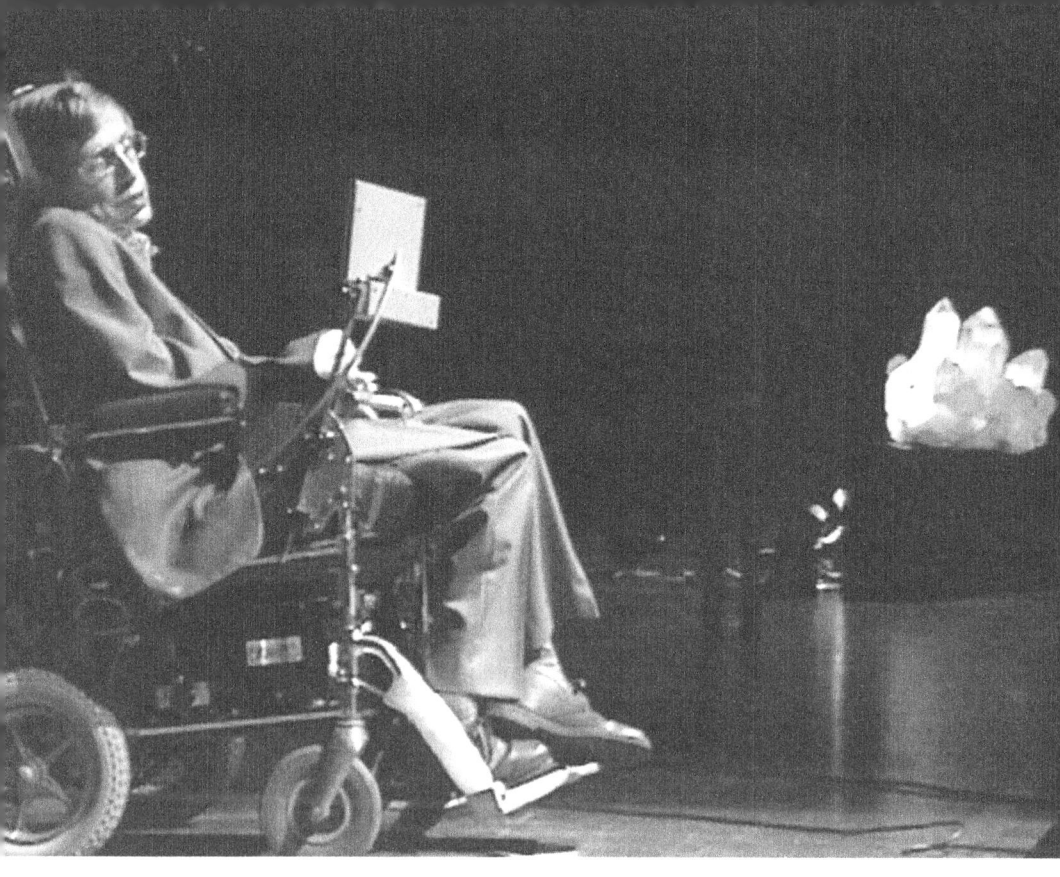

Stephen and ancient quartz

In his work to unify Einstein's relativity with quantum theory Hawking brought the physics at the smallest scale, in quantum mechanics, to the workings of the world at the largest, cosmological scale. Hawking theorized that tiny black holes emitted a type of radiation – now named after him, 'Hawking Radiation' – and would someday explode. "Because it's proportional to one over the mass, it loses energy in this funny way. The smaller it is the hotter it gets and the faster it radiates. So you get this runaway acceleration and eventually the black hole explodes."

In subsequent years, Hawking has noted with irony that he was working to disprove a substantial part of the work that first made his name in science. He came to believe that there was no single Big Bang beginning to the universe. His focus then became to see if he could show whether the collapse of black holes results in the elimination of "information" – meaning the energy and matter taken in by the gravitational sinkhole. "Other physicists don't like the idea," he continued. The information-loss issue became the subject of a famous bet with Hawking and his colleague Kip Thorne and another Cal Tech physicist, John Preskill. Hawking maintained that the information was lost irretrievably, Thorne and Preskill offered that the information reemerged as Hawking Radiation.

Hawking dispelled some of the more fanciful notions connected to black-hole theory. For instance, the idea that they might be a means of traveling through space. He said no one could survive the turmoil inside one. "Black holes might be useful for getting rid of garbage or some of one's friends," he said. "However, they would be a country from which no traveler can ever return."

At the end of the lecture Hawking remarked: "I hope for a day when a theory of the universe is understandable in broad principle by everyone, not just a few scientists. Then we shall all, philosophers, scientists and just ordinary people, be able to take part in the discussion of why it is that we and the universe exist."

After the lecture, as Stephen emerged into the reception in the lobby of the Seattle Opera House, he was mobbed by a group of students from the Seattle School District. These students – some with disabilities – had wanted their own session with Hawking like the one at Seattle University. Meeting him at the reception had become the default alternative. Although I had largely forgotten about this, they certainly had not. So the first fifteen minutes of Stephen's time at the reception was spent with them – about ten feet from where he had entered the lobby.

Then he just broke loose, zipping around the reception, visiting with this group and that – photos and questions first here and then there.

44

Precedents – The Socratic Turn and The Pragmatic Turn

American Pragmatist Charles Sanders Peirce once remarked that when you think you have found something really important to say, if it has not been said before, you are most likely incorrect. In graduate school I expressed a certain frustration to fellow grad student, Scott Borg, that virtually everything we were thinking was most important had been said before. Scott surprised me by responding, 'Of course. All the deepest insights have been formulated in earlier and other civilizations. Our task is to provide a new better articulation, relevant to the current issues of the modern cultural milieu.'

There is a consensus among historians and philosophers of science that what we call Modern Science – beginning with the iconic figures of Copernicus, Galileo, Kepler and Newton – is a re-introduction of the Ancient Science with roots at least 2000 years ago. The blossoming of the Ancient Scientific Research Program is well-represented at the beginning of the Golden Age of Ancient Greece in the 5th and 6th centuries BCE and is identified with the iconic 'first scientists': Thales, Anaximander, Anaxagoras, Pythagoras, Parmenides and Heraclitus, among others.

Modern Science and Ancient Greek Science share the same fundamental commitment to the Scientific Hypothesis – that all phenomena in the universe are governed by One universal order. It should not be too surprising, then, to find that the modern search for a post-scientific superseding More General Theory also has a precedent in Ancient Greece.

The Socratic Turn

There was, indeed, an ancient Paradigm Shift from the detached, determin-istic Scientific Research Program to a proposed superseding More General Participant framework. The Paradigm Shift is often referred to as The Socratic Turn. (By far the best modern presentation of The Socratic Turn, in my opinion, is to be found in *Masters of Greek Thought*, by Professor Robert Bartlett, Boston College (The Great Courses, course 4460)). The most striking matching theme between the ancient and modern Paradigm Shifts is captured in Socrates' reflections on his life. He points to his en-counter with complementarity as a crucial step leading him to make The Turn.

Socrates tells us that as a youth he was drawn to the tradition of scien-tific inquiry into nature.

Phaedo 96a/b "When I was young, Cebes, I had a prodigious desire to know that department of philosophy which is called the investigation of nature; to know the causes of things, and why a thing is and is created or destroyed appeared to me to be a lofty profession; and I was always agitating myself with the consideration of questions such as these."

As he carefully studied the different schools of thought (viz. successful research programs) he was unable to discern any uniquely correct approach to such inquiries. He came to recognize the perennial disputes.

"Those who pride themselves most on their discussion of these points differ from each other, as madmen do… One sect has discovered that Being is one and indivisible. Another that it is infinite in number. If one proclaims that all things are in a continual flux, another replies that nothing can pos-sibly be moved at any time. The theory of the universe as a process of birth and death is met by the counter theory, that nothing ever could be born or ever will die."

In making The Turn – what he literally refers to as his 'second sailing' (viz. second effort at inquiry into the nature of reality) – Socrates resolves to think 'outside' and beyond the Ancient Scientific Research Program. "At last I concluded myself to be utterly and absolutely incapable of [resolving] these [complementary scientific] enquiries."

Socrates recounts his new (second sailing) exploration leading to his Turn to a new *type* of research program. Socrates gradually recognizes that he possesses a certain sort of wisdom, emphasizing that it is a wisdom such as may be attained by any man, distinct from the superhuman wisdom that some suppose they possess.

Socrates tells the story of the impetuous Chaerephon who asked the Oracle at Delphi whether anyone was wiser than Socrates. The Pythian prophetess answered that there was no man wiser. Socrates tells of his perplexity when he heard this. So he began an inquiry, seeking out the reputedly wisest men in politics, poetry, and practical arts, that he might find one wiser and refute the oracle. "And I said to myself, go I must to all who appear to know, and find out the meaning of the oracle."

After one query he remarks: "So I left him, saying to myself, as I went away: Well, although I do not suppose that either of us knows anything really beautiful and good, I am better off than he is, for he knows nothing, [and yet] thinks that he knows. I neither know nor think that I know. In this latter particular, then, I seem to have slightly the advantage of him."

"Is not this ignorance of a disgraceful sort, the ignorance which is the [ideological] conceit that a man knows what he does not know? And in this respect only I believe myself to differ from men in general, and may perhaps claim to be wiser than they are: that whereas I know but little of the world below, I do not suppose that I know."

The Socratic Method

The practical expression of the embrace of complementarity can be seen in Socrates' famous method of questioning. Socrates questions ideologues in such a way so as to lead them to recognize, first, that their ideology is incomplete. Second, Socrates leads them to see that they not only believe what they thought they believed but that they also believe the opposite – its complement. For instance, those who profess to know that reality is universally competitive are led to realize that they also believe – reminiscent of Oakeshott and the soccer game – that reality is also irreducibly cooperative.

In the first stage Socrates' questioning seems to 'defeat' each ideologue's claim to 'objective', 'one right answer' knowledge. The defeated ideologues often imagine at that point that Socrates must have some greater knowledge – some greater, more comprehensive, 'objective' knowledge. How else could he have 'known' that their ideology was incomplete? "And I am called wise, for my hearers always imagine that I myself possess the wisdom which I find wanting in others." When they insist that he reveal this greater knowledge, he responds that he doesn't have any such knowledge.

This second stage of Socratic questioning is often characterized by scholars and commentators as leading his interlocutors to a state of 'aporia', the Greek word for puzzlement. This is when the ideologue realizes that he actually believes both sides of perennial issues. Aporia arises from the enigmatic, often extremely uncomfortable, realization of the complementary character of (all) meaningful beliefs – like the realization in modern science of multiple 'one right answer' objectivities.

At that little café, drinking tea, in the outer terminal of the Eugene Airport, having wrestled for two days with the same questions about competition and cooperation, selfishness and selflessness, Jonathan, the nurses, Hawking and I, had our moment of revelatory aporia – the recognition that there was no 'one right answer', that the question was somehow formally undecidable. The recognition is not simply of one's personal inability to resolve the issue, or even of humanity's inability to resolve the issue. It is the revelation that there 'really' is no resolution – and that this says something about us and about the nature of reality. But what does it say? What's next? Hawking's reflective comment – "I felt for him" – led me forward in search of a more general understanding.

The Socratic method of questioning reminds me of my own dialogues with friends and colleagues on both the political right and left, where I was accused of being a right-wing capitalist by leftists and a left-wing socialist by right-wingers. Since I have come to appreciate the more general middle-ground position, my self-conscious aim in introductory dialogues is now to lead my interlocutors to aporia – to an appreciation of an intermediate, dynamically balancing, middle-ground, political position, and then, hopefully, toward an appreciation of the Parliamentary Attitude.

The embrace of complementarity in quantum theory – and arguably in Relativity – brought about a state of aporia in the modern Scientific Research Program. The embrace of complementarity in the rebels' challenge to the Positivist philosophy of science brought about, in parallel, a state of aporia in modern philosophy of science. Niels Bohr expressed the new sense of aporia in both science and philosophy, "The opposite of a correct statement is a false statement. But the opposite of a profound truth may well be another profound truth." In particular situations one or the other 'reading' of reality might be clearly the best, the most appropriate, allowing one to make 'a correct (local) statement' about a competitive aspect. But the competitive and cooperative 'readings' of reality remain complementary 'profound truth(s)' – locally competitive, while globally compatible.

The aporia arising from the embrace of complementarity is not an end state but a stimulus to seek a More General Understanding, a stimulus to make the Turn to a new Middle Way, to a new Middle Way Research and Development Program. The aporia arising from the embrace of complementarity is a stimulus to seek a new, more general *type* of questioning, to seek a new *type* of understanding of our natural questioning, to seek a new, self-referentially coherent representation of inquiry, to seek a new self-referentially coherent representation of the nature of reality and our place in it.

The actual undecidability of the perennial controversies about the objectivity of either competition or cooperation allows us to make sense of the failure to arrive at any 'objective' resolution in the Avinash Dialogues. It also provides clues to understand the phenomenon of lateral conversions – the flip from one belief system framework (and its ideological attitude) to the, equally profound, complementary opposite.

The Socratic program might be understood as bringing people to understand Oakeshott's political theme that all real social systems have (and should have) both competitive and cooperative organizational and operational components.

Xenophon's Recollection of the Turn

Besides Plato, Socrates' other commentator, and close friend, was Xenophon, a celebrated Greek solider, statesman, historian and philosopher. Xenophon tells of the very day that Socrates made The Turn.

Xenophon, in his essay, *Oeconomica*, tells of Socrates making The Turn to a new way of philosophizing – to begin a new *type* of inquiry. The conversation that day was about the management of the small farming estate. ('Oeconomica' translates as 'the skilled household manager.') The Socratic Turn is from asking, as if from a detached position, 'how the world works', to asking, from a participant position, 'how to work in the world' – 'how to make the world work better'. The new type of question, the new understanding of inquiry is about 'how we should live' – about how to bring about a more desirable future.

The Socratic Turn is to this new *type* of questioning and, concomitantly, to a new *type* of understanding of the importance and meaningfulness of questioning itself. The Socratic Turn is to this new *type* of understanding of inquiry, learning and knowledge, relevant to an exploratory Participant Research and Development Program in a developing reality.

By contrast, in the detached Spectator representation, the notion of real, meaningful inquiry cannot possibly be made sense of. The real, practical value of inquiry cannot be made sense of in a fully deterministic space-time invariant representation of the nature of reality.

There is additional support of the hypothesis that the ancient Socratic Turn is an earlier version of the modern Participant Paradigm Shift. Socrates explicitly addresses the 'detached' inquirers ('speculators'), recognizable as modern advocates of 'pure/basic' research, who tell us that they have no 'official', 'overt' interest in or expectation of any practical benefit from their sought-after knowledge of how the universe works.

Xenophon recalls how Socrates addressed these 'detached' inquirers in an ironic tone. According to Xenophon: "Socrates' questioning on the merits of these speculators sometimes took another form. The student of human learning expects, Socrates said, to make something of his studies for the benefit of himself or others, as he likes. Do these [detached] explorers into divine operations hope that when they have discovered by what forces the various phenomena occur, they will create winds and waters at will and fruitful seasons? Will they manipulate these and the like to suit their needs? Or has no such notion perhaps ever entered their heads, and will they be content simply to know how such things come into existence?" (*Memoribilia* (61))

With the embrace of complementarity a qualitatively new *type* of open-ended questioning becomes meaningful: How should we live? How should we work in the world? How might we make the world work better? These are not meaningful questions in a fully deterministic, mechanical, time-space invariant, clockwork reality.

Socrates suggests that the most important questions necessarily involve an irreducible evaluative component – how we *should* choose, how we *should* act, how we *should* inquire, how we *should* attempt to progressively develop reality, to bring about a more desirable, a more valuable future. The initial 'household management' context of these most important questions naturally expands to concerns as to how (according to what methods) we should manage our lives, how we should develop ourselves, our society and our world. The questions concerning household management naturally scale up to questions of how to better govern a city-state, addressed in Plato's famous dialogue, *The Republic*. There is a notable thematic continuity to the new type of developmental questioning – scaling up to modern concerns. As Participants, we are naturally concerned with the Earth's biosphere. As Participants, embedded, embodied parts of the biosphere, we are naturally concerned with the household management of this our 'not so small', farming estate. The self-inclusive Participant questions of how to manage a socio-politico-economic system in *The Republic* are about how to manage, how to organize, how to design our lives. These are engineering questions in the modern context, these are questions about how we should live; these are problems of design.

Of crucial importance in understanding the modern manifestations of the Socratic Turn is that the new context of inquiry, of research and development is necessarily experimental and exploratory. Participant inquiry and action, by their very nature, have an open-ended potential for improving how reality – including us – works.

The Socratic Turn supersedes scientific knowledge, incorporating scientific knowledge, albeit understanding it in a new way, incorporating it in a new More General Research and Development Program.

Xenophon notes Socrates saying: "Doubtless, skill in carpentering, building, smithying, farming, of the art of governing men, together with the theory of these processes, and the sciences of arithmetic, economy, strategy, are affairs of study, and within the grasp of human intelligence." *Memoribilia* (30)

Xenophon recalls Socrates famously ironic response to a question of whether his new more general understanding rejected the value of practical knowledge, "as though a man should inquire, "Am I to choose an expert driver as my coachman, or one who has never handled the reins?" "Shall I appoint a mariner to be skipper of my vessel or a landsman?" And so with respect to all we may know by numbering, weighing, and measuring. To seek advice from Heaven on such matters was [to Socrates] a sort of profanity." *Memoribilia* (41)

Xenophon recalls Socrates saying: "Our duty is plain…where we are permitted to work through our natural faculties, there let us by all means apply them. But in things which are hidden [to be discovered], let us seek to gain knowledge from above, by divination; for the gods," he added, "grant signs to those to whom they will be gracious." *Memoribilia* (41) Plato's dialogue, *Ion*, explores in more detail the sort of divine rhapsody involved in creativity and innovation.

A modern expression of similar guidance for exploration and experimentation in the face of irreducible uncertainty comes from *The Little Prince* (Antoine de Saint Exupery, Chapter XXI): "Here is my secret. It is very simple: It is only with the heart that one can see rightly. What is essential is invisible to the eye."

As to the nature of the essential dialogue involved in the new type of questioning, Xenophon tells us of Socrates: "He himself never wearied of discussing human topics. What is piety [virtue and goodness]? What is impiety? What is the beautiful? What the ugly? What the noble? What the base? What are meant by just and unjust? What by sobriety and madness? What by courage and cowardice? What is a state? What is a statesman? What is a ruler over men? What is a ruling character? and other like problems, the knowledge of which, as he put it, conferred a patent of nobility on the possessor, whereas those who lacked the knowledge might deservedly be stigmatized as slaves."

The 'slaves' reference could be understood in the modern existentialist literature as the 'inauthentic', those who pretend to themselves that they do not have freewill and so are not responsible for their choices (viz. often supposing they are victims of their genes, personal history and external forces).

The 'enlightened' dialogue on questions of justice and beauty is an essential methodological component in the creative exploration and development of the Middle Way.

The Socratic Turn doesn't reject but moves beyond the diversity of scientific knowledge and the practical arts (how to do things), subsuming and superseding, understanding them in a new way, making sense of them in a More General Participant context. The Socratic Turn supersedes the scientific search for a universal, participant-independent, objective theory, supersedes the scientific 'one right description' of reality. Concomitantly, the Socratic Turn is a move beyond the presumed Religious Research Program that is searching for the ideological 'one right value system', for a universal, time-space invariant, moral-actor-independent, 'one right prescription' for how we should live. The Socratic Turn doesn't reject prior advances in moral understanding, in the understanding of what is valuable, but subsumes and supersedes them in an ongoing open-ended exploratory enterprise to discover and bring about a better, a more desirable future.

The Socratic Turn from a Spectator to a Participant perspective provides a renewed understanding of our natural role in the evolution of human society, our natural role in the evolution of the Earth's biosphere and our natural role in the evolution of the cosmos as a whole.

The Pragmatic Turn

The Socratic Turn has a modern correlate referred to as The Pragmatic Turn. American Pragmatism was founded by Charles Sanders Peirce, and further developed by William James, Josiah Royce, John Dewey and many others. The inspiration for pragmatism can be traced back to the 18th century German Philosophical Enlightenment associated with the iconic figures of Immanuel Kant, Johann Fichte, Friedrich Schelling and George Hegel. According to James, the pragmatic way of thinking begins with the acceptance that we have freewill. In his famous essay, "Pragmatism: A New Name for Some Old Ways of Thinking," James embraces complementarity in all choices and ways of reasoning – and then moves toward the Middle Way. Another distinguishing feature of American Pragmatism is the presupposition that the universe is evolving – maturing. Pragmatism is not an attempt to *describe* reality simply in terms of a better superseding 'theory' as if it were within a Spectator Research Program. Rather, The Pragmatic Turn involves a Paradigm Shift – a Problem Shift – to a superseding, developing More General Participant enterprise concerned with 'method', concerned with *how* we should live. Dewey argues that 'an advance in knowledge is an advance in method.' Pragmatism is a Participatory, methodologically Middle Way representation of inquiry, understood in the context of the more general question of how we should live, of how to make the world, including ourselves, work better.

Both the Socratic Turn and the Pragmatic Turn emerge, through hindsight, from a critique of the Spectator's fixed objectivist, observer/actor-independent framework leading to a point of aporia. Pragmatism is an early modern exploration attempting to better articulate a post-scientific More General Theory. I have argued elsewhere that Pragmatism is a Philosophy of Engineering (cf. terrybristol.org). Like the engineer, the pragmatist is a Participant problem solver attempting to move from the current state of affairs to a more desirable future state of affairs – to a world, self-referentially including ourselves, that is better, that works better.

In the Pragmatist's Participant framework learning is about the creative discovery and implementation of innovative new ideas, about new ways (methods) of doing things, always in the context of seeking to bring about a better future. Pragmatic inquiry seeks knowledge of how to actualize and develop both value and potential value. In order for the Pragmatist program of inquiry to be successful it must be possible to improve the nature of reality, the organizational structure and function of reality. The Pragmatist representation of the Participant's Research and Development Program views successful inquiry as cumulative, qualitatively progressive, self-enabling and conceptually emergent.

In terms of The Parallel Hypothesis the self-inclusive Pragmatic representation of embodied inquiry and innovative methods points to a similarly progressively developing reality that learns recursively – boot-strapping. Yet the direction of learning and development is inherently under-determined by the present and open-ended. There can be no fixed, pre-conceivable final end-point, no final specifiable 'objective' structure or way of operating, way of living. The 'direction of development' is both constrained and qualitatively and conceptually emergent – by its very nature. Like the Taoist theme, the real path forward cannot be pre-conceived or pre-specified, can't be 'stated' – it emerges creatively through the Middle Way.

There is a tacit realism suggested in the Participant framework. The human enterprise is, by its very nature, 'a part of' One, unfolding, universal enterprise seeking to actualize value, seeking to bring about a better universe. The embodied enterprise – the nature of reality including us – is seeking – by its very nature – to bring about a reality that progressively increases its ability to make reality (itself) better.

45

'Essentially Contested Concepts' and Carnot's Epiphany

Back in Berkeley, in Sproul Plaza during the Free Speech Movement, Joan Baez had offered our rebellious justice-seeking enterprise, with its growing animosity to our opponents, the simple wisdom: "Do this thing with love in your hearts." On reflection, over the years, I have come to understand this 'very Berkeley wisdom in the 1960s' as an expression of the Parliamentary Attitude. The Regents of the University of California were to be treated as our loyal opposition. Baez was leading us away from ideological confrontation toward the puzzling middle ground – to an aporia. Love your enemy. The path forward, to a more desirable future, was to be discovered through a creative dialogic middle way.

The wisdom of the Socratic and Pragmatic Turns is not the result of any greater 'one right answer' knowledge, either theoretical or practical. In recognizing that he possessed a certain sort of wisdom, Socrates had emphasized that it is a wisdom such as may be attained by anyone. What the Socratic wisdom offers is a methodological strategy about how to proceed once you have recognized the complementarity of the perennial disputes. Pragmatist W.B. Gallie, in his book, *Peirce and Pragmatism*, pointed to the fundamental status of the perennial oppositions – what I have called complementary concepts. Gallie brilliantly characterizes these as 'essentially contested concepts.' William Connolly, a commentator on Gallie's thesis, remarked that it is only when people realize that they are dealing with inherently unresolvable perennial issues, with essentially contested concepts, that 'enlightened dialogue' can begin – the middle way. What is justice? What is virtue? What is beauty? What is reality?

When I came to Berkeley to major in Astronomy, one of my formative concerns arose from my encounter with the litany of the Scientific Research Program. I was genuinely surprised to find an enforced resistance to critical self-reflection. The universal validity of the presuppositions defining the Scientific Research Program were not to be questioned. Only much later, on reflection, I realized that such 'outside the box' questions literally could not be addressed from *within* the Scientific Research Program. Such critically self-reflective questions were scientifically nonsensical, could not be made sense of in the conceptual framework defined by the Scientific Hypothesis. Eventually, I realized that these formative concerns applied to the un-self-critical litanies of all 'one right answer' ideologies. Popper's original concern with the Marxist ideologues had to do with their resistance to self-critical reflection. Popper had hoped that sensitivity to the 'scientific evidence' would allow us to distinguish between pseudo-scientific ideologues and how he imagined 'real', supposedly self-critical, science worked. The Surprising Answer to Popper's Question was that the 'evidence' for the limit of each ideology couldn't be seen, couldn't be understood from *within* the conceptual framework of the ideology. The critical evidence demonstrating the limit of an ideology can only come from outside, from the recognition of the success of a complementary perspective.

Only with the embrace of complementarity and the entailments of the Parliamentary Attitude does a self-referentially critical understanding of 'real', genuinely progressive learning make sense. The Parliamentary Attitude accepts that other points of view that don't make sense in terms of your current understanding have, nonetheless, captured an essential aspect of the understanding of reality. It is because reality is not conceptually uniform that progressive, emergent learning is even possible. In the Island thought experiment, if you put a group of Republicans (or Democrats) on an island they naturally separate over time into Republicans and Democrats. However, if one of these groups, seeking 'rational' uniformity, subsequently manages to eliminate the opposition, the complementary type, they undermine the potential of the island community to learn. The result would be what I refer to as the iterative, self-defeating, self-destructive Closing Phenomena. It is not surprising to find that most wisdom traditions counsel the value of your 'apparent' opponents. It is by working with your opponents – taking the path of the (win-win) middle way – that 'real', otherwise unpredictable types of advances are realized. Without self-referentially critical feedback 'real' learning is not possible. The very idea of learning is not self-referentially coherent in a logico-mathematically or mechanically uniform reality. Ideologies – by their very nature – presuppose a uniformity of reality, tacitly suppose that they have the 'one right way' to understand reality. They don't need to listen to or tolerate other views. Complementarity leads us to an understanding of reality that can make sense of questioning, inquiry and the embodied development of a better reality.

The More General Theory cannot be understood as a new superseding objectivity. It cannot be understood as a new superseding ideology. There is no 'one right way' to understand reality, in classical scientific terms, as a uniform 'objective' reality, 'out there'. However, the More General Theory does allow us to say something about reality. The More General Theory leads to a new *type* of realism. Since the paradigm shift is also a problem shift, a shift to a new more general type of problem, to a new more general type of question, to a new more general understanding of embodied inquiry, what can be said and how it is said are different from, subsuming and superseding, the classical 'scientific' descriptions. The context defined by the more general understanding of inquiry is about 'how we should live'. In the Participant engineering interpretation this question is embodied in the naturally self-critical Research and Development Program.

Instead of offering us a new scientific answer to the classical Spectator question as to the nature of objective reality 'out there', the embrace of complementarity presents us with an unresolvable problem, an unresolvable uncertainty. What first appears *within* the Spectator enterprise as this 'unresolvable problem' is newly, better understood in the superseding More

General Theory as an open-ended opportunity to problem solve, to actualize value, to bring about a more desirable future.

In the Spectator representation of inquiry learning should have been understandable as logico-mathematical, something that could be turned over to mechanical computers. In an extreme, Laplacean-like interpretation one should have been able to logico-mathematically reason future theories from current or past theories. In the Spectator representation of inquiry 'learning' was a convergence from an 'unexplained' state of ignorance to a complete and consistent understanding of the time-space invariant 'clockwork' nature of reality. In contrast, in the rebel's Participant representation, 'real' learning is inherently problematic, requiring discovery through a searching, experimental and exploratory strategy. In the Participant representation, 'real' learning is an embodied emergence, a qualitatively unfolding Research and Development enterprise. In the Participant representation advances in innovative methods (viz. Participant knowledge) are, concomitantly, advances in (new types of) ignorance, where ignorance is thought of in terms of meaningful questions. New learning generates new questions, new types of questions. Innovative new methods (viz. ways of observing and acting) constitute new freedoms, enabling novel, experimental explorations. Ignorance doesn't decline to certainty and choice doesn't disappear – they both develop and expand.

In the debate over the nature of scientific method, Paul Feyerabend argued that there simply wasn't One, universal, scientific method – one way to learn. In the Participant Paradigm there are distinct research programs – diverse ways of living and learning. And how we learn changes as we live. Learning develops, progressively opening qualitatively diverse opportunities for further experimental exploration. There isn't just one 'conceivable' path ('one right path') of learning and development. The design of an inquiring system should embrace controversy and should even encourage the loyal opposition. Feyerabend argued that a society's 'official' method of inquiry should be qualitatively diverse, embodying and encouraging an irreducible element of logico-mathematical, scientific anarchy. Lakatos, characterizing himself oppositely as a scientific fascist countered, insisting on the enforcement of standards of practice. I had opportunity to ask both Feyerabend and Lakatos, who were close friends and colleagues, about the reasonableness of their extreme positions. Feyerabend said that 'of course' his anarchy was 'enlightened' enough to allow for periods of fascism, and Lakatos said that 'of course' his fascism was 'enlightened' enough to allow for periods of anarchy. Lakatos emphasized that in his occasional embrace of Feyerabend's anarchy he would still need 'to be empowered to distinguish between the geniuses and the cranks.' Feyerabend reluctantly acknowledged the need to distinguish genius from crank but insisted that history showed that the established 'scientific (fascist) authority' was ill-equipped to make that judgment. This deliberately enigmatic embrace, by Feyerabend and Lakatos, of mutually interpenetrating, incommensurable opposites, was a sort of ongoing theatrical presentation, illustrating both their loyal opposition and still problematic nature of that opposition. They were pointing to a problematic character of the middle ground while at the same time – illustrated in their friendship – the path of the middle way. As I understood it, this oppositional drama was to be the central 'enlightened' message of their planned joint publication – scuttled by Lakatos's untimely premature death.

There is much to be said and much more to be discovered about the new realism, about this new post-scientific, post-objective realism. However, except for the following brief remarks a more in-depth treatment of the realism of the More General Participant perspective must await a future publication (see terrybristol.org for updates).

In one of his memorable, initially puzzling, insights Lakatos pointed at the 'naturalness' of inquiry: 'scientists don't need a theory of science in order to do science anymore than fish need a theory of hydrodynamics in order to swim.' After several years of reflections on this, I came up with a simple extension: children don't need a theory of how to ask questions in order to ask questions anymore than fish need a theory of hydrodynamics in order to swim. 'Why Daddy?' 'Yes, but why?' 'Yes, but why?'… We don't teach children to ask questions. Questioning – inquiry – is in our Participant natures. Inquiry is an irreducible aspect of the recursively developing nature of reality.

The nature of reality from the perspective of the Participant is the same as the nature of the Participant. In the *Timaeus* Plato offers the evolution of 'the eye' as an exemplar of the types of things – structures and functions – that 'we', the Imperfect Master Craftsmen, have brought forth.

Systems Nature of The Participant Perspective in Reality

Royce's Criterion of Self-Referential Coherence argued that whatever understanding of reality that you come up with must be able to make sense of itself – must be able to make sense of the problem solving involved in coming up with that understanding. Royce's argument entails that 'problem solving' (viz. natural questioning) must be an irreducible aspect of any self-referentially coherent understanding of the nature of reality. The Spectator representation tacitly presents us with One, idealized Spectator learning about a unified objective reality governed by One, universal, time-space invariant order. Royce's argument naturally suggests reflexively that there are 'other' problem solvers in reality. In the Participant representation the embodied Participant inquirer is embedded in a world of diverse Participant inquirers (viz. 'parts' of the Imperfect Master Craftsman). In the Participant representation reality is not one evolving learner learning about another fixed learner. And it is not simply that the Participant is attempting to learn about other Participants that are attempting to learn about him. In the Participant's realty we don't learn in detached isolation by ourselves or in relation to a universally fixed reality. Per hypothesis, the Participant reality is a sort of 'web of learning' – progressively learning, progressively developing. A representative image is that of the evolving biosphere, composed of partially autonomous learners and learning sub-systems. Fritjof Capra suggested this 'web of life' metaphor to supersede the classical clockwork in his book, *The Web of Life: A New Scientific Understanding of Living Systems (1997)*. More recently in his book, *The Systems View of Life: A Unifying Vision,* Capra has further developed his argument that the Systems View of reality properly subsumes and supersedes the inherently limited Mechanical View.

The Systems perspective also serves to answer a common critical query about the engineering agenda. As represented here and in the *Timaeus*, it seems as if seeking the 'good' in engineering is only for the human good, to the possible detriment of other 'parts' of society, the biosphere and the cosmos. However, all real problems for all Participants are systems problems about making the system better – and the most general system is the universe itself. Because of the ubiquity of complementarity and the quantum choice, individuals, human society, the biosphere and the cosmos all have the same 'One' common agenda, one common historical narrative. Although it cannot be pre-conceived, predicted or spoken the common agenda is about bringing about a better future.

University of Exeter Philosopher of Biology John Dupre, countering the widespread neo-Darwinian mantra that humans are no 'better' than other animals, plants or house flies, suggests that humans are, indeed, the single most advanced creation so far in the history of Earth's evolution. 'Advanced' here might be understood as most potentially beneficial in the context of the Participant's evolutionary agenda. Humans are like concentrated bundles of (imperfectly intelligent) freedom, the most enabled, the most empowered actors in the ongoing global engineering enterprise. Another partial response to the critical query about the 'direction' of the evolving engineering agenda is George Bugliarello's inspired insistence on the narrative continuity of modern human engineering with the previous 3.7B years of biological evolution. 'Modern engineers should be taught that their activities are a natural, narrative, extension and expansion of biological evolution.'

Perhaps the single most definitive characteristic that distinguishes the dynamic under-determined, evolving Participant representation of reality from the fully determined classical clockwork representation of reality, is *feedback*. The feedback characteristic of Participant reality is a *natural*, holistic self-reflexive feedback. This cannot be properly understood as analogous to simple feedback mechanisms like those that control household temperature. The feedback of the Participant reality is due to the dynamic, progressive interplay of diverse complementary processes. These must somehow systemically balance through an ongoing, under-determined – 'critical' – feedback, thereby serving to keep the system on the constructive path of the middle way. The under-determined 'direction' of evolution is constrained by the ongoing need to maintain a 'reasonable' balance. By analogy what I can do in the world is constrained by my need to maintain various homeostatic properties of my metabolic system. The homeostasis enables, but under-determines, my actions. Per hypothesis, the dynamic, interpenetrating interplay of complementary processes is the middle way dance of the Taoist yin and yang.

'Real' learning involves the discovery and practical implementation of novel, mutually beneficial, win-win middle ground relationships, ways (methods) of living – arrived at through the middle way. Evolution is not a convergent adaptation but is a recursively enabling constructive emergence.

Carnot's Epiphany

The Participant cannot be understood as a fully determined cog in a fixed mechanical clockwork universe. The Participant cannot be understood as an engineer operating in a fully determined, mechanical clockwork universe. The expression of the engineer's self-referentially inclusive understanding of reality comes from Sadi Carnot and his father, Lazere Carnot, both associated with the first modern Engineering University, the École Polytechnique in Paris. Lazere Carnot was one of the founders of the university during the French Revolution in 1794. Based on comments by Sadi Carnot, what I have come to refer to as Carnot's Epiphany – is that 'we are engineers in a world of engineering.'

The embrace of complementarity has led us, per hypothesis, to an understanding of how a holistic system like the universe can develop and emerge self-critically. The Participant-engineer, seeking to make the world work better, is embedded and embodied in a universe that is seeking to make itself work better. And although constrained, since what is 'better' emerges, there is no script, no mechanical plan, no pre-conceivable plan, no ideologically expressible plan. The path forward isn't universally left or universally right, or universally anything pre-specifiable. Real learning – what is 'better' – can't be understood in terms of any universal, 'one right answer', ideological conception of the rational or the good.

46

Hawking's Heart and Peirce's
Evolutionary Love

So who is the real Stephen Hawking? I have argued that Stephen Hawking is better understood, more fully understood, as a Participant in a developing universe. Indeed, since the Spectator representation of scientific inquiry presupposes a fully deterministic universe, the understanding of Stephen Hawking as a scientist can only actually be made sense of as a limited, idealized representation from within a More General Participant understanding of reality. The understanding of Hawking, as classical scientist, as a detached Spectator, seeking to discover the universal deterministic laws governing the universe, is subsumed and superseded by the new more general understanding of Stephen Hawking the Participant, seeking to bring about a more desirable future. Hawking, of course, in this telling, is symbolic. The question of who is the real Stephen Hawking scales to the question of who is the real scientist. I, for one, certainly thought of myself as a 'scientist' seeking the universal laws, the One order, governing all phenomena. And I was notably shocked when I came to the conclusion that I couldn't understand my actions as doing that because the new physics had demonstrated that there wasn't One order governing all phenomena. I couldn't have been doing that even though that was what I thought I was doing – albeit largely unreflectively, uncritically. It was the embrace of complementarity that first led me to the beginning of my aporia – my puzzlement. This puzzlement has also been the tacit condition of 20th-21st century Modern Science. The realization that comes to awareness – perhaps only slowly and begrudgingly to all ideologues – is that there isn't just 'one right answer'. There isn't just one rational path of inquiry and development. It is this aporia – about how we should inquire, about how we should act, about how we should live – that has stimulated the search for a post-scientific, post-ideological More General Theory.

What remains is to try to capture a sense of the value system that Hawking embodies. Stephen's role as physicist is transparent to the public. However, what I have seen, while hosting him on these visits, is less commonly appreciated. I now hasten to correct the occasional, uninformed fan who suggests that Stephen must lead the life of a detached, cloistered, Spectator physicist. On the contrary, without qualification, I have never met anyone who leads such an active life, appreciative, certainly, of his science but also of the aesthetic, in art, music, fine food and time with friends and family. ("I don't have time to do all the things I can do, so it doesn't seem important to worry about the things I can't do.")

I think Hawking understands himself most fully as a Participant. Despite his leadership in a scientific community, often associated with determinism, Hawking answers the question about determinism by saying: "I have noticed that even people who claim everything is predetermined and that we can do nothing to change it, look before they cross the road."

In his recent book, *The Grand Design*, Hawking eschews ambitions to arrive at a scientific Theory of Everything, defaulting to the still scientifically enigmatic 'model-based realism' – whatever models work to some extent capture to some extent some aspect of reality. The value of a model is solely in its use.

The single word – 'Don't' – on the Oregon Coast was a moral act. One can argue from a detached Spectator perspective that it was a selfless act of charity, or that somehow it was a selfish act – perhaps to assure him of a prompt lunch. But such interpretations are ideological stretches, ideological rationalizations. Stephen's later reflection on his motive – 'I felt for him' – was more revealing and much more challenging to understand 'scientifically'. The idea of empathy is a start: the ability to understand someone else's feelings or difficulties. Sympathy is closer: the ability to enter into, to understand, or share somebody else's situation and feelings. Being the philosopher, let me offer an understanding of empathy and sympathy that relates to the broader themes of this essay. Aristotle, with his characteristically profound insight, offered a definition of friendship: 'a friend is like a second self.' Anecdotally, if you found your best friend's wallet/purse on the street, full of cash, you wouldn't even think of not returning it, or of taking out the cash for yourself before returning it. You naturally presuppose a common interest, a common agenda. You understand your friend's success as your success. Hawking was treating the waiter as he would treat a friend.

Who was this waiter at the restaurant at the Inn at Spanish Head? Who was he to Hawking? Most likely they will never interact directly again. And the waiter never knew what Hawking had done in encouraging us to rethink our retributive justice agenda. Furthermore, an easily overlooked aspect of Hawking's 'Don't' was its long-term impact on the rest of us at the table. Stephen was tacitly offering himself as a role model. And with this telling – here – there will be an impact on a wider scale.

What was revealed on the Oregon Coast was something quite general about Hawking's attitude to the world and about Hawking's more general understanding of his place in reality, in a universe with an irreducible moral aspect.

Hawking's meetings with the young students with disabilities further reveals the breadth of his moral attitude. He understood his special, unique opportunity to make a positive, progressive difference in their lives – and, expanding, in all the lives they touched. He took time from his scientific work and his family and his closest friends to try to move the universe from its current state of affairs to a more desirable future state of affairs. Stephen's response to the student's question at Science World in Vancouver, B.C., about how he was able to relate to others when he was so disabled, was enlightening. Stephen's answer: 'Everyone has disabilities. Some are just more obvious than others.' His moral attitude is completely inclusive – compassionate. His efforts weren't just for people with 'obvious' disabilities. They were for all of us for all time. Stephen Hawking tries to serve as best he can as a moral role model, as a moral role model for everyone. Stephen sees himself in others and others in himself. I am sorry that in Vancouver he wasn't able to connect with the Dali Lama, the more commonly recognized master of universal compassion.

Such leadership considerations of compassion and morality make no sense in a fully deterministic framework. In the Participant Paradigm every choice you make, every inquiry you undertake, every problem you try to solve, every activity in which you engage has irreducible moral content.

Throughout the period covered by the four-city lecture tour, Hawking's answer to the inevitable 'God question' was always the same: "It's not for me to say." There was always an audible 'rise' from the audience with this answer. 'Appropriate' – coming from a scientist, coming from someone representing the Scientific Research Program. That was my interpretation. However, Hawking's outwardly respectful and sympathetic expressions have changed over the years. His rejection of formal religious institutions and their litany has become quite explicit. Hawking, I believe, wants to encourage people 'to think for themselves.'

Hawking's life is such an inspiration to so many, and the man I encountered was so exceptional in both his attitude and his actions; the question of how one might characterize his spiritual worldview in a way that might illuminate his 'strength' and 'resolve' remains.

There is a poem that I heard recited a number of times in my youth, whenever my father was asked about his religious beliefs. I don't presume to put these words into Hawking's mouth. I have never asked him about the poem. However, somewhat surreptitiously I did ask two of Hawking's closest long-term nurses if they agreed with me – and they did – that the poem at least begins to capture Hawking's heart.

Abou Ben Adhem (may his tribe increase!)
Awoke one night from a deep dream of peace,
And saw, within the moonlight in his room,
Making it rich, and like a lily in bloom,
An angel writing in a book of gold:—

Exceeding peace had made Ben Adhem bold,
And to the Presence in the room he said
"What writest thou?"—The vision raised its head,
And with a look made of all sweet accord,
Answered "The names of those who love the Lord."
"And is mine one?" said Abou. "Nay, not so,"
Replied the angel. Abou spoke more low,
But cheerly still, and said "I pray thee, then,
Write me as one that loves his fellow men."
The angel wrote, and vanished. The next night
It came again with a great wakening light,
And showed the names whom love of God had blessed,
And lo! Ben Adhem's name led all the rest.

James Henry Leigh Hunt (1784-1859)

Niels Bohr tells of losing his Christian faith and of a subsequent conversation with his father. Bohr had tried hard to believe in salvation and to understand what that might mean. But he lost the struggle and 'realized with complete conviction that the truths of Christian dogma were not true.' When he confessed this to his atheist Lutheran father 'he received a sage smile.' Bohr writes to Margrethe, his soon to be wife, "That smile… told me a lot. My courage roared so wildly, wildly, for I knew then that I too could think." Berkeley Professor of History, John Heilbron, suggests 'the approving smile of the man [Bohr] admired most in the world taught him that he belonged among the few who could reason their way free from the standard beliefs of their class and culture, place and time.' Heilbron goes on: 'He [Bohr] would repeat the performance when recognizing that ordinary mechanics represented the truth of the microworld no better than conventional religious belief accorded with the meaning of life.'

Heilbron goes on: "The primary payoff of his engagement with quantum physics for his wider philosophy was the discovery that multiple truths come… in complementary pairs." Citing newly available correspondence with his fiancée, Margrethe Norlund, Heilbron notes that Bohr discusses the different truths expressed in sermons, great works of literature, and science. Bohr wrote, in the spirit of the new superseding merger: "It's something I feel very strongly about, I can almost call it my religion, that I think that everything that is of value is true."

I don't know if I am really surprised by the implication of the More General Theory that Hawking and I and all purported scientists are really engaged in what is better understood in this more general perspective as an emergent, value-actualizing enterprise. Bohr's notion that 'everything of value is true' does seem to makes sense in the More General Participant understanding. Reality, as currently manifest, is the result of the Participant's (viz. Imperfect Master Craftsmen's) creative, historically cumulative Research and Development Program. Reality is the result of the discovery and implementation of (better) innovative ways of living (methods). The Participant's recursively enabling problem solving advances reality by creating value – progressively unfolding a more desirable future. Creative problem solving is 'really' novel value actualization. Similarly, in pragmatism, truth isn't something that is fixed and discovered. What is true has been created, has creatively emerged. The *discovery* of a novel truth is concomitantly *a creative innovation*, requiring a new, better way of observing, involving a new better way of acting; implementing a new better way of living. As Wheeler suggested in his Participatory Anthropic Principle, we create the future. New truths, new values, are creatively discovered through middle way experimentation and exploration.

One implication of the Participant Paradigm is that the best design for a socio-economic learning system is the best design for a moral learning system. Paul Romer's Paradigm Shift to New Growth Economics, subsuming and superseding scientific (supply-demand equilibrium) economics, recognizes that the path to a better socio-economic system is through the discovery and implementation of new, better ideas – new, better innovative methods, new better ways of living (viz. 'tools and rules'). Romer emphasizes the global benefits of what begins as local innovation. The computer is an easily understood example. Although 'I didn't do anything' to directly contribute to the innovations, I benefited from the many innovations of computer technology. The moral lesson is that it is 'in my interest that others discover and develop new ideas.' It is 'in my interest that others succeed.' Reciprocally, 'it is in the interests of others that I succeed.' Hawking 'didn't do anything' directly to invent or develop the specific engineering advances that have allowed him to survive and thrive.

Romer's New Growth Economics understanding of economics as a natural web of innovation constitutes a new way of understanding, subsuming and superseding the perennial controversy between competitive and cooperative ideologies. Speculating, the global inclusiveness of balancing reciprocity of oppositions, wherein 'it is in my interest that you (we) succeed', reminds me of the classical moral wisdom of the Golden Rule and Immanuel Kant's Categorical Imperative. These are methodological, organizing principles – engineering design principles – concerned with how we should live. But like the path of the Tao 'that cannot be spoken', they don't offer a globally pre-conceivable course of action for specific situations. And yet, they do suggest a defining constraint on the 'direction' toward a balanced, progressive emergence, toward a non-ideological middle way. Similarly, the Ancient Greeks pointed to their most fundamental insight as 'Nothing too much. Everything in its measure.' Speculating again, along the same lines, the lesson of the Parliamentary Attitude is to seek new, progressive relationships with 'apparent' opponents, with others with discontinuous perspectives, through the formation of new, innovative relationships – ways of living.

Earlier explorers of the Participant Paradigm, from Plato and Socrates to Kant, Hegel and Dewey (and many others), speculated that a more comprehensive understanding of the 'value aspect' necessarily involves the aesthetic – 'the good is beautiful.' Kant, in his third critique, *The Critique of Judgment*, argues that decisions of practical reason necessarily involve an *aesthetic judgment* – consideration of the beauty of the solution. The experimental exploration of how to bring about a more desirable world is concomitantly an experimental exploration of aesthetics, of the beautiful. The engineer's problem of design is – by its very nature – a problem of how to bring about a more beautiful design. In his TED-talk, legendary Silicon

Valley product design engineer Donald Norman, reflected the problem of design. He has now realized in his later years that the most important characteristic of a good design is its beauty.

The Participant Paradigm, embracing complementarity, accepts the ubiquity of uncertainty. Choice is understood in the Participant engineering context as having an irreducible exploratory and experimental aspect, as being under-determined by one's nature, nurture or circumstance – as naturally free. The Spectator's concept of ignorance is subsumed and superseded, understood in a new way, in the Participant perspective. Our 'ignorance' is newly understood in terms of our freedom: To be ignorant is to be free. Both the Participant Paradigm and traditional Pragmatism presuppose freewill as essential to making sense of 'real' learning. By learning, we are seeking increasing freedom, increasing ability to explore and develop the universe, increasing freedom to learn. Participant learning is not seeking to converge to a point of no ignorance. Since learning is expansive and emergent, seeking to learn is, in a curious sense, seeking novel ignorance, better questions. In seeking expansive knowledge we are seeking novel forms of ignorance – novel forms of freedom. Seeking to learn is seeking to realize the opportunity to make more types of better choices. The (quantum) uncertainty doesn't decline; it develops, it evolves, emerging as better questions, offering better choices.

In the earlier, initial development of American Pragmatism – what I now take to be the earlier modern articulation of a Philosophy of Engineering – William James embraced the complementarity of the perennial controversies of rationality. James also recognized that what was then being called 'pragmatism' was 'a new name for some old ways of thinking.' (See James's famous essay, *Pragmatism: a new name for some old ways of thinking*.)

In 1893, Pragmatist Charles Sanders Peirce, citing numerous ancient precedents, noted that progressive philosophies and theologies, from the earliest times seemed to have a common vision of the nature of our evolving reality. Peirce offered his speculative impression of the direction of cosmic evolution, in his famous essay entitled *Evolutionary Love*. Fundamental to the new Participant Paradigm is the middle way search to discover, understand and implement progressive win-win relationships – ways of living and learning.

Afterword – The Way Forward

I once asked Paul Feyerabend why he was so focused on the issues in physics whereas the inadequacies of the scientific worldview were so much more apparent in the biological, socio-economic and political realms. I don't recall his exact words but the essence was that until it was clear how to move beyond the classical scientific worldview in physics the overall cultural dominance of the scientific mentality was likely to remain. We need to understand how to move to a post-scientific physics.

This book has been limited primarily to re-thinking basic assumptions about the nature of the problems that arose with the 'new physics' and the 'new philosophy of science'. One of the main impediments to 'thinking outside the box' seeking a new post-scientific understanding of the universe and our role in it, has been the continuing dominance of 'rationalizing' attempts to resolve the paradoxes while retaining the defining presuppositions of the Spectator's classical mechanical science and Positivist's logico-mathematical philosophy of science.

In keeping with Feyerabend's theme I think it is unlikely that the physics community will be able to move forward until a post-scientific physics has been more clearly articulated. In discussing this with my friend, Columbia University physicist, Brian Greene I posited that *at the very least* we needed a new post-mechanical, post-Boltzmannian thermodynamics. Brian responded that, indeed, 'its *all* about the thermodyanmics'.

I have always been fond of Greene's reflection on (what Kuhn would have called his indoctrination) learning about Loschmidt's Paradox of Entropic Reversibility: "When I first encountered this idea many years ago, it was a bit of a shock. Up until that point, I had thought I understood the concept of entropy fairly well, but the fact of the matter was that, following the approach of textbooks I'd studied, I'd only ever considered entropy's implications for the future. And, as we've just seen, while entropy applied toward the future confirms our intuition and experience, entropy applied toward the past just as thoroughly contradicts them. It wasn't quite as bad as suddenly learning that you've been betrayed by a longtime friend, but for me, it was pretty close." (*Fabric of the Cosmos*, page 168).

One item on the research agenda of the More General Theory Research Program is to develop a post-mechanical thermodynamics. For some time I have been attracted to 'engineering thermodynamics' and Carnot's insight that no 'actions' are ever completely mechanical, perfectly efficient. There is always an incommensurable 'loss' – a mechanically discontinuous division, a bifurcation. That's a start. But there are larger issues in trying to make sense of how the 'interplay of opposites' generates a recursively expanding thermodynamics so that we can make sense of the considerable evidence that the universe is winding up (viz. increasing capacity to perform work) rather than winding down to a heat death. For those working in this area you might be interested in my 2014 Linus Pauling Memorial Lecture posted on YouTube (search: Life Ascendant). There I focus on the observation of what I refer to as 'Lovelock's Problem': that the atmosphere of the Earth is optimized for power extraction and has been in thermodynamic disequilibrium for at least 3 billion years. See also my forthcoming book: *Rethinking the Second Law*.) Per hypothesis, the interplay of complementary opposites must be historically net constructive – of the universe. The complementary 'interplay of opposites' is non-mechanically symmetric and therefore must result in this net 'product'.

Another outstanding problem has to do with the emergence of probability and statistics in the history of science. David Lindley, in his excellent book, *Uncertainty: Einstein, Heisenberg, Bohr and the Struggle for the Soul of Science*, suggests that the entry of complementarity into physics (and other sciences) begins with a variety of pre-quantum theory observations of chance-governed phenomena. The adoption of Born's Rule in quantum theory has seemed to entail, as Einstein put it, that god plays dice with the universe. My general theme in this book that 'you can't make sense of the irrational in terms of the rational.' See my forthcoming book about a new way to understand chance-governed phenomena in the middle-ground: *Of Clouds and Clocks*.

And finally, there is a huge literature of what is referred to as 'the quantum measurement problem'. Much of the confusion derives from the Spectator presuppositions of what we are to understand as an 'observation' or 'measurement'. 'Observation of objective reality' in the Spectator representation is tacitly presupposed to be a one-way flow *to* the Spectator *from*, what is tacitly presupposed to be, an isolated system – self-contained, by its very nature. This is another case where the Spectator representation of inquiry is not self-referentially coherent. What is called for, and needs to be articulated, is a Participant representation of observations and inquiry that somehow resolves the confusion of 'the measurement problem.' Dewey offers some clues in his 'transactional' model of observer-observed relations, moving us to a middle-ground away from completely passive and completely active models of observation. However, it is my suspicion that any acceptable resolution must understand observation and measurement as involving 'the performance of work', and so consequently, requiring, as indicated above, a post-mechanical thermodynamic understanding of observation and measurement.

All this will need to be in a framework that is able to understand the universe, self-reflexively and self-coherently, as emerging through some sort of recursively enabling experimental and exploratory evolutionary process.

For further and ongoing updates visit my website at terrybristol.org.

*Jon Wood, Stephen and the author
celebrating successful 2012 event*

Photo Acknowledgements

Cover
  Hubble Deep Field (courtesy of NASA)
  Cover design by Lenny Naar lennynaar.com
Title Page
  Hawking weightless (preparing for space flight) (courtesy of NASA)
Prologue
  NASA jets (courtesy of NASA)
Chapter One
  Paul Feyerabend (courtesy of copyright © creator, Grazia Borrini-Feyerabend)
  Thomas Kuhn (photo copyright © Stanley Rowin, stanstudio.com)
Chapter Six
  Aaron's moment of insight (by permission Michael Lloyd, The Oregonian)
Chapter Twelve
  Sir Karl Popper (from London School of Economics Collections)
Chapter Thirteen
  Imre Lakatos (from London School of Economics Collections)
  Vera Rubin (Archives and Special Collections, Vasser College)
Chapter 22
  Bohr's Family Coat of Arms (Niels Bohr Archives)

Chapter 27
   Stephen Jay Gould (Harvard University Library)
Chapter 28
   John Barrow (by permission copyright © Serafino Amato)
Chapter 30
   John Archibald Wheeler (University of Texas)
Chapter 34
   Niels Bohr (courtesy of Niels Bohr Archive)
Chapter 37
   Sean Carroll (courtesy Ken Weingart)
   Kip Thorne (California Institute of Technology)

Author

Terry Bristol initially attended University of California at Berkeley in the mid-1960s to major in Astronomy. Finding the disciplinary narrowing too constraining he switched to the 'all sciences' field of Philosophy of Science. Paul Feyerabend was his Honors Thesis Advisor at Berkeley and encouraged him to attend the University of London for graduate work in History and Philosophy of Science, where he worked with Imre Lakatos and the 'Karl Popper Group'. In recent years he has been a leading-edge contributor to the emerging Paradigm Shift from the Scientific Worldview to the new broader, superseding, post-scientific Engineering Worldview.

Bristol has held teaching positions at University of California at Santa Barbara, Linfield College, Portland State University, Marylhurst University and Portland Community College. Since 1987, as President of the Institute for Science, Engineering and Public Policy, affiliated with Portland State University, he has run the award-winning Linus Pauling Memorial Lecture Series.

Intermittently over a forty-year period he has engaged in biomedical research projects at Oregon Health and Science University. His work on the biological effects of DMSO was published in the Annual of the New York Academy of Sciences. He also undertook a multi-year research project in cryobiology and served, for over twenty years, as scientific editor of the newsletter of Scleroderma International Foundation.

Terry Bristol served as President of the Columbia Willamette Chapter of Sigma Xi (Research Society of America) for several years. He has a number of both scientific and philosophical publications and has made numerous conference presentations.

He is an active member of the Philosophy of Science Association, the History of Science Society, the Society for the History of Technology, the Society for the Philosophy of Technology, the Forum on Philosophy, Engineering and Technology, Sigma Xi, The American Philosophical Association, the American Physical Society, and the AAAS.

Lightning Source UK Ltd.
Milton Keynes UK
UKOW06n1854210316

270614UK00011B/200/P